Lecture Notes in Chemistry

Edited by G. Berthier, M. J. S. Dewar, H. Fischer
K. Fukui, H. Hartmann, H. H. Jaffé, J. Jortner
W. Kutzelnigg, K. Ruedenberg, E. Scrocco, W. Zeil

17

Hermann Gerhard Hertz

Electrochemistry

A Reformulation of the Basic Principles

Springer-Verlag
Berlin Heidelberg New York 1980

Author

Hermann Gerhard Hertz
Institute of Physical Chemistry and Electrochemistry
University of Karlsruhe, Kaiserstr. 12
7500 Karlsruhe/West Germany

Library of Congress Cataloging in Publication Data. Hertz, Hermann Gerhard. Electrochemistry. (Lecture notes in chemistry ; 17) Bibliography: p. Includes index. 1. Electrochemistry. I. Title. QD553.H53 541.3'7 80-17798

ISBN-13: 978-3-540-10008-9 e-ISBN-13: 978-3-642-86534-3
DOI: 10.1007/ 978-3-642-86534-3

Introduction

In this book a presentation of a phenomenological theory of electrochemistry is given. More precisely, it should be stated that only one part of the whole field of electrochemistry is developed. It is the purpose of this treatment to describe the interconnection between the electric current in a composite thermodynamic system and the rate of production of a certain substance on the one side, the rate of depletion of another substance on the other side, and the work per unit time which has to be delivered to or is supplied by the system. The last part of this programme leads to the computation of the electric potential or the electromotive force of a typical arrangement called a galvanic cell. It will only be the electric current J_2 which is considered, not the change of the electric current per unit time, i.e. dJ_2/dt. The variation of J_2 with time would have to be the subject of the second part of this new treatment of electrochemistry.

Although on the following pages macroscopic systems will be dealt with almost exclusively the stimulus for the undertaking to reformulate the basic principles of electrochemistry arose from an argument connected with the microscopic picture of a material system. In order to give an expression for the velocity cross-correlation coefficients between different ions in electrolyte solutions in terms of the corresponding mutual diffusion coefficient, the author found it necessary to revise in a certain sense Arrhenius' ionic theory and to reintroduce in the microscopic representation of an electrolyte solution the concept of a salt molecule; a NaCl molecule, for instance [1-5]. Of course, proper and suitable definition is required. The comparison of this modified picture of an electrolyte solution with the conventional one immediately led the author into the field of electrochemistry because many of the quantities characterizing electrolyte solutions - for instance the transport numbers - were obtained by electrochemical methods. The author then being faced with the conceptual construction of the electrochemical potential did not feel satisfied with the basic tenets of the current theory. It is well known that Gibbs[6] and Guggenheim[7,8] found the concept of the chemical potential of a single ion to be connected with severe intrinsic difficulties. Guggenheim then introduced the concept of the electrochemical potential[8] and he stated the condition under which a separation of the chemical potential into an electrostatic and a chemical part should be permissible. Then, subsequently the concept of the electrochemical potential found ample application certainly in many cases without the necessary precautions required by the special physical situation. However, since the end-formulas only contained observable i.e. measurable quantities, these treatments were generally accepted in spite of the fact that the starting point of the derivation and various steps in the development lacked proper basis.

In summary, the critical reader when making acquaintance with the construction of electrochemistry very soon comes to the conclusion that the system of conventional arguments does not withstand a penetrating questioning; and moreover, if he tries to obtain more clarity by asking his colleagues working in the field of electrochemistry, he realizes that they too have a feeling of uncertainty and sometimes even perplexity. Four examples can be given where the weaknesses of the classical theory becomes particularly apparent. The problem of the continuation of the electric field of a double layer between an electrolyte solution and a metal surface outside the material system when the latter arrangement does not form a galvanic cell. The same question arises for the electric field thought to be connected with a non-uniform electrolyte solution which leads to the diffusion or liquid junction potential. Then, closely related to this problem is the explanation of the effectiveness of the salt bridge, and finally, the understanding of the membrane potentials. Referring to the salt bridge the situation may be typically characterized by the following sentences, taken from a very recent textbook of physical chemistry[9)]: "The reason for the success of the salt bridge is obscure. The high concentration of KCl in the bridge ensures that the K^+ and Cl^- ions carry most of the current in the cell irrespective of the direction of current flow, and they also carry equal shares of the current. Furthermore, the bridge introduces two junction potentials, and it is both hoped and thought that these tend to cancel".

The method presented in this book avoids all these difficulties. The central concept in the treatment is the electric current, then, for the determination of the work supplied to the system during a change of its material composition, the concept of the electrical potential is also introduced. However, the concept of the electrical charge is not needed and consequently does not occur. Also, the electric field does not play any role in the treatment, and in particular, concepts like the electrical potential or the electrical field strength inside matter are rejected. This then, of course, has the consequence that the electrochemical potential is entirely eliminated from the theory and also the concept of ionic species is no longer of fundamental significance. Ionic species may be introduced or not as one pleases, our treatment is more general and leaves this question to the convenience of the user of the theory which in most cases will be determined by historical developments during the past hundred years.

In the following pages, in chapters 2 - 6, the methodic basis of the treatment will be outlined, thereafter, in chapters 8 - 12 it will be demonstrated with a fairly large number of examples, that in all cases the method is able to give the results which are known from experiments and classical theory, with however the distinctions which are typical for the two types of approach.

References

1) H.G. Hertz, Ber.Bunsengesellschaft Phys.Chem., 1977, 81, 656

2) H.G. Hertz, K.R. Harris, R. Mills, and L.A. Woolf, Ber.Bunsengesellschaft Phys.Chem., 1977, 81, 664

3) H.G. Hertz and R. Mills, J. Phys.Chem., 1978, 82, 952

4) H.G. Hertz, in: "Protons and Ions Involved in Fast Dynamic Phenomena" 1978, Elsevier Scientific Publishing Co., Amsterdam

5) A. Geiger and H.G. Hertz, J.C.S. Faraday I, 1980, 76, 135

6) J.W. Gibbs, "Collected Works", Vol.I, 1957 Yale University Press, New Haven

7) E.A. Guggenheim, "Thermodynamics", 1977, North Holland Publ. Co. Amsterdam

8) E.A. Guggenheim, J.Phys.Chem., 1929, 33, 842

9) P.W. Atkins, "Physical Chemistry", 1978, Oxford University Press

CONTENTS page

Chapter 1

Description of the multicomponent electrolyte solution in the equilibrium state.

1.1 Components, constitutents, salt molecules and ions.

In order to be reasonably general we start off with a mixture of four salts with water. Let the four salts be the 1-1 electrolytes, KCl, NaI, NaCl, KI, for example. Any of these salts may be replaced by another 1-1 electrolyte. The description would not be much simplified if we used general symbols with suitable indices for the solution components. The author feels that reading is easier if the derivations given later are connected with real chemical species. The corresponding equations for polyvalent electrolytes will only occasionally be added. Their derivation is straightforward in all cases.

We refrain from a detailed analysis of what an "electrolyte solution" is meant to be. We take this concept here as denoting any non-metallic liquid mixture in which an electric current can exist. In a limiting situation the total number of components is one.

The choice of a five-component electrolyte as a starting point is suggested by the construction of a galvanic cell in which two different binary electrolytes make contact through a liquid junction. Such systems will be treated below.

Now we say that the mixture has five components, or, in order to emphazise the operational aspect more, we say that the system is prepared or constructed from five components. By this procedure we can also uniquely define five molecular species to be present in the solution. The masses of the components are m_{H_2O} , m_{KCl} , m_{NaI} , m_{NaCl}, and m_{KI} , the total volume of the system is V , thus the partial densities are ρ_{H_2O}, ρ_{KCl} , ρ_{NaI}, ... , and finally, the molar concentrations of the molecular species are:

$$c_{H_2O} = \frac{\rho_{H_2O}}{M_{H_2O}} , \quad c_{KCl} = \frac{\rho_{KCl}}{M_{KCl}} , \quad c_{NaI} = \frac{\rho_{NaI}}{M_{NaI}} , \quad \ldots\ldots$$

where the M_i's are the molar masses. We can say that these molecular concentrations are analytically well-defined quantities. They are defined by the addition of the respective substances to the system. We are aware of the fact that usually such a definiton is not made. This, however, has been for historical reasons and not so much as a consequence of the absence of a logical justification.

Next we define constituent masses by the relations

$$
\begin{aligned}
m_{Na} &= \frac{M_{Na}}{M_{NaI}} m_{NaI} + \frac{M_{Na}}{M_{NaCl}} m_{NaCl} \\
m_{Cl} &= \frac{M_{Cl}}{M_{NaCl}} m_{NaCl} + \frac{M_{Cl}}{M_{KCl}} m_{KCl} \\
m_{K} &= \frac{M_{K}}{M_{KCl}} m_{KCl} + \frac{M_{K}}{M_{KI}} m_{KI} \\
m_{I} &= \frac{M_{I}}{M_{KI}} m_{KI} + \frac{M_{I}}{M_{NaI}} m_{NaI}
\end{aligned}
\tag{1}
$$

where M_{Na}, M_K, M_{Cl}, M_I, are the corresponding atomic weights. For the solvent water the constituent mass has been taken to be the same as the component mass. This has been done for simplification. Thus we refrain from an inclusion of "acid-base equilibria" in this treatment. This would require a separate development. Thus, in the following sections, in all mass transformation relations we shall omit the fifth constituent, i.e. water. We have seen that the system is uniquely defined by the set of five component masses through the manner of construction. As a consequence, the constituent masses m_{Na}, m_{Cl}, . . . are also uniquely defined through the linear transformations (1). In the same way constituent concentrations are well-defined quantities

$$
C_{Na} = \frac{m_{Na}}{M_{Na}} V^{-1}, \quad C_{Cl} = \frac{m_{Cl}}{M_{Cl}} V^{-1}, \quad \ldots\ldots
\tag{1a}
$$

Inspection of eq.(1) tells us that the two component masses m_{KI} and m_{NaI} occurring in the fourth line have already been used to determine m_K and m_{Na}. Thus the constituent mass m_I cannot be varied independently of the other constituent masses. We divide the equation containing the m_i by M_i, i = Na, K, Cl, I, and add the resulting equations corresponding to the metal and the halide constituent, respectively, pairwise. The result is that the interdependence of the constituent masses can be expressed as

$$
\frac{m_{Na}}{M_{Na}} + \frac{m_K}{M_K} = \frac{m_{Cl}}{M_{Cl}} + \frac{m_I}{M_I}
\tag{2}
$$

or

$$c_{Na} + c_K = c_{Cl} + c_I \tag{2a}$$

These equations are the representation of the construction constraint that the constituents can only be added pairwise to the system. The extension of these relations to cases where the aggregation during addition is different is straightforward. For example we would have

$$2c_{Ca} + c_K = c_{Cl} + c_I \tag{2b}$$

when the system is constructed from Ca salts instead of Na salts.

The validity of eq.(2) has the consequence that eqs.(1) cannot be reversed. Given a system which is characterized by a set of four constituent masses, then a unique set of four component masses m_{KCl}, m_{NaI}, . . . from which the system is to be constructed does not exist. Only three component masses can be determined uniquely. In other words: a multicomponent electrolyte solution which is physically identical with that described by eq.(1) can also be constructed by the addition of only three components to the solvent, such as KCl, NaI, and NaCl.

Now assume, we have added these three components to water. Then, analyzing the composition of the vapour or the system of salts which crystallizes out from the solution: we find that these phases also contain (or may contain) KI. Thus there must have been a chemical reaction in the liquid phase:

$$\text{KCl} + \text{NaI} \longrightarrow \text{NaCl} + \text{KI} \qquad \text{(I)}$$

The appearance of a given salt in the crystalline phase depends on the relative stabilities of the solid components. At most only three salts will appear in the equilibrium product which crystallizes out.

The fact that the chemical reaction (I) occurs is implied in the definition of the constituent masses and constituent concentrations given in eqs.(1) and (1a). Moreover, the reaction time τ for (I) must be very much shorter than the "observation time", that is, the time during which the system is observable as a macroscopic system $\tau \gtrsim 1$ s say.

Now, if m_{KCl}, m_{NaI} and m_{NaCl} are the component masses which have been used for the construction of the system, then the transformation relations to the three independent constituent masses can be given as:

$$m_{Na} = \frac{M_{Na}}{M_{NaCl}} m_{NaCl} + \frac{M_{Na}}{M_{NaI}} m_{NaI} \tag{3}$$

$$m_{Cl} = \frac{M_{Cl}}{M_{NaCl}} m_{NaCl} + \frac{M_{Cl}}{M_{KCl}} m_{KCl} \tag{4}$$

$$m_{K} = \frac{M_{K}}{M_{KCl}} m_{KCl} \tag{5}$$

m_I is given by eq.(2).
So far eqs.(3),(4),(5),(2), and (1a) represent the definition of the constituent concentrations C_{Na} , C_K . . . . with the following meaning. To each set of quantities $m_{KCl}, m_{NaI}, m_{NaCl}$, a set of quantities m_{Na}, m_{Cl}, m_K, can be assigned uniquely. However, they are also, as a rule, fairly well-defined quantities via suitable analytical operations performed on the system itself or on samples extracted from the solution. For example, spectroscopic techniques to determine the amount of sodium in the sample, can be used to measure C_{Na}.

We return to eqs.(3),(4),(5). <u>The molecular</u> concentrations C_{NaCl}, C_{KCl} , C_{NaI}, C_{KI} , are connected with the constituent concentrations as under:

$$C_{Na} = C_{NaCl} + C_{NaI} \tag{6}$$

$$C_{Cl} = C_{NaCl} + C_{KCl} \tag{7}$$

$$C_{K} = C_{KCl} + C_{KI} \tag{8}$$

Thus we have only three independent equations to determine the four molecular concentrations C_{NaCl}, C_{KCl}, C_{NaI}, C_{KI}. Molecular concentrations are always well-defined if $m_{NaI} = 0$. In particular, the salt molecule concentration is well-defined for a binary electrolyte solution. For example $C_{NaCl} = m_{NaCl}/M_{NaCl}V$. This manner of description is only unusual when one connects the idea of an "internal rigidity" with the concept of a molecule. In fact, in the gas phase we do have rigidity of the NaCl molecule. Thus, it is easy and unique to use the addition operations NaCl + H_2O as the definition of the concept "NaCl molecule". Further, the concept of a salt molecule is preferable to the concept of an ionic particle, which will be introduced shortly, because certain misinterpretations are avoided. Some examples will be given below. In the general case, where the system is constructed from KCl, NaI, and NaCl, we may choose any arbitrary molecular concentration which does not contradict eqs.(6)-(8) as being fixed. This corresponds to our original procedure mentioned in connection with eq.(1).

However, it is important to explain another more general method of assigning molecular concentrations here. In order to make a statement about all four molecular concentrations we need one further equation

$$\varphi(c_{NaCl}, c_{NaI}, c_{KCl}, c_{KI}) = 0 \tag{9}$$

apart from eqs.(6)-(8). The most usual relation is given by the requirement that the "distribution" of Na among Cl and I, apart from a constant factor f, is the same as the distribution of K among Cl and I:

$$\frac{c_{NaCl}}{c_{NaI}} = f \cdot \frac{c_{KCl}}{c_{KI}} \tag{10}$$

Eq.(10) is equivalent to the simple law of mass action. Then the combination of eqs.(6)-(8) with eq.(10) yields:

$$c_{NaCl} = -\frac{c_K - c_{Cl} + f(c_{Na} + c_{Cl})}{2(1-f)} \pm \left\{ \frac{f}{1-f} c_{Na} c_{Cl} + \frac{1}{4}\left[\frac{c_K - c_{Cl} + f(c_{Na} + c_{Cl})}{1-f}\right]^2 \right\}^{1/2}$$

$$f > 0, \quad f \neq 1$$

For f = 1 the situation is particularly simple

$$\begin{aligned} c_{NaCl} &= \frac{c_{Na} c_{Cl}}{c_K + c_{Na}} \\ c_{NaI} &= \frac{c_{Na}(c_K + c_{Na} - c_{Cl})}{c_K + c_{Na}} \\ c_{KCl} &= \frac{c_K c_{Cl}}{c_K + c_{Na}} \\ c_{KI} &= \frac{c_K(c_K + c_{Na} - c_{Cl})}{c_K + c_{Na}} \end{aligned} \tag{11}$$

Thus we have

$$c_{NaCl} = c_{NaCl}(c_{Na}, c_K, c_{Cl})$$
$$c_{KCl} = c_{KCl}(c_{Na}, c_K, c_{Cl})$$
$$\cdots\cdots\cdots\cdots\cdots$$

i.e. the NaCl, KCl, ... molecular concentrations as a function of the independent constituent concentrations. As an example let us assume that we have

$$c_{Na} = 7$$
$$c_K = 4$$
$$c_{Cl} = 5$$
$$c_I = 6$$

Then we find from eqs.(11)

$$c_{NaCl} = \frac{7\cdot 5}{4+7} = 3,\ldots \qquad c_{KCl} = \frac{4\cdot 5}{4+7} = 2-0,\ldots$$
$$c_{NaI} = \frac{7(11-5)}{4+7} = 4-0,\ldots \qquad c_{KI} = \frac{4(11-5)}{4+7} = 2,\ldots$$

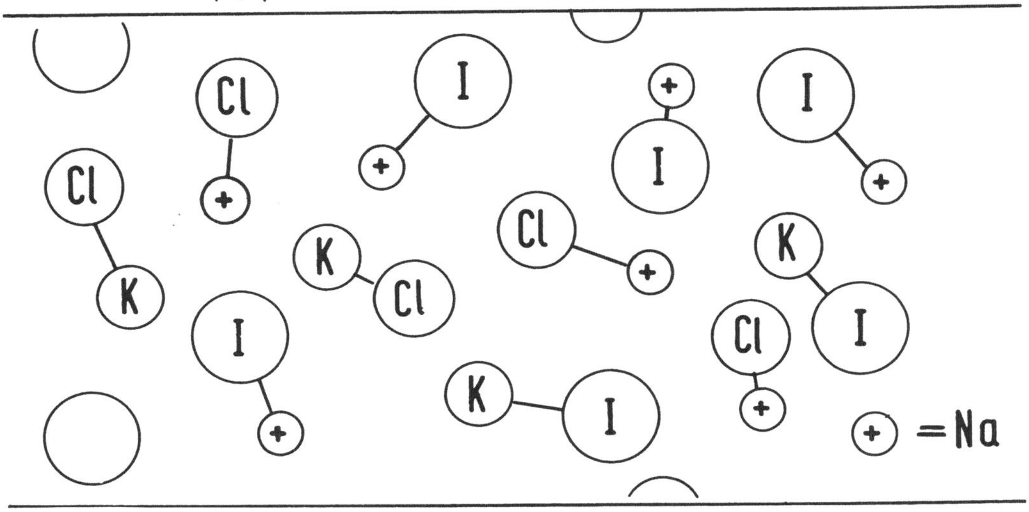

Fig.1 A scheme of connecting lines which together with eq.(10) and f = 1 define the salt molecules in a multicomponent electrolyte solution.

In Fig.1 we have drawn as many lines between the corresponding constituent particles as there are molecules. We have connected those pairs of constituents which have the smallest distances between them. Of course, this is not essential for the concept of the salt molecule. As already mentioned, eq.(9) may be any arbitrary equation leading to a well-behaved function for c_{ij} , for instance:

$$c_{NaCl} = \phi(c_{Na}, c_K, c_{Cl}) \tag{12}$$

such that ϕ = 0 for c_{Na} = 0, and c_{Cl} = 0 and $\phi \leq c_{Cl}$. Having such an equation, then all other concentrations, c_{KCl} , c_{NaI}, . . . are well-defined. In particular, we may have any experimental quantity

$$\alpha = \alpha(C_{NaCl}, C_{Na}, C_K, C_{Cl}) \tag{13}$$

A relation of this type explicitly or implicitly imposes a condition for the configurational arrangements of the constituent particles which has to be satisfied.

Next we perform an important step in extending eqs.(6),(7),(8) in the following way

$$C_{Na} = C_{NaCl} + C_{NaI} + C_{Na^+} \tag{6a}$$

$$C_{Cl} = C_{NaCl} + C_{KCl} + C_{Cl^-} \tag{7a}$$

$$C_K = C_{KCl} + C_{KI} + C_{K^+} \tag{8a}$$

and we have also

$$C_I = C_{NaI} + C_{KI} + C_{I^-} \tag{8b}$$

where C_{Na^+}, C_{K^+}, C_{Cl^-}, C_{I^-}, are the free ion concentrations. These are particles which in the microscopic representation are definitely not allowed to have a connecting line drawn between them as in Fig.1. Thus the $\pm$ sign added to the constituent symbols means that a connecting line must not be drawn. Comparison of the sum of eqs.(6a) and (8a) with the sum of eqs.(7a) and (8b) and applying eq.(2a) yields

$$C_{Na^+} + C_{K^+} = C_{Cl^-} + C_{I^-} \tag{14}$$

which is the well-known electroneutrality relation. We have derived this equation from the restriction found experimentally to be imposed on the manner of adding constituents to the system (pairwise addition in the form of components NaCl, KCl, . . .). Thus it should be kept in mind that eq.(2a) is more general than eq.(14); it holds for all systems of reacting components containing appropriate constituent pairs.

In order to obtain well-defined free ion concentrations C_{Na^+}, C_{K^+}, C_{Cl^-}, C_{I^-}, we need three further equations of type (10), (12) or (13). These equations may serve to define the free ionic species or they may be deduced from some intrinsic physical properties of the electrically charged particles. These details need not be discussed here. If we extend our treatment to polyvalent electrolytes, then, what we have given is a short description of the study of complex equilibria in ionic solutions. The only difference is that usually one starts from the free ion picture whereas for reasons of higher generality we started from the molecular concentration. Let us for a moment adopt the conventional standpoint and consider the solution to be made up only of free ions and water molecules, i.e. in eqs.(6a)-(8a) we set $c_{NaCl} = c_{KCl} = c_{NaI} = c_{KI} = 0$. Then our constituent concentrations are the same as the free ion concentrations, and thus it is irrelevant whether we call them ions or constituents. The fact that the ions are electrically charged has not been discussed yet. In the present treatment the concept of the electrical charge is not needed and therefore will be avoided. The definition of the signs $\pm$ will be given below. The statement that the free ion concentration in the absence of any other solute species is identical with the constituent concentration is of great importance. To demonstrate this let us consider a binary electrolyte solution, a NaCl solution say. Then it follows from eq.(3), (1a) and (6a)

$$c_{Na^+} = c_{Na} = \frac{1}{M_{NaCl}} \cdot \frac{m_{NaCl}}{V} \tag{15}$$

and as a consequence of eq.(14)

$$c_{Cl^-} = \frac{1}{M_{NaCl}} \cdot \frac{m_{NaCl}}{V}$$

The meaning of this equation is as follows. Whenever we postulate that there is a finite ionic concentration c_{Na^+} then there must be a finite component mass m_{NaCl} from which the system is constructed, thus the operation of adding NaCl to H_2O defines the ionic concentration.

Usually in electrochemistry, free ionic species are understood to be entities in themselves, and as we will see below, this can lead to misinterpretations; in particular, when non-equilibrium systems are considered. To be correct, the ionic concentration as expressed by eq.(15) has to be understood as being a transformed representation of the solute (construction) component. The solution is not a mixture of Na^+ ions, Cl^- ions, and water because the concept "mixture" is defined by the operations of addition and - at least in principle - separation of the components. These operations do not exist with respect to Na^+ and Cl^-.

Now we release the condition that only free ions are present which means that molecular species appear. However, in all our systems a real neutralization of electrical charges by "electric currents" between the ions of opposite charge never occurs. Always we are left with the electric dipole moments of the molecular species, or at least local electric dipole moments. In fact, all questions concerning the number of neutral or ionic particles present are only a matter of configurational classification and may be answered in a number of different ways. Therefore, one way of procedure, which is formally entirely correct is to take the constituents to be always ionic species wherever they occur. However we have not yet defined the $\pm$ sign characterizing the nature of the ions; this we shall do later. Still, our transformation relations eqs.(3)-(5) are fully valid. For this reason we shall now exclusively use the concept "constituent"; this concept being based in an operational sense on eqs.(3),(4), and (5).

1.2 Coordinate system transformations.

We have seen that we may construct a multicomponent electrolyte solution from the three component masses m_{KCl}, m_{NaI}, m_{NaCl}, and the fourth component water. Then, considering only the solute, we obtain the four constitutent masses:

$$m_{Na} = \frac{M_{Na}}{M_{NaI}} m_{NaI} + \frac{M_{Na}}{M_{NaCl}} m_{NaCl} \tag{3}$$

$$m_{Cl} = \frac{M_{Cl}}{M_{NaCl}} m_{NaCl} + \frac{M_{Cl}}{M_{KCl}} m_{KCl} \tag{4}$$

$$m_{K} = \frac{M_{K}}{M_{KCl}} m_{KCl} \tag{5}$$

$$m_{I} = \frac{M_{I}}{M_{NaI}} m_{NaI} \tag{5a}$$

These equations can be reversed : If the three constituent masses m_{Na}, m_{Cl}, and m_K are given and if

$$\frac{m_K}{M_K} + \frac{m_{Na}}{M_{Na}} - \frac{m_{Cl}}{M_{Cl}} > 0$$

(see eq.(2)), then, regarding the system as the result of construction from the components NaCl, KCl, NaI, the three component masses are given by the equations:

$$m_{KCl} = M_{KCl} \frac{m_K}{M_K} \tag{16}$$

$$m_{NaCl} = M_{NaCl}\left(\frac{m_{Cl}}{M_{Cl}} - \frac{m_K}{M_K}\right) \tag{17}$$

$$m_{NaI} = M_{NaI}\left(\frac{m_{Na}}{M_{Na}} + \frac{m_K}{M_K} - \frac{m_{Cl}}{M_{Cl}}\right) \tag{18}$$

If the three constituent masses m_K, m_{Na}, and m_I (instead of m_{Cl}) are known, then

$$m_{KCl} = M_{KCl}\frac{m_K}{M_K} \tag{16a}$$

$$m_{NaI} = M_{NaI}\frac{m_I}{M_I} \tag{17a}$$

$$m_{NaCl} = M_{NaCl}\left(\frac{m_{Na}}{M_{Na}} - \frac{m_I}{M_I}\right) \tag{18a}$$

Of course, all three component masses must be positive quantities, and thus we must have

$$\frac{m_{Cl}}{M_{Cl}} - \frac{m_K}{M_K} \geqslant 0 \tag{17b}$$

or

$$\frac{m_{Na}}{M_{Na}} - \frac{m_I}{M_I} \geqslant 0 \tag{18b}$$

If these requirements are not fulfilled then the system cannot be constructed from the three components NaCl, NaI, and KCl. Rather the construction components are KI, NaCl, and NaI.

Now eq.(3) remains unchanged, but eqs.(4),(5),(5a) are modified. We have

$$m_{Na} = \frac{M_{Na}}{M_{NaI}}m_{NaI} + \frac{M_{Na}}{M_{NaCl}}m_{NaCl}$$

$$m_I = \frac{M_I}{M_{NaI}}m_{NaI} + \frac{M_I}{M_{KI}}m_{KI}$$

$$m_K = \frac{M_K}{M_{KI}}m_{KI}$$

$$m_{Cl} = \frac{M_{Cl}}{M_{NaCl}}m_{NaCl}$$

The reverse relations are:

$$m_{NaCl} = M_{NaCl}\frac{m_{Cl}}{M_{Cl}}$$

$$m_{KI} = M_{KI}\frac{m_K}{M_K}$$

$$m_{NaI} = M_{NaI}\left(\frac{m_{Na}}{M_{Na}} - \frac{m_{Cl}}{M_{Cl}}\right)$$

or

$$m_{NaCl} = M_{NaCl}\frac{m_{Cl}}{M_{Cl}}$$

$$m_{KI} = M_{KI}\frac{m_K}{M_K}$$

$$m_{NaI} = M_{NaI}\left(\frac{m_I}{M_I} - \frac{m_K}{M_K}\right)$$

Thus we obtain the result. If the inequalities

$$\frac{m_{Na}}{M_{Na}} - \frac{m_{Cl}}{M_{Cl}} \geqslant 0 \tag{19}$$

and

$$\frac{m_I}{M_I} - \frac{m_K}{M_K} \geqslant 0 \tag{20}$$

are fulfilled, then the system can be constructed from the components NaCl, KI and NaI. However, both sets of inequalities i.e. (17b),(18b) and (19),(20) may be fulfilled simultaneously. Then one has to specify the set of construction components which is to be chosen. For instance, we may state that we consider a given system to be constructed from the components KCl, NaCl, and NaI. Having done this, the system is uniquely defined by the three component masses m_{KCl}, m_{NaCl}, and m_{NaI}. It may occur that the constituent masses satisfy the inequalities

$$\frac{m_K}{M_K} - \frac{m_{Cl}}{M_{Cl}} \geqslant 0$$

$$\frac{m_I}{M_I} - \frac{m_{Na}}{M_{Na}} \geqslant 0$$

or

$$\frac{m_{Cl}}{M_{Cl}} - \frac{m_{Na}}{M_{Na}} \geqslant 0$$

$$\frac{m_K}{M_K} - \frac{m_I}{M_I} \geqslant 0$$

which are the reversed inequalities (17b), (18b),(19),(20). In such a situation the system is constructed from the components NaI, KCl, KI, and NaCl, KI, KCl, respectively.

If the system only contains three constituents, e.g. K, Na, Cl, then, in eq.(3) we have $m_{NaI} = 0$, eq.(5a) disappears and eq.(5) becomes the dependent one. Now the system is only constructed from two components, namely KCl and NaCl, and as such it is uniquely defined. Finally, if we have a binary solution, then eqs.(5) and (5a) disappear and eq.(4) is no longer an independent one, the system is uniquely constructed from one component, NaCl say.

When we have so far described our system in two different "coordinate systems", or "property spaces", i.e. the construction or component coordinate system and the constituent coordinate system, we were dealing only with uniform systems. For non-uniform systems we divide all our quantities by V, the total volume of the system. Then we have the formal partial densities, e.g.

$$\rho^*_{NaCl} = m_{NaCl}/V$$
$$\rho^*_{KCl} = m_{KCl}/V$$
$$\rho^*_{NaI} = m_{NaI}/V$$

characterizing the local composition in the construction space, and

$$\rho_{Na} = m_{Na}/V, \quad \rho_K = m_K/V, \quad \rho_{Cl} = m_{Cl}/V,$$
$$\rho_I = m_I/V$$

giving the local composition in the constituent coordinate system. The transformation relations between the two coordinate systems referring to the local quantities are the same as given for the total masses, only a division by V is needed. As an example we have in place of eqs.(3)-(5), and (16)-(18)

$$\rho_{Na} = \frac{M_{Na}}{M_{NaI}}\rho^*_{NaI} + \frac{M_{Na}}{M_{NaCl}}\rho^*_{NaCl} \tag{21}$$

$$\rho_{Cl} = \frac{M_{Cl}}{M_{NaCl}}\rho^*_{NaCl} + \frac{M_{Cl}}{M_{KCl}}\rho^*_{KCl} \tag{22}$$

$$\rho_K = \frac{M_K}{M_{KCl}}\rho^*_{KCl} \tag{23}$$

$$\rho^*_{KCl} = M_{KCl}\frac{\rho_K}{M_K} \tag{21a}$$

$$\rho^*_{NaCl} = M_{NaCl}\left(\frac{\rho_{Cl}}{M_{Cl}} - \frac{\rho_K}{M_K}\right) \tag{22a}$$

$$\rho^*_{NaI} = M_{NaI}\left(\frac{\rho_{Na}}{M_{Na}} + \frac{\rho_K}{M_K} - \frac{\rho_{Cl}}{M_{Cl}}\right) \tag{23a}$$

These local partial densities then yield the instrument to treat non-equilibrium systems as will be shown in chapter 2.

When in the preceding paragraphs we reduced the number of constituents stepwise, then, similarly the number of independent equations of type (3),(4),(5), . . . was reduced until finally we had only one equation, for instance

$$m_{Na} = \frac{M_{Na}}{M_{NaCl}} m_{NaCl}$$

There is still one dependent equation, e.g. that giving the Cl constituent mass:

$$m_{Cl} = \frac{M_{Cl}}{M_{NaCl}} m_{NaCl}$$

Now we wish to rearrange our equation such that in the last step we only have one equation involving one independent variable. Of course, in the binary case this is the solute NaCl. So we introduce a new coordinate system to describe the multicomponent electrolyte system in the following way: Given that the system is constructed from a certain set of components, e.g. from the masses m_{NaCl} , m_{NaI} , and m_{KCl}, then we may describe the solution in a unique manner by the total solute mass m_s, and two constituent masses, for example m_{Na} and m_{Cl} . The transformation relations are

$$m_s = m_{KCl} + m_{NaCl} + m_{NaI}$$

$$m_{Na} = \frac{M_{Na}}{M_{NaCl}} m_{NaCl} + \frac{M_{Na}}{M_{NaI}} m_{NaI}$$

$$m_{Cl} = \frac{M_{Cl}}{M_{NaCl}} m_{NaCl} + \frac{M_{Cl}}{M_{KCl}} m_{KCl}$$

In order to get local quantities we introduce the total solute density

$$\rho_s = m_s / V$$

leading to the relations

$$\rho_s = \rho^*_{NaCl} + \rho^*_{KCl} + \rho^*_{NaI} \tag{24}$$

$$\rho_{Na} = \frac{M_{Na}}{M_{NaCl}} \rho^*_{NaCl} + \frac{M_{Na}}{M_{NaI}} \rho^*_{NaI} \tag{25}$$

$$\rho_{Cl} = \frac{M_{Cl}}{M_{NaCl}} \rho^*_{NaCl} + \frac{M_{Cl}}{M_{KCl}} \rho^*_{KCl} \tag{26}$$

The meaning of these equations is the following: We know that the system is constructed from a given set of components - and we should perhaps write $\rho_s^{(NaCl,\,KCl,\,NaI)}$, however, we omit this superscript for simplicity - then ρ_s is given only by a mass determination, an operation with a balance. However, the total solute has still two degrees of freedom, the two constituent partial mass densities ρ_{Na} , and ρ_{Cl} , and they have to be determined by analytical operations which also specify the kind of material. With decreasing complexity of the system these internal degrees of freedom are reduced stepwise until we arrive at the situation where the total solute density has no "internal" degrees of freedom, NaCl, for example.

Returning to the four constituent systems, we have

$$\rho_s = \rho_{Na} + \rho_{Cl} + \rho_K + \rho_I \tag{24a}$$

and as a consequence, ρ_K and ρ_I may be calculated when ρ_s, ρ_{Na}, and ρ_{Cl} are known:

$$\rho_K = \frac{\rho_s - \rho_{Na}(1 + M_I/M_{Na}) - \rho_{Cl}(1 - M_I/M_{Cl})}{(1 + M_I/M_K)} \tag{27}$$

$$\rho_I = M_I \left(\frac{\rho_{Na}}{M_{Na}} + \frac{\rho_K}{M_K} - \frac{\rho_{Cl}}{M_{Cl}} \right) \tag{28}$$

Now let us assume we have a system which is characterized by the set of three parameters ρ_s, ρ_{Na}, ρ_{Cl}, where the three quantities fulfil the requirement that $\rho_K > 0$, $\rho_I > 0$, according to eqs.(27) and (28). Then this mode of description may be transformed to the construction space:

$$\rho^*_{KCl} = \frac{M_{KCl}}{M_{KI}} \left[\rho_s + (M_I - M_{Cl}) \frac{\rho_{Cl}}{M_{Cl}} - M_{NaI} \frac{\rho_{Na}}{M_{Na}} \right] \tag{29}$$

$$\rho^*_{NaCl} = \frac{M_{NaCl}}{M_{KI}} \left[M_{KCl} \frac{\rho_{Cl}}{M_{Cl}} + M_{NaI} \frac{\rho_{Na}}{M_{Na}} - \rho_s \right] \tag{30}$$

$$\rho^*_{NaI} = \frac{M_{NaI}}{M_{KI}} \left[\rho_s + \frac{\rho_{Na}}{M_{Na}} (M_K - M_{Na}) - M_{KCl} \frac{\rho_{Cl}}{M_{Cl}} \right] \tag{31}$$

These equations have been derived by inverting eqs. (24),(25),(26).

We give some simple examples of the application of this formula. Assume we find that

$$\rho_{Na} = \frac{M_{Na}}{M_{NaCl}} \rho_s \; ; \qquad \rho_{Cl} = \frac{M_{Cl}}{M_{NaCl}} \rho_s$$

then it follows from eqs.(29)-(31)

$$\rho^*_{KCl} = 0 \, , \quad \rho^*_{NaCl} = \rho_s \, , \quad \rho^*_{NaI} = 0$$

and likewise, if we find that

$$\rho_{Cl} = \frac{M_{Cl}}{M_{KCl}} \rho_s \quad ; \quad \rho_{Na} = 0$$

it follows

$$\rho^*_{KCl} = \rho_s \ , \quad \rho^*_{NaCl} = \rho^*_{NaI} = 0$$

and finally for

$$\rho_{Na} = \frac{M_{Na}}{M_{NaI}} \rho_s \quad ; \quad \rho_{Cl} = 0$$

we have

$$\rho^*_{KCl} = \rho^*_{NaCl} = 0 \ , \quad \rho^*_{NaI} = \rho_s$$

For a solution which is only constructed from two components and having three constituents, for instance the components KCl and NaCl, one derives the following transformation relations between the component space and the total solute constituent space:

$$\rho_s = \rho^*_{KCl} + \rho^*_{NaCl} = \rho_{KCl} + \rho_{NaCl}$$

$$\rho_{Na} = \frac{M_{Na}}{M_{NaCl}} \rho^*_{NaCl} = \frac{M_{Na}}{M_{NaCl}} \rho_{NaCl}$$

Here we can write $\rho^*_{NaCl} = \rho_{NaCl}$, because the molecular partial densities are the same as those from which the system is constructed. The reverse transformation relations are

$$\rho_{NaCl} = \rho^*_{NaCl} = M_{NaCl} \frac{\rho_{Na}}{M_{Na}} \tag{32}$$

$$\rho_{KCl} = \rho^*_{KCl} = \rho_s - M_{NaCl} \frac{\rho_{Na}}{M_{Na}} \tag{33}$$

For completeness we give the transformation relations from the full constitutents space to the total solute-two constituents space with the reverse relation eq.(27).

$$\rho_s = \rho_{Na}\left(1 + \frac{M_I}{M_{Na}}\right) + \rho_K\left(1 + \frac{M_I}{M_K}\right) + \rho_{Cl}\left(1 - \frac{M_I}{M_{Cl}}\right)$$

$$\rho_{Na} = \rho_{Na} \ , \quad \rho_{Cl} = \rho_{Cl} .$$

1.3 The chemical potentials.

Our liquid multicomponent electrolyte solution may be part of a more extended system consisting also of the gas phase and a number of solid phases. The number of solid phases can be as high as four: ice and three salts; however in general it will be smaller. In equilibrium there are the six quantities T, p, μ^*_{KCl}, μ^*_{NaCl}, μ^*_{NaI}, μ^*_w which are constant throughout the total system, that is, independent of position. T is the temperature, p is the pressure, the μ_i^* are the specific chemical potentials

$$\mu_i^* = \left(\frac{\partial G}{\partial m_i}\right)_{T,p,m_{j\neq i}}$$

where G is the Gibbs-free energy, and $w \equiv H_2O$. Here we have assumed that the set of stable solid phases is really NaCl, KCl, and NaI. Clearly the validity of this special statement is immaterial for our argument.

There is always a gas phase (superscript $^{(g)}$) coexisting with the liquid phase and we have

$$\mu_w^* = \mu_w^{*(g)} \; ; \quad \mu_{NaCl}^* = \mu_{NaCl}^{*(g)} \; ; \quad \mu_{NaI}^* = \mu_{NaI}^{*(g)}$$

$$\mu_{KCl}^* = \mu_{KCl}^{*(g)}$$

The quantities without superscript refer to the liquid phase. The μ_i^*'s are the specific chemical potentials of the components, that is, they are taken with respect to the <u>added amounts</u> of masses m_i . In principle the amounts of salt present in the gas phase may be measurable at some suitable temperature; the same may be true for the vibrational and rotational properties of the gaseous salt molecule. Then the $\mu_i^{(g)}$ could be calculated at any desired temperature and this would define the μ_i^*'s in the solution.

However, since the gas phase only contains extremely small salt concentrations the vapour pressure p is due only to the component water. Then to determine the μ_i^*'s for the salts we have a set of differential equations which follow from the Gibbs-Duhem relations:

$$\left.\begin{aligned}
-m_w\left(\frac{\partial \mu_w^*}{\partial m_{NaCl}}\right)_{T,p,m_{NaI},m_{KCl}} &= \sum m_i\left(\frac{\partial \mu_i^*}{\partial m_{NaCl}}\right)_{T,p,m_{NaI},m_{KCl}} \\
-m_w\left(\frac{\partial \mu_w^*}{\partial m_{NaI}}\right)_{T,p,m_{KCl},m_{NaCl}} &= \sum m_i\left(\frac{\partial \mu_i^*}{\partial m_{NaI}}\right)_{T,p,m_{NaCl},m_{KCl}} \\
-m_w\left(\frac{\partial \mu_w^*}{\partial m_{KCl}}\right)_{T,p,m_{NaI},m_{NaCl}} &= \sum m_i\left(\frac{\partial \mu_i^*}{\partial m_{KCl}}\right)_{T,p,m_{NaI},m_{NaCl}}
\end{aligned}\right\} \; i = NaCl,\; NaI,\; KCl$$

These coupled differential equations may be integrated, the left-hand sides are known from vapour pressure measurements. The constants of integration can be obtained from the study of the solid-liquid equilibrium, for which we have

$$\mu_{NaCl}^{\gamma} = \mu_{NaCl}^{*(s)}\ ;\quad \mu_{NaI}^{\gamma} = \mu_{NaI}^{*(s)}\ ;\quad \mu_{KCl}^{*} = \mu_{KCl}^{*(s)}$$

where the superscript (s) means. solid state.

So we see that the specific chemical potentials of the liquid mixtures as a function of the amounts of salt from which the mixture is prepared (constructed) are quantities which can be determined experimentally. Essentially they represent a projection of the multi-phase behaviour on one single phase, here the liquid phase.

Then the Gibbs free energy of the system is

$$G = m_W \mu_W^{*} + m_{NaCl}\mu_{NaCl}^{*} + m_{NaI}\mu_{NaI}^{*} + m_{KCl}\mu_{KCl}^{*} \qquad (34)$$

In this equation, the three variables m_{NaCl}, m_{NaI}, and m_{KCl} are given by our operation to construct the system. Consequently, eq.(34) represents the Gibbs free energy expressed in the component or construction space. Now it is easy to transform eq.(34) to the constituent space. The μ_i^{*}'s in eq.(34) are functions of m_W , m_{NaCl} , m_{NaI} , and m_{KCl}. These variables may be substituted by the aid of the transformation relations (16),(17),(18). Then we obtain from eq.(34)

$$\begin{aligned} G &= m_W\mu_W^{*} + \mu_{NaCl}^{*} M_{NaCl}\left(\frac{m_{Cl}}{M_{Cl}} - \frac{m_K}{M_K}\right) \\ &\quad + \mu_{NaI}^{*} M_{NaI}\left(\frac{m_{Na}}{M_{Na}} + \frac{m_K}{M_K} - \frac{m_{Cl}}{M_{Cl}}\right) \\ &\quad + \mu_{KCl}^{*} M_{KCl}\frac{m_K}{M_K} \\ &\equiv m_W\tilde{\mu}_W + m_{Na}\tilde{\mu}_{Na} + m_K\tilde{\mu}_K + m_{Cl}\tilde{\mu}_{Cl} \end{aligned} \qquad (35)$$

with

$$\tilde{\mu}_W = \mu_W^{*} \qquad (36)$$

$$\tilde{\mu}_{Na} = \frac{1}{M_{Na}}\mu_{NaI}^{*} M_{NaI} \qquad (37)$$

$$\tilde{\mu}_K = \frac{1}{M_K}\left(\mu_{NaI}^{*} M_{NaI} + \mu_{KCl}^{*} M_{KCl} - \mu_{NaCl}^{*} M_{NaCl}\right) \qquad (38)$$

$$\tilde{\mu}_{Cl} = \frac{1}{M_{Cl}}\left(\mu_{NaCl}^{\gamma} M_{NaCl} - \mu_{NaI}^{*} M_{NaI}\right) \qquad (39)$$

We may also introduce mole numbers instead of the masses, then we get the (partial) chemical potentials $\mu_i = M_i \tilde{\mu}_i$, also $\mu_i = M_i \mu_i^*$

$$G = n_W \mu_W + n_{Na} \mu_{Na} + n_K \mu_K + n_{Cl} \mu_{Cl} \tag{40}$$

This equation has formally a similar appearence to the conventional expression of the Gibbs free energy containing single ion chemical potentials. However, the physical significance is quite different here. The μ_i's are linear transformations of the μ_i^*, i = NaCl, KCl, NaI, in the construction space, and the latter represent the projection of a multi-phase structure on a single phase. An important and characteristic difference is given by the fact that in our treatment for the constituent I, a chemical potential is not defined and does not exist. If we wish to have a μ_I defined, then, by the aid of the inequalities (19) and (20) we would have to examine whether the system can also be constructed from the components NaCl, KCl, and KI. In the positive case transformation from the construction space to the constituent space is analogous to that given above. However, then a μ_{Cl} does not exist for the system. The corresponding reasoning can also be applied when the system is to be described in such a way that only μ_{Na} or μ_K exists, however both, μ_{Cl} and μ_I are physically defined.

Next we wish to give an expression for the Gibbs free energy in the "total solute - constituents" coordinate system. We multiply eqs.(29)-(31) by the volume of the system. This gives the component masses in terms of the total solute mass and the two constituent masses m_{Na} and m_{Cl}. These expressions we insert in eq.(34) to get:

$$G = m_W \mu_W^* + m_s \frac{1}{M_{KCl}} \left\{ \mu_{KCl}^* M_{KCl} - \mu_{NaCl}^* M_{NaCl} + \mu_{NaI}^* M_{NaI} \right\}$$

$$+ \frac{m_{Na}}{M_{Na}} \frac{M_{NaI}}{M_{KI}} \left\{ \mu_{NaCl}^* M_{NaCl} - \mu_{KCl}^* M_{KCl} + \mu_{NaI}^* (M_K - M_{Na}) \right\}$$

$$+ \frac{m_{Cl}}{M_{Cl}} \frac{M_{KCl}}{M_{KI}} \left\{ \mu_{NaCl}^* M_{NaCl} + \mu_{KCl}^* (M_I - M_{Cl}) - \mu_{NaI}^* M_{NaI} \right\}$$

This can be rewritten as

$$G = m_W \mu_W^* + m_s \mu_s^* + m_{Na} \mu_{Na}^* + m_{Cl} \mu_{Cl}^* \tag{41}$$

with

$$\mu_S^* = \frac{1}{M_{KI}}\left\{\mu_{KCl}^* M_{KCl} - \mu_{NaCl}^* M_{NaCl} + \mu_{NaI}^* M_{NaI}\right\} \tag{41a}$$

$$\mu_{Na}^* = \frac{1}{M_{Na}}\frac{M_{NaI}}{M_{KI}}\left\{\mu_{NaCl}^* M_{NaCl} - \mu_{KCl}^* M_{KCl} + \mu_{NaI}^*(M_K - M_{Na})\right\} \tag{41b}$$

$$\mu_{Cl}^* = \frac{1}{M_{Cl}}\frac{M_{KCl}}{M_{KI}}\left\{\mu_{NaCl}^* M_{NaCl} + \mu_{KCl}^*(M_I - M_{Cl}) - \mu_{NaI}^* M_{NaI}\right\} \tag{41c}$$

where μ_S^* is the specific chemical potential with respect to the total solute mass and μ_{Na}^*, μ_{Cl}^* are the specific chemical potentials with respect to the constituent masses of Na and Cl, respectively. Again μ_{Na}^* and μ_{Cl}^* are definitely not "ionic" chemical potentials; this becomes quite obvious from eq.(41) because this relation contains only two quantities referring to the constituent masses, whereas there are four constituents in the system. Again, the variables m_{Na} and m_{Cl} should be thought of as being internal parameters specifying the nature of the total solute mass m_S in more detail.

For completeness we add the corresponding formulas for a system which is only constructed from H_2O, NaCl and KCl. Now in eq.(34), $m_{NaI} = 0$. In order to write G in the constituent space, m_{NaCl} and m_{KCl} are taken from eqs.(3) and (5). Thus we obtain

$$G = m_W\mu_W^* + m_{Na}\frac{1}{M_{Na}}\mu_{NaCl}^* M_{NaCl} + m_K\frac{1}{M_K}\mu_{KCl}^* M_{KCl}$$

$$= m_W\tilde{\mu}_W + m_{Na}\tilde{\mu}_{Na} + m_K\tilde{\mu}_K$$

with

$$\tilde{\mu}_{Na} = \frac{1}{M_{Na}}\mu_{NaCl}^* M_{NaCl} \; ; \quad \tilde{\mu}_K = \frac{1}{M_K}\mu_{KCl}^* M_{KCl}$$

Another possibility is to write the analog of eq.(40)

$$G = n_W\mu_W + n_{Na}\mu_{Na} + n_K\mu_K$$

with

$$\mu_{Na} = M_{NaCl}\mu_{NaCl}^* \; , \quad \mu_K = M_{KCl}\mu_{KCl}^*$$

In order to obtain G in the total solute-constituents coordinate system we insert the eqs.(32),(33) in eq.(34) (after they have been multiplied by V). The result is

$$G = m_w \mu_w^* + m_{Na} \frac{M_{NaCl}}{M_{Na}} (\mu_{NaCl}^* - \mu_{KCl}^*) + m_s \mu_{KCl}^*$$

Thus we have

$$G = m_w \mu_w^* + m_s \mu_s^* + m_{Na} \mu_{Na}^* \tag{42}$$

with

$$\mu_s^* = \mu_{KCl}^* \tag{42a}$$

$$\mu_{Na}^* = \frac{1}{M_{Na}} M_{NaCl} (\mu_{NaCl}^* - \mu_{KCl}^*) \tag{42b}$$

Chapter 2

The multicomponent electrolyte solution in the non-equilibrium situation

2.1 The local mass conservation in a non-equilibrium system

Assume that at t = 0 the mixing of the components H_2O, KCl, NaCl, and NaI is completed such that the system is uniform. This means that we have

$$\frac{\partial \varrho^*_{NaCl}}{\partial x} = \frac{\partial \varrho^*_{KCl}}{\partial x} = \frac{\partial \varrho^*_{NaI}}{\partial x} = 0$$

where x is the position coordinate in the system. For simplicity we assume that the cross-section A of the vessel containing the solution is constant. So it suffices to take the spatial coordinate as the variable x. We have also

$$\frac{\partial \varrho^*_{NaCl}}{\partial t} = \frac{\partial \varrho^*_{KCl}}{\partial t} = \frac{\partial \varrho^*_{NaI}}{\partial t} = 0$$

and likewise all the composition variables in the constituent coordinate system and the total solute-constitutents coordinate system have vanishing derivatives with respect to the position and the time. However the concentrations of molecular species present in the solution may change with time. In our system we have certainly the chemical reaction

$$NaI + KCl \underset{k_2}{\overset{k_1}{\rightleftharpoons}} NaCl + KI \qquad \text{(I)}$$

where k_1 and k_2 are the rate constants for the reactions indicated. Molecular species are defined in the sense of eqs.(9)-(13). Of course, in the usual practical situation the reaction will be unmeasurably fast, however for the following it will be of great importance to clarify the physical situation in principle. Then in an uniform concentration distribution ($\partial c_i / \partial x$ = 0, i = NaI, KCl, NaCl, KI) deviating from equilibrium with respect to the molecular concentrations we have

$$\begin{aligned} \frac{\partial c_{NaI}}{\partial t} &= -k_1 c_{NaI}\left(c_{KCl} - c^0_{KCl}(c_{NaCl}, c_{NaI}, c_{KI})\right) \\ &\cdots\cdots\cdots\cdots \\ \frac{\partial c_{NaCl}}{\partial t} &= -k_2 c_{KI}\left(c_{NaCl} - c^0_{NaCl}(c_{KCl}, c_{NaI}, c_{KI})\right) \\ &\cdots\cdots\cdots\cdots \end{aligned} \qquad (43)$$

The c^0_j are the equilibrium concentrations of the molecular species j which is connected with the given concentrations c_k, c_l, c_m, of the other species. The equilibrium concentrations are related to each other by the quantity

$$f^* = \left[\frac{c_{NaCl} \cdot c_{KI}}{c_{NaI} \cdot c_{KCl}}\right]_{equilibr.} \tag{44}$$

(See, for instance, eq.(10) and the following formulas)
With eqs.(44), eq.(43) may be rewritten as

$$\frac{\partial c_{NaI}}{\partial t} = -k_2 \left(f^* c_{KCl} c_{NaI} - c_{KI} c_{NaCl}\right)$$

$$\cdots\cdots\cdots\cdots\cdots \tag{45}$$

$$\frac{\partial c_{NaCl}}{\partial t} = -k_2 \left(c_{KI} c_{NaCl} - f^* c_{NaI} c_{KCl}\right)$$

$$\cdots\cdots\cdots\cdots\cdots$$

Next we consider a system for which the distribution of molecular concentrations is no longer uniform, i.e. we have $\partial c_i/\partial x \neq 0$ (i = NaCl, KCl, NaI, KI). Then we observe that with a known set of quantities k_2 and f^* the $\partial c_i/\partial t$ (i = NaCl, KCl, . . .) are no longer given by eqs.(45). The difference between the observed concentration change and the one predicted by eqs.(45) we assign to the divergence of a molecular flux $\bar{j}_i$, i.e. we set

$$\left(\frac{\partial c_i}{\partial t}\right)_{obs.} - \left(\frac{\partial c_i}{\partial t}\right)_{eq(45)} = -\,div\,\bar{j}_i \qquad i = \text{NaCl, NaI, KCl, KI}$$

Then the total set of equations describing the change of molecular concentrations is

$$\frac{\partial c_{NaI}}{\partial t} = -\,div\,\bar{j}_{NaI} - k_2\left(f^* c_{KCl} c_{NaI} - c_{KI} c_{NaCl}\right) \tag{46a}$$

$$\frac{\partial c_{KCl}}{\partial t} = -\,div\,\bar{j}_{KCl} - k_2\left(f^* c_{KCl} c_{NaI} - c_{KI} c_{NaCl}\right) \tag{46b}$$

$$\frac{\partial c_{NaCl}}{\partial t} = -\,div\,\bar{j}_{NaCl} - k_2\left(c_{KI} c_{NaCl} - f^* c_{NaI} c_{KCl}\right) \tag{46c}$$

$$\frac{\partial c_{KI}}{\partial t} = -\,div\,\bar{J}_{KI} - k_2\left(c_{KI}\,c_{NaCl} - f^{*}c_{NaI}\,c_{KCl}\right) \qquad (46d)$$

Let us assume that our system is closed, then at the boundaries we have

$$\bar{J}_{NaI} = \bar{J}_{KCl} = \bar{J}_{NaCl} = \bar{J}_{KI} = 0 \qquad (47)$$

The fluxes occurring in eqs.(46a-47) are molecular fluxes. We may also work with molecular mass fluxes:

$$J_i = M_i\bar{J}_i \qquad i = \text{NaCl, KCl, NaI, KI,}$$

which gives, instead of eqs.(46)

$$\frac{\partial \rho_{NaI}}{\partial t} = -\,div\,J_{NaI} - k_2 M_{NaI}\left(f^{*}c_{KCl}\,c_{NaI} - c_{KI}\,c_{NaCl}\right)$$

$$\cdots\cdots\cdots \qquad (48)$$

$$\frac{\partial \rho_{NaCl}}{\partial t} = -\,div\,J_{NaCl} - k_2 M_{NaCl}\left(c_{KI}\,c_{NaCl} - f^{*}c_{NaI}\,c_{KCl}\right)$$

$$\cdots\cdots\cdots$$

$$J_{NaCl} = J_{NaI} = \ldots = 0 \quad \text{at the boundaries} \qquad (49)$$

Eqs. (46),(47) and eqs.(48),(49) represent the definition of the molecular and molecular mass fluxes, respectively in terms of the time dependent molecular concentration field. In practical situations the second term on the right-hand side of eqs.(46) and (48) can very often be neglected because the chemical reaction is very slow ($k_2 \rightarrow 0$) or very fast ($f^{*}c_i c_j - c_k c_l \rightarrow 0$) compared with the diffusion process. In the latter case the fourth equation, eq.(46d), is no longer an independent one. Now experience shows that the mass fluxes (or molecular fluxes) vanish wherever we have $\partial c_i/\partial x$ = 0. Therefore the fluxes may be written as linear functions of concentration gradients:

$$\bar{J}_i = -\sum_j \bar{D}_{ij}\frac{\partial c_j}{\partial x} \qquad i,j = \text{NaI, KCl, NaCl....} \qquad (50)$$

$$J_i = -\sum_j D_{ij} \frac{\partial \rho_j}{\partial x} \tag{51}$$

The concentration gradient of water is not independent because we have

$$1 = c_{NaCl}\bar{v}_{NaCl} + c_{KCl}\bar{v}_{KCl} + c_{NaI}\bar{v}_{NaI} + c_{KI}\bar{v}_{KI}$$

or

$$1 = \rho_{NaCl}\bar{V}_{NaCl} + \rho_{KCl}\bar{V}_{KCl} + \rho_{NaI}\bar{V}_{NaI} + \rho_{KI}\bar{V}_{KI}$$

where $\bar{v}_i$ and $\bar{V}_i$ are the partial molar volumes or partial specific volumes, respectively. Combination of eqs.(46) and (50) or (48) and (51) gives the set of general diffusion equations governing the multicomponent mixing process with the inclusion of the chemical reaction (I). All these equations involve our definition of the molecular species, either as a formal relation (eq.(10)) or in the form of an appropriate experimental operation. The chemical reaction terms on the right-hand side of eqs.(46) and (48) are characterized by the two parameters k_2 and f^* which are well-defined in a uniform system. Of course, such a uniform system is a closed one and consequently we have internal mass conservation during the process. This can be applied to our eqs. (46) and (48). If the partial mass density of one of the molecular species changes, for example $\partial\rho_{NaI}/\partial t > 0$, then the increase of NaI-mass per volume element is supplied partly by the mass flux $J_{NaI}(x)$ through the walls of the volume element and partly by the material within the volume element. In fact, part of the molecular species NaCl and KI are used up in this way. Thus, all processes contributing to the conservation of mass for a given process occur within the volume element, in other words, they are of short range character. So we may speak of the occurrence of a local mass conservation during the process of production of a (or several) molecular species. We shall see below that a delocalized mass conservation can also occur in the system when suitable conditions are established. The demonstration of this fact was the main purpose for writing eqs.(46) and (48).

2.2 Description of the diffusion process in a multicomponent electrolyte solution.

Now we add eqs.(46a) and (46c), eqs.(46b) and (46c), and eqs.(46b) and (46d), the result is:

$$\frac{\partial c_{Na}}{\partial t} = -\operatorname{div}\bar{J}_{Na} = -\operatorname{div}(\bar{J}_{NaCl} + \bar{J}_{NaI})$$

$$\frac{\partial c_{Cl}}{\partial t} = -\operatorname{div}\bar{J}_{Cl} = -\operatorname{div}(\bar{J}_{NaCl} + \bar{J}_{KCl}) \tag{52}$$

$$\frac{\partial c_{K}}{\partial t} = -\operatorname{div}\bar{J}_{K} = -\operatorname{div}(\bar{J}_{KCl} + \bar{J}_{KI})$$

These equations give the change with time of the constituent concentrations c_{Na}, c_{Cl}, c_K. The change with time of the fourth constituent concentration is given by the time derivative of eq.(2a)

$$\frac{\partial c_{Na}}{\partial t} + \frac{\partial c_K}{\partial t} = \frac{\partial c_{Cl}}{\partial t} + \frac{\partial c_I}{\partial t}$$

$\bar{J}_{Na}$, $\bar{J}_{Cl}$, $\bar{J}_K$ are the molar constituent fluxes. Of course, now the chemical reaction terms have dropped out, indeed, the chemical reaction merely recombines the constituents; however it does not change the amount of constituent present in the volume element. The constituent fluxes are connected with the gradients of the constituent concentrations:

$$-\bar{J}_i = \frac{M_{Na}}{M_i} D^*_{iNa}\frac{\partial c_{Na}}{\partial x} + \frac{M_K}{M_i} D^*_{iK}\frac{\partial c_K}{\partial x} + \frac{M_{Cl}}{M_i} D^*_{iCl}\frac{\partial c_{Cl}}{\partial x} \tag{53}$$

i= Na, K, Cl

The number of terms is now reduced from four to three because the gradient of the fourth constituent is not independent. We have

$$\frac{\partial c_{Na}}{\partial x} + \frac{\partial c_K}{\partial x} = \frac{\partial c_{Cl}}{\partial x} + \frac{\partial c_I}{\partial x}$$

The D^*_{ij} are the constituent mutual diffusion coefficients, they represent a 3 x 3 matrix. Of course, due to the addition operation by which they were obtained, they may be expressed as functions of the molecular mutual diffusion coefficients and these functions also involve the relations defining the molecular species. However since the constituent concentrations are themselves well-defined observables, the matrix has a physical meaning itself.

It is sometimes more convenient to treat diffusion problems in terms of specific mass densities, thus we multiply eq.(52) and (53) by the atomic masses to get:

$$\frac{\partial \rho_i}{\partial t} = -\, div\, j_i \qquad i = \text{Na, K, Cl} \quad (54)$$

and

$$\frac{\partial \rho_i}{\partial t} = div\left(D^*_{iNa}\frac{\partial \rho_{Na}}{\partial x} + D^*_{iK}\frac{\partial \rho_K}{\partial x} + D^*_{iCl}\frac{\partial \rho_{Cl}}{\partial x}\right) \qquad (54a)$$

$$i = \text{Na, K, Cl}$$

where j_{Na}, j_K, and j_{Cl} are the three independent constituent mass fluxes; they are defined by eq.(54), together with the boundary conditions

$$j_{Na} = j_K = j_{Cl} = 0 \qquad \text{at} \qquad x = 0,\; x = \ell$$

This definition is based on the fact that the ρ_i's are observable quantities. Eq.(53) has been written in such a form that a redefinition of the diffusion coefficients is not necessary.

Eqs.(54) represent the set of diffusion equations for a four consitutents electrolyte solution. In addition to the three eqs.(54) one other equation could be written for the solvent. However this fourth equation is not independent as long as the volume of the system remains constant during the diffusion process. This leads to the requirement that the partial specific volumes do not depend on the composition of the solution. Of course this will never be fulfilled strictly. However for our treatment the discussion of the volume change on mixing can be omitted.

2.3 Description of multicomponent electrolyte diffusion in various property spaces.

We are now ready to apply the coordinate transformations defined in chapter 1 to our diffusion equations (54). First we transform to the component or construction space. We apply eqs.(21a)-(23a) and find

$$\frac{\partial \rho^*_{KCl}}{\partial t} = \frac{M_{KCl}}{M_K}\frac{\partial \rho_K}{\partial t} = -\, div\, \tilde{j}_{KCl} = -\, div\, \frac{M_{KCl}}{M_K} j_K \qquad (55)$$

$$\frac{\partial \rho^*_{NaCl}}{\partial t} = M_{NaCl}\left(\frac{1}{M_{Cl}}\frac{\partial \rho_{Cl}}{\partial t} - \frac{1}{M_K}\frac{\partial \rho_K}{\partial t}\right)$$

$$= -\, div\, \tilde{j}_{NaCl} = -\, div\left(\frac{M_{NaCl}}{M_{Cl}} j_{Cl} - \frac{M_{NaCl}}{M_K} j_K\right) \qquad (56)$$

$$\frac{\partial \rho^*_{NaI}}{\partial t} = M_{NaI}\left(\frac{1}{M_{Na}}\frac{\partial \rho_{Na}}{\partial t} + \frac{1}{M_K}\frac{\partial \rho_K}{\partial t} - \frac{1}{M_{Cl}}\frac{\partial \rho_{Cl}}{\partial t}\right)$$
$$= -\operatorname{div}\tilde{j}_{NaI} = -\operatorname{div}\left(\frac{M_{NaI}}{M_{Na}} j_{Na} + \frac{M_{NaI}}{M_K} j_K - \frac{M_{NaI}}{M_{Cl}} j_{Cl}\right) \quad (57)$$

The three fluxes $\tilde{j}_{KCl}$, $\tilde{j}_{NaCl}$, and $\tilde{j}_{NaI}$ are the component mass fluxes and they are defined by the left-hand sides of eqs.(55)-(57) and the requirement that these fluxes vanish at the boundaries. The physical significance of the set of observables $\partial \rho^*_{NaCl}/\partial t$, $\partial \rho^*_{KCl}/\partial t$, and $\partial \rho^*_{NaI}/\partial t$ is as follows. Add to a solution (ρ^*_{NaCl}, ρ^*_{KCl}, ρ^*_{NaI}) such amounts of the three components NaCl, KCl, and NaI that the increase of the partial component mass densities is the same as $d\rho^*_{NaCl}$ = $(\partial \rho^*_{NaCl}/\partial t)dt$, ... according to eqs.(55)-(57). This construction operation defines the local composition change of the solution. Of course, also an analytical field in the construction space is defined, and now the component mass fluxes can be connected to the gradients in this space.

$$-\tilde{j}_i = D_{iNaCl}\frac{\partial \rho^*_{NaCl}}{\partial x} + D_{iNaI}\frac{\partial \rho^*_{NaI}}{\partial x} + D_{iKCl}\frac{\partial \rho^*_{KCl}}{\partial x}$$

i = NaCl, NaI, KCl.

In the limit $\rho^*_{NaI} \rightarrow 0$ (i.e. $\rho_I \rightarrow 0$) we must have

$$D_{NaI,NaCl} = D_{NaI,KCl} = 0$$

The diffusion coefficients are the mutual diffusion coefficients in the component space; they are linear functions of the diffusion coefficients in the constituent space:

$$D_{NaCl,NaCl} = \frac{M_{Na}}{M_{Cl}} D_{Cl,Na} - \frac{M_{Na}}{M_K} D_{K,Na} + D_{Cl,Cl} - \frac{M_{Cl}}{M_K} D_{K,Cl}$$
$$D_{NaCl,NaI} = \frac{M_{NaCl}}{M_{NaI}}\left(\frac{M_{Na}}{M_{Cl}} D_{Cl,Na} - \frac{M_{Na}}{M_K} D_{K,Na}\right)$$
$$D_{NaCl,KCl} = \frac{M_{NaCl}}{M_{KCl}}\left(D_{Cl,Cl} - \frac{M_{Cl}}{M_K} D_{K,Cl} + \frac{M_K}{M_{Cl}} D_{Cl,K} - D_{K,K}\right)$$

$$D_{NaI,NaCl} = \frac{M_{NaI}}{M_{NaCl}}\left(D_{Na,Na} + \frac{M_{Na}}{M_K} D_{K,Na} - \frac{M_{Na}}{M_{Cl}} D_{Cl,Na} + \right.$$

$$+ \frac{M_{Cl}}{M_{Na}} D_{Na,Cl} + \frac{M_{Cl}}{M_K} D_{KCl} - D_{Cl,Cl} \Big)$$

$$D_{NaI,NaI} = D_{Na,Na} + \frac{M_{Na}}{M_K} D_{K,Na} - \frac{M_{Na}}{M_{Cl}} D_{Cl,Na}$$

$$D_{NaI,KCl} = \frac{M_{NaI}}{M_{KCl}} \Big(\frac{M_{Cl}}{M_{Na}} D_{Na,Cl} + \frac{M_{Cl}}{M_K} D_{K,Cl} - D_{Cl,Cl}$$

$$+ \frac{M_K}{M_{Na}} D_{Na,K} + D_{K,K} - \frac{M_K}{M_{Cl}} D_{Cl,K}$$

$$D_{KCl,NaCl} = \frac{M_{KCl}}{M_{NaCl}} \Big(\frac{M_{Na}}{M_K} D_{K,Na} + \frac{M_{Cl}}{M_K} D_{K,Cl} \Big)$$

$$D_{KCl,NaI} = \frac{M_{KCl}}{M_{NaI}} \cdot \frac{M_{Na}}{M_K} D_{K,Na}$$

$$D_{KCl,KCl} = \frac{M_{Cl}}{M_K} D_{K,Cl} + D_{K,K}$$

The situation becomes much simpler if we consider only a mixture of NaCl, KCl, and water. Now the partial density of the constituent I is exactly zero. This is a ternary diffusion problem; it has two solute components and three solute constituents. We have

$$\frac{\partial \rho^*_{NaCl}}{\partial t} = - \operatorname{div} \tilde{j}_{NaCl} = - \operatorname{div} \frac{M_{NaCl}}{M_{Na}} j_{Na}$$

$$\frac{\partial \rho^*_{KCl}}{\partial t} = - \operatorname{div} \tilde{j}_{KCl} = - \operatorname{div} \frac{M_{KCl}}{M_K} j_K$$

and

$$-\tilde{j}_i = D_{i,NaCl} \frac{\partial \rho^*_{NaCl}}{\partial x} + D_{i,KCl} \frac{\partial \rho^*_{KCl}}{\partial x}$$

i = NaCl, KCl

Further, the transformation relations of the mutual diffusion coefficients are

$$D_{NaCl,NaCl} = D_{Na,Na} \; ; \qquad D_{NaCl,KCl} = \frac{M_{NaCl}}{M_{KCl}} \frac{M_K}{M_{Na}} D_{Na,K}$$

$$D_{KCl,NaCl} = \frac{M_{KCl}}{M_{NaCl}} \frac{M_{Na}}{M_K} D_{K,Na} \; ; \qquad D_{KCl,KCl} = D_{K,K}$$

In Chapter 1 we introduced a third type of property space within which the composition of the system can be described. This is the total solute-constituents space. Of course, the diffusion process can also be treated in this representation. We shall use this mode of description very often in the following chapters.

From equations (24a) and (2) we have

$$\frac{\partial \rho_s}{\partial t} = \left(1 + \frac{M_I}{M_{Na}}\right)\frac{\partial \rho_{Na}}{\partial t} + \left(1 + \frac{M_I}{M_K}\right)\frac{\partial \rho_K}{\partial t} + \left(1 - \frac{M_I}{M_{Cl}}\right)\frac{\partial \rho_{Cl}}{\partial t} \tag{58}$$

where $\partial \rho_s / \partial t$ is the local change with time of the total solute mass density. This leads to the definition of the total solute mass flux:

$$\frac{\partial \rho_s}{\partial t} = -\operatorname{div} j_s \tag{59}$$

$$j_s = 0 \qquad \text{at } x = 0 \text{ and } x = x_l \tag{60}$$

According to eq.(58), j_s is connected with the three constituent mass fluxes

$$j_s = j_{Na}\left(1 + \frac{M_I}{M_{Na}}\right) + j_K\left(1 + \frac{M_I}{M_K}\right) + j_{Cl}\left(1 - \frac{M_I}{M_{Cl}}\right)$$

It follows that j_K and j_I are also known when j_s, j_{Na}, and j_{Cl} are given:

$$j_K = \frac{j_s - j_{Na}(1 + M_I/M_{Na}) - j_{Cl}(1 - M_I/M_{Cl})}{(1 + M_I/M_K)} \tag{61}$$

$$j_I = j_s - j_{Na} - j_K - j_{Cl} \tag{62}$$

In the total solute-constituents space a 3 x 3 square matrix of the mutual diffusion coefficients exists also. We have

$$j_i = \tilde{D}_{i,s}\frac{\partial \rho_s}{\partial x} + \tilde{D}_{i,Na}\frac{\partial \rho_{Na}}{\partial x} + \tilde{D}_{i,Cl}\frac{\partial \rho_{Cl}}{\partial x} \tag{63}$$

$$i = s, \text{Na}, \text{Cl}$$

These diffusion coefficients are linear functions of the mutual diffusion coefficients in other coordinate systems; for example, the relations between $D_{i,j}$ and $\tilde{D}_{i,j}$, i,j = Na, K, Cl are:

$$\tilde{D}_{S,S} = \frac{\beta_1}{\beta_2} D_{Na,K} + D_{K,K} + \frac{\beta_3}{\beta_2} D_{Cl,K}$$

$$\tilde{D}_{S,Na} = \beta_1 \left(D_{Na,Na} - \frac{\beta_1}{\beta_2} D_{Na,K} - \frac{\beta_3}{\beta_2} D_{Cl,K} - D_{K,K} \right) + \beta_2 D_{K,Na} + \beta_3 D_{Cl,Na}$$

$$\tilde{D}_{S,Cl} = \beta_3 \left(- \frac{\beta_1}{\beta_3} D_{Na,K} + D_{Cl,Cl} - D_{K,K} - \frac{\beta_3}{\beta_2} D_{Cl,K} \right) + \beta_1 D_{Na,Cl} + \beta_2 D_{K,Cl}$$

$$\tilde{D}_{Na,S} = \frac{D_{Na,K}}{\beta_2} \quad ; \quad \tilde{D}_{Na,Na} = D_{Na,Na} - \frac{\beta_1}{\beta_2} D_{Na,K}$$

$$\tilde{D}_{Na,Cl} = D_{Na,Cl} - \frac{\beta_3}{\beta_2} D_{Na,K}$$

$$\tilde{D}_{Cl,S} = \frac{D_{Cl,K}}{\beta_2} \quad ; \quad \tilde{D}_{Cl,Na} = D_{Cl,Na} - \frac{\beta_1}{\beta_2} D_{Cl,K}$$

$$\tilde{D}_{Cl,Cl} = D_{Cl,Cl} - \frac{\beta_3}{\beta_2} D_{Cl,K}$$

with

$$\beta_1 = 1 + \frac{M_I}{M_{Na}} \quad ; \quad \beta_2 = 1 + \frac{M_I}{M_K} \quad ; \quad \beta_3 = 1 - \frac{M_I}{M_{Cl}}$$

From eqs.(59) and (63) we obtain the complete set of diffusion equations in the total solute-constituents space.

$$\frac{\partial \rho_S}{\partial t} = \mathrm{div} \left(\tilde{D}_{S,S} \frac{\partial \rho_S}{\partial x} + \tilde{D}_{S,Na} \frac{\partial \rho_{Na}}{\partial x} + \tilde{D}_{S,Cl} \frac{\partial \rho_{Cl}}{\partial x} \right) \qquad (64)$$

$$\frac{\partial \rho_{Na}}{\partial t} = \mathrm{div} \left(\tilde{D}_{Na,S} \frac{\partial \rho_S}{\partial x} + \tilde{D}_{Na,Na} \frac{\partial \rho_{Na}}{\partial x} + \tilde{D}_{Na,Cl} \frac{\partial \rho_{Cl}}{\partial x} \right) \qquad (65)$$

$$\frac{\partial \rho_{Cl}}{\partial t} = \mathrm{div} \left(\tilde{D}_{Cl,S} \frac{\partial \rho_S}{\partial x} + \tilde{D}_{Cl,Na} \frac{\partial \rho_{Na}}{\partial x} + \tilde{D}_{Cl,Cl} \frac{\partial \rho_{Cl}}{\partial x} \right) \qquad (66)$$

If the diffusion coefficients are constant, i.e. $\partial \tilde{D}_{i,j} / \partial x = 0$, then eqs.(64)-(66) are the generalized Fick's second law equations. Now let us consider the limit $\rho_{Cl} \longrightarrow 0$. In this situation we must have $\tilde{D}_{Cl,S} = \tilde{D}_{Cl,Na} = 0$.

On the other hand, if $\rho_I \longrightarrow 0$, then, since $j_I \longrightarrow 0$ the coupling between j_I and $\partial\rho_{Na}/\partial x$ and $\partial\rho_{Cl}/\partial x$ must vanish. Then application of eqs.(24a) and (27) together with eqs.(64)-(66) yields

$$\frac{\partial\rho_I}{\partial t} = \frac{1}{M_{KCl}}\left[M_K\tilde{D}_{S,S} + (M_{Na}-M_K)\tilde{D}_{Na,S} - M_{KCl}\tilde{D}_{Cl,S}\right]\frac{\partial^2\rho_S}{\partial x^2}$$

and the diffusion coefficients must fulfil the condition

$$M_K\tilde{D}_{S,Na} + (M_{Na}-M_K)\tilde{D}_{Na,Na} - M_{KCl}\tilde{D}_{Cl,Na} = 0$$

$$M_K\tilde{D}_{S,Cl} + (M_{Na}-M_K)\tilde{D}_{Na,Cl} - M_{KCl}\tilde{D}_{Cl,Cl} = 0$$

if

$$\rho_I \longrightarrow 0$$

If ρ_I is exactly zero, then we have only a ternary system, i.e. the solution is constructed from the two solutes NaCl and KCl and water. Now the diffusion equations are

$$\frac{\partial\rho_S}{\partial t} = -\,\mathrm{div}\,j_S = \mathrm{div}\left(\tilde{D}_{S,S}\frac{\partial\rho_S}{\partial x} + \tilde{D}_{S,Na}\frac{\partial\rho_{Na}}{\partial x}\right) \qquad (67)$$

$$\frac{\partial\rho_{Na}}{\partial t} = -\,\mathrm{div}\,j_{Na} = \mathrm{div}\left(\tilde{D}_{Na,S}\frac{\partial\rho_S}{\partial x} + \tilde{D}_{Na,Na}\frac{\partial\rho_{Na}}{\partial x}\right) \qquad (68)$$

$$\rho_S = \rho_{Na} + \rho_{Cl} + \rho_K$$

$$j_K = \left[j_S - j_{Na}\left(1 + \frac{M_{Cl}}{M_{Na}}\right)\right]\left(1 + \frac{M_{Cl}}{M_K}\right)^{-1}$$

$$j_{Cl} = j_S - j_K - j_{Na}$$

The transformation relations of the diffusion coefficients from the constituents coordinate system are

$$\tilde{D}_{S,S} = \frac{\beta_1'}{\beta_3'}D_{Na,K} + D_{K,K}$$

$$\tilde{D}_{S,Na} = \beta_1' D_{Na,Na} - \frac{\beta_1'^2}{\beta_2'}D_{Na,K} + \beta_2' D_{K,Na} - \beta_1' D_{K,K}$$

$$\tilde{D}_{Na,S} = \frac{1}{\beta_2'}D_{Na,K}$$

$$\tilde{D}_{Na,Na} = D_{Na,Na} - \frac{\beta_1'}{\beta_2'}D_{Na,K}$$

with

$$\beta_1' = 1 + \frac{M_{Cl}}{M_{Na}} ; \beta_2' = 1 + \frac{M_{Cl}}{M_K}$$

A very important experiment which may be mentioned briefly here is the measurement of the tracer or self-diffusion coefficient. For this case we replace ρ_K by ρ_{Na^*}, where ρ_{Na^*} is the sodium tracer partial mass density. Then, in an ordinary experiment we have $\rho_{Na^*} \to 0$ and consequently $D_{Na^*,Na} = 0$. Thus the diffusion equations in the constituent space are

$$\frac{\partial \rho_{Na}}{\partial t} = D_{Na,Na} \frac{\partial^2 \rho_{Na}}{\partial x^2} + D_{Na,Na^*} \frac{\partial^2 \rho_{Na^*}}{\partial x^2}$$

$$\frac{\partial \rho_{Na^*}}{\partial t} = D_{Na^*,Na^*} \frac{\partial^2 \rho_{Na^*}}{\partial x^2}$$

whereas in the ρ_s, ρ_{Na} representation the above transformation relations together with eqs.(67) and (68) yield:

$$\frac{\partial \rho_s}{\partial t} = D_{Na^*,Na^*} \frac{\partial^2 \rho_s}{\partial x^2} + \beta_1' \left(D_{Na,Na} - D_{Na^*,Na^*} - \frac{\beta_1'}{\beta_2'} D_{Na,Na^*} \right) \frac{\partial^2 \rho_{Na}}{\partial x^2}$$

$$\frac{\partial \rho_{Na}}{\partial t} = \frac{1}{\beta_2'} D_{Na,Na^*} \frac{\partial^2 \rho_s}{\partial x^2} + \left(D_{Na,Na} - \frac{\beta_1'}{\beta_2'} D_{Na,Na^*} \right) \frac{\partial^2 \rho_{Na}}{\partial x^2}$$

with

$$\beta_1' = 1 + \frac{M_{Cl}}{M_{Na}} ; \quad \beta_2' = 1 + \frac{M_{Cl}}{M_{Na^*}}$$

D_{Na^*,Na^*} is the tracer diffusion coefficient of Na in a NaCl solution; it will be seen that it also approximately governs the rate of change of the total solute mass density.

2.4 The binary diffusion process and some additional remarks

In the next step we consider the ordinary binary diffusion problem. Let us assume that we have the system NaCl + H_2O. Then the total solute partial mass density is equal to the NaCl density and we have

$$\frac{\partial \rho_s}{\partial t} = div \left(D_s \frac{\partial \rho_s}{\partial x} \right)$$

with $\rho_s = \rho_{NaCl}$. The constituent mass densities are of course dependent quantities

$$\rho_{Na} = \frac{M_{Na}}{M_{NaCl}} \cdot \rho_s = \frac{M_{Na}}{M_{NaCl}} \rho_{NaCl}$$

$$\rho_{Cl} = \frac{M_{Cl}}{M_{NaCl}} \cdot \rho_s = \frac{M_{Cl}}{M_{NaCl}} \rho_{NaCl}$$

So far we have described coordinate transformations only in an abstract space which we have called property space; the properties in question being various observables defining the composition of the system. However in the literature the coordinate system transformations which are treated in connection with diffusion problems refer to ordinary space and the various coordinate systems are distinguished by their velocities relative to one another.

In this sense all the mutual diffusion coefficients $\mathcal{D}_{ij}$ given so far are diffusion coefficients in the coordinate system given by the measuring apparatus, i.e. they are cell-fixed quantities. If there is no volume change during diffusion they are at the same time the diffusion coefficients in the so-called[1] volume-fixed reference frame. As already mentioned, for simplicity, we shall specialize our treatment to the case of negligible volume change. In this situation the change of the partial density of the solvent with time, $\partial\rho_w/\partial t$ is completely determined by our diffusion equations, for instance by the three eqs. (64)-(66). Then we have for the mass flux of the solvent:

$$\frac{\partial \rho_w}{\partial t} = - \operatorname{div} j_w$$

and

$$- j_w = \mathcal{D}_{w,s} \frac{\partial \rho_s}{\partial x} + \mathcal{D}_{w,Na} \frac{\partial \rho_{Na}}{\partial x} + \mathcal{D}_{w,Cl} \frac{\partial \rho_{Cl}}{\partial x}$$

where, as may be shown the diffusion coefficients $\mathcal{D}_{w,i}$ are given by the relations:

$$-\mathcal{D}_{w,i} = \frac{\bar{V}_s}{\bar{V}_w} \mathcal{D}_{s,i} + \frac{\bar{V}_{Na}}{\bar{V}_w} \mathcal{D}_{Na,i} + \frac{\bar{V}_{Cl}}{\bar{V}_w} \mathcal{D}_{Cl,i}$$

$$i = s, \text{Na}, \text{Cl}$$

where the $\bar{V}_i$ are the partial specific volumes with respect to the variables m_s, m_{Na}, m_{Cl}. It might be briefly mentioned here that occasionally mixed solvents have been used for the study of electrolyte diffusion problems. Such solvents have been used for transference number measurements with the intention of obtaining information about preferential solvation of ions. The inclusion of a second solvent in our diffusion equations should be a straightforward procedure.

It is unnecessary within the framework of the present treatment to discuss any details of transformation to those coordinate systems which have been of considerable interest in the past. These are the coordinate systems which are at rest relative to the local centre of mass, and the other ones which are at rest relative to the solvent (we have already mentioned that the third "classical" coordinate system, the reference frame at rest relative to the local centre of volume is identical with the cell-fixed system when the volume change can be neglected). It is only of interest here to give a definition for the velocity of the local centre of mass (or the local mean velocity)

$$u = \frac{j_s + j_w}{\rho} \tag{69}$$

where ρ is the total mass density at a given point in the system, $\rho = \rho_w + \rho_s$. Since the mass fluxes j_s and j_w according to the equations given above are functions of the diffusion coordinate x and the time, we have $u = u(x,t)$.

During the diffusion process the centre of mass of the system moves and the walls of the container exert an external force on the system. The local mean velocity u as defined by eq.(69) has finite values in those regions where the concentration gradients differ from zero, and thus, as a consequence of the equation of motion [1])

$$\rho \frac{du}{dt} = - \mathrm{Div}\, \sigma \tag{70}$$

the stress tensor σ is also a function of x and t which leads to finite forces at the boundaries. Of course these forces are very small.

Finally we mention that the analytical field expressed, for example, by the three variables $\rho_s(x,t)$, $\rho_{Na}(x,t)$ and $\rho_{Cl}(x,t)$ can also be transformed to a description which involves the specific chemical potentials. In fact, by application of equations (41) the gradients $\partial\rho_i/\partial x$ can be transformed to the gradients $\partial\mu_i^*/\partial x$ (i = s,Na,Cl). Then a redefinition of the diffusion coefficients is required which then are very often denoted by the symbols Ω_{ij}, and they are termed phenomenological coefficients, although the original use of this name has another source. Details of this treatment need not be given here.[1])

In concluding this chapter we give two schematic examples of a multicomponent electrolyte diffusion process. Fig. 2a - c describes the general situation where at each position, all four constituents are present. At x = 0 and x = 1 we have vanishing gradients of the quantities characterizing the analytical field, and also on both sides of the system all three components are present. In Fig.2a the diffusion process is described in component or construction space. The arrows give a schematic representation of the various component mass fluxes. These fluxes are local quantities and we have chosen them at the position x_o as indicated by the dashed line. It may be seen that there is a NaI (component) mass flux from the right to the left, the KCl (component) mass flux also is directed from the right to the left, however the NaCl (component) mass flux is in the opposite direction. Of course, if other compositions on the left- and right-hand side had been chosen, then the fluxes could have other directions. In Fig.2b the same system is depicted, however now the description occurs in the constituents space defined by the three constituents Na, K, Cl. It will be seen that the constituent mass fluxes j_{Cl} and j_{Na} are both directed from left to right, the constituent mass of K flows in the opposite direction. If we are interested in the "ionic" fluxes, then the directions are the same as the constituent mass fluxes, however the magnitudes would differ according to the different atomic weights. In Fig.2c the total solute mass flux is given instead of the K constituent mass flux. It will be seen that it has the same direction as the K mass flux, however the magnitude differs.

In Fig. 3a-c a special situations exists. Here at t = 0 the left-hand part only contains the component KCl, the right-hand part only contains NaI. Inspection of the relations (17b) and (18b) tells us that at t = 0 and $t \rightarrow \infty$ the system can be considered to be constructed from NaI, NaCl, and KCl; here the equalities are valid. But the construction scheme schould also be applicable during the mixing process. The question is in which kind of components space can the system be described for times $0 < t < \infty$? In Fig. 3a we have drawn a distribution of the component NaCl which implies that eqs.(17b) and (18b) are fulfilled, i.e. the construction from NaI, KCl, and NaCl is acceptable. Should this assumption be incorrect, then the NaCl distribution around $x = x_o$ has to be replaced by a "KI distribution". For large times the NaCl distribution (or the KI distribution respectively) disappears again. So we see here that the mapping of a non-equilibrium system on an equilibrium system is not a priori unique. If the constituent mutual diffusion

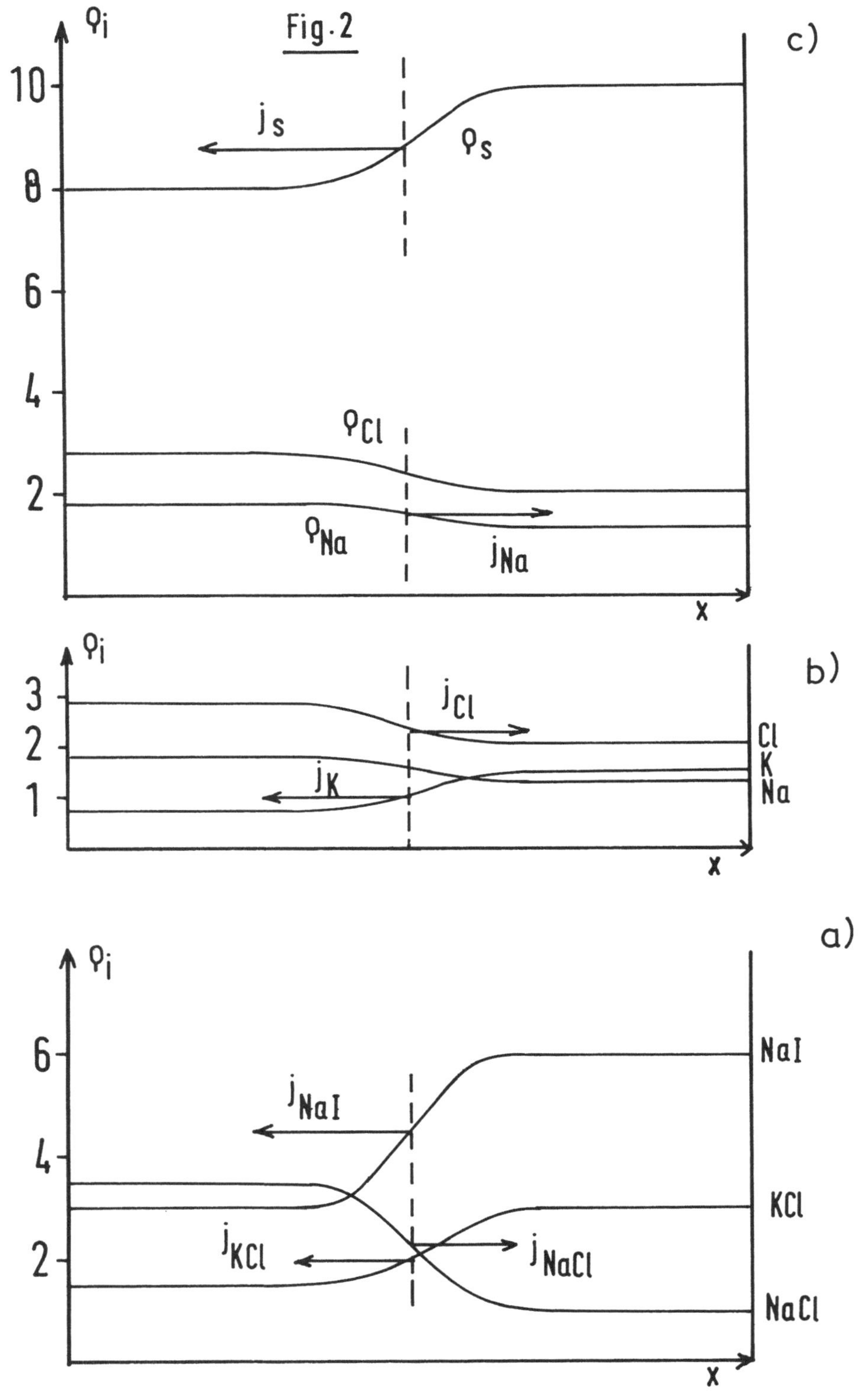
Fig. 2
c)
ρi
10
8
6
4
2
js
ρs
ρCl
ρNa
jNa
x
b)
ρi
3
2
1
jCl
jK
Cl
K
Na
x
a)
ρi
6
4
2
jNaI
jKCl
jNaCl
NaI
KCl
NaCl
x

Fig. 3

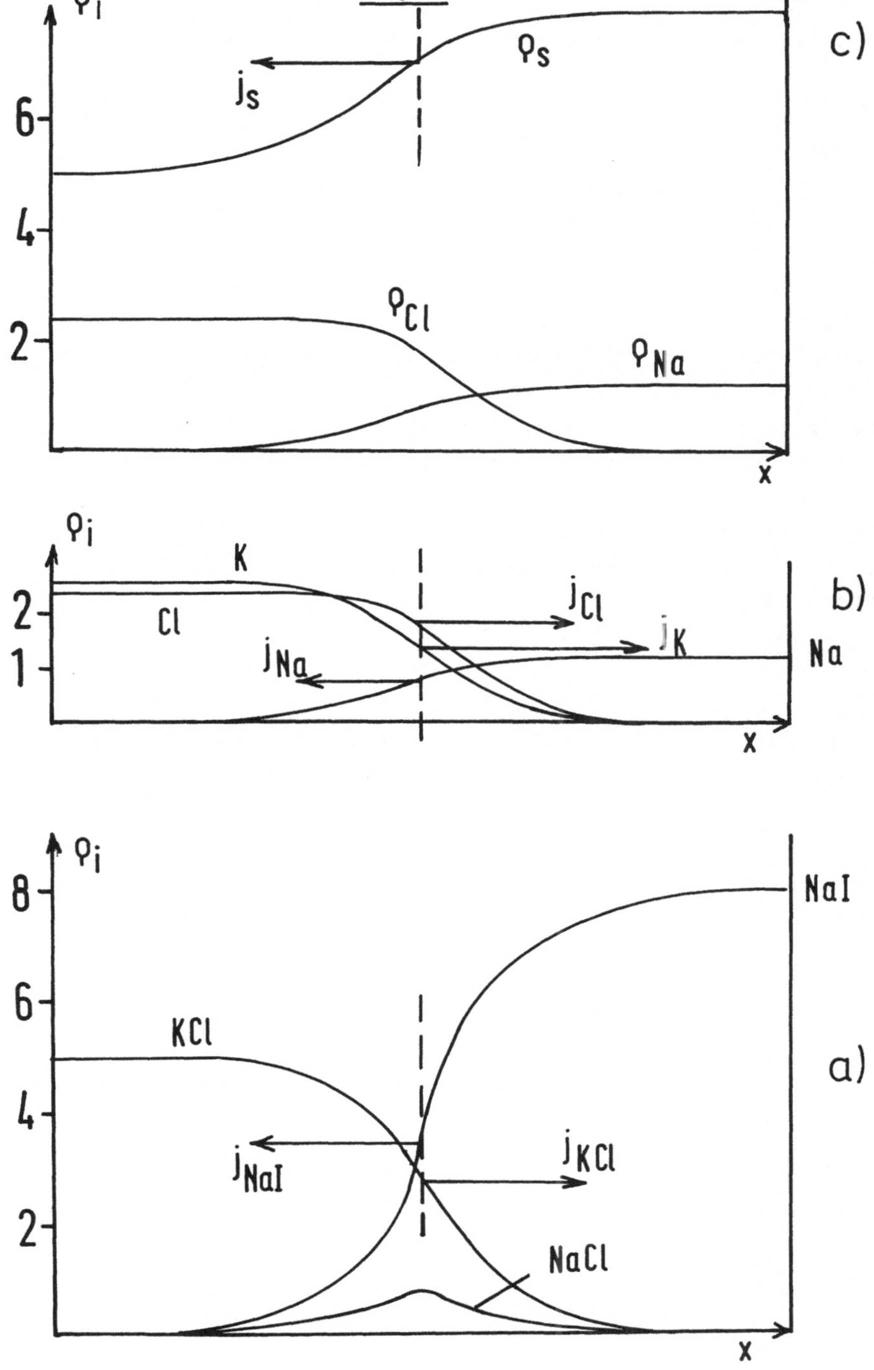

ρi
6
4
2
js
ρs
ρCl
ρNa
x
c)
ρi
K
2
1
Cl
jCl
jK
jNa
Na
x
b)
ρi
8
6
4
2
KCl
NaI
jNaI
jKCl
NaCl
x
a)

Fig. 2: Schematic representation of a diffusion process in multicomponent electrolyte solutions.
a) Representation in construction space, b) in constituents space, and c) in total solute-two constituents space.

Fig. 3: The diffusive mixing of the two binary electrolyte solutions KCl and NaI. The process is described a) in the construction space, b) in the constituents space, and c) in the total solute-two constituents space.

coefficients are known, then they can be transformed to the NaI, KCl, NaCl or to the NaI, KCl, KI construction scheme. The particular coordinate system, for which all partial component densities retain non-negative values throughout the total diffusion process, is the correct one for describing diffusive mixing. In Fig.3b and 3c the same diffusion process is shown in the constituents and total solute-constituents coordinate system, respectively. Here the complication of the non-uniqueness of the representation does not occur. ρ_s is uniquely given if the independent constituent partial densities and the kind of the fourth constituent are known.

Reference

1) See any text book of irreversible thermodynamics, for example D.D. Fitts, Nonequilibrium Thermodynamics, MCGraw-Hill, New York 1962

Chapter 3

The diffusion system in the presence of an electric current.

3.1 Local rates of change of composition in the presence of an electric current.

First we consider a system which has no electrodes but contains an electrolyte solution in which an electric current can exist. This in effect means that the electrodes are a very great distance away from the cell which we are observing. Alternatively we could have a closed loop of electrolyte solution, parts of which we can move through a static magnetic field. As we shall see below, the characterization of the system as a closed loop is a more fundamental picture than a cell construction where the electrodes are at very large distances from the section being observed. Assume now that we have switched on an electric current $\mathcal{J}_z$ by some device. The electric current can be identified by a magnetic field around the cell. When the electric current is switched on let it be constant in time. Since electric work is performed on the system, in order to keep the temperature constant, heat has to be continuously delivered to the surroundings. Of course, the electric current is constant along the total system. For simplicity we assume that the cross-section A of the tube containing the electrolyte solution is constant. j_z is the electric current density and we have

$$\mathrm{div}\, j_z = 0 \tag{71}$$

Let the electrolyte solution be the same as previously in that it contains the four constitutents, Na, K, Cl, and I, and as before, the composition is not constant everywhere along the coordinate x which gives the location in the cell. The positive x direction is the same as the positive direction of the vector j_z . For the description of the composition we apply the total solute-constituents coordinate system.

Now we observe that in the presence of an electric current density the three analytical field quantities

$$\rho_s(x,t)$$

$$\rho_{Na} = \rho_{Na}(x,t) = M_{Na}\, c_{Na}(x,t)$$

$$\rho_{Cl} = \rho_{Cl}(x,t) = M_{Cl}\, c_{Cl}(x,t)$$

are no longer described correctly by eqs.(64)-(65), that is, we no longer have

$$\frac{\partial \rho_s}{\partial t} = -\operatorname{div} j_s$$

$$\frac{\partial \rho_{Na}}{\partial t} = M_{Na}\frac{\partial c_{Na}}{\partial t} = -\operatorname{div} j_{Na} = -M_{Na}\operatorname{div}\tilde{j}_{Na}$$

$$\frac{\partial \rho_{Cl}}{\partial t} = M_{Cl}\frac{\partial c_{Cl}}{\partial t} = -\operatorname{div} j_{Cl} = -M_{Cl}\operatorname{div}\tilde{j}_{Cl}$$

where the solute mass and the constituent mass fluxes are given by the gradients of the respective partial densities. Rather, in the presence of the electric current there are excess variations of the solute and constituent partial densities. These excess variations we write as primed quantities:

$$\frac{\partial \rho_s'}{\partial t} \;;\quad \frac{\partial \rho_{Na}'}{\partial t} = M_{Na}\frac{\partial c_{Na}'}{\partial t} \;;\quad \frac{\partial \rho_{Cl}'}{\partial t} = M_{Cl}\frac{\partial c_{Cl}'}{\partial t}$$

Then the total observed partial density changes with time are

$$\left(\frac{\partial \rho_s}{\partial t}\right)_{total} = \frac{\partial \rho_s}{\partial t} + \frac{\partial \rho_s'}{\partial t} \tag{72}$$

$$\left(\frac{\partial \rho_{Na}}{\partial t}\right)_{total} = \frac{\partial \rho_{Na}}{\partial t} + \frac{\partial \rho_{Na}'}{\partial t} \tag{73}$$

$$\left(\frac{\partial \rho_{Cl}}{\partial t}\right)_{total} = \frac{\partial \rho_{Cl}}{\partial t} + \frac{\partial \rho_{Cl}'}{\partial t} \tag{74}$$

Instead of $\partial \rho_s/\partial t$ we could also have measured the change with time of the third constituent concentration, $(\partial c_K/\partial t)_{total}$, say. Now, the total variation with time of the constituent concentrations are well-defined analytical quantities, thus, as a consequence of eq.(2a) we must have

$$\frac{\partial c_{Na}}{\partial t} + \frac{\partial c_{Na}'}{\partial t} + \frac{\partial c_K}{\partial t} + \frac{\partial c_K'}{\partial t} = \frac{\partial c_{Cl}}{\partial t} + \frac{\partial c_{Cl}'}{\partial t} + \frac{\partial c_I}{\partial t} + \frac{\partial c_I'}{\partial t}$$

The unprimed quantities are those occurring in the ordinary diffusion equations, see for instance eqs.(52) and (53), and as such satisfy the relation

$$\frac{\partial c_{Na}}{\partial t} + \frac{\partial c_K}{\partial t} = \frac{\partial c_{Cl}}{\partial t} + \frac{\partial c_I}{\partial t} \tag{75}$$

and consequently we must also have

$$\frac{\partial c'_{Na}}{\partial t} + \frac{\partial c'_K}{\partial t} = \frac{\partial c'_{Cl}}{\partial t} + \frac{\partial c'_I}{\partial t} \tag{75a}$$

Returning to the ρ_s, ρ_{Na}, ρ_{Cl} description, it is clear that for $\partial\rho'_s/\partial t$ the relation holds:

$$\frac{\partial \rho'_s}{\partial t} = M_{Na}\frac{\partial c'_{Na}}{\partial t} + M_K\frac{\partial c'_K}{\partial t} + M_{Cl}\frac{\partial c'_{Cl}}{\partial t} + M_I\frac{\partial c'_I}{\partial t} \tag{76}$$

3.2 Excess constituent mass fluxes and generalized transport numbers.

Next we define the concepts of an excess total solute mass flux

$$\frac{\partial \rho'_s}{\partial t} = -\operatorname{div} j'_s \tag{77}$$

and of the excess constituent mass fluxes

$$\frac{\partial \rho'_{Na}}{\partial t} = -\operatorname{div} j'_{Na} \tag{78}$$

$$\frac{\partial \rho'_{Cl}}{\partial t} = -\operatorname{div} j'_{Cl} \tag{79}$$

When one uses eqs.(75a), (76) and (77) then the divergences of the two other excess constituent mass fluxes are also given

$$\operatorname{div} j'_K = \frac{\operatorname{div} j'_s - (1 + M_I/M_{Na})\operatorname{div} j'_{Na} - (1 - M_I/M_{Cl})\operatorname{div} j'_{Cl}}{1 + M_I/M_K}$$

$$\operatorname{div} j'_I = \frac{M_I}{M_{Na}}\operatorname{div} j'_{Na} + \frac{M_I}{M_K}\operatorname{div} j'_K - \frac{M_I}{M_{Cl}}\operatorname{div} j'_{Cl}$$

As a consequence of eqs.(72)-(74) and (77)-(79) we obtain for the total partial mass density changes in the presence of an electric current

$$\left(\frac{\partial \rho_s}{\partial t}\right)_{total} = -\operatorname{div} j_s - \operatorname{div} j'_s \tag{80}$$

$$\left(\frac{\partial \rho_{Na}}{\partial t}\right)_{total} = -\operatorname{div} j_{Na} - \operatorname{div} j'_{Na} \tag{81}$$

$$\left(\frac{\partial \rho_{Cl}}{\partial t}\right)_{total} = -\operatorname{div} j_{Cl} - \operatorname{div} j'_{Cl} \tag{82}$$

where, as before, the fluxes j_i (i = s, Na, Cl) are linear functions of the partial density gradients.

We note that so far only the divergences of the excess mass fluxes are defined. This is due to the fact that the boundary conditions are not yet specified. Thus our excess mass fluxes are only defined apart from an additive constant. Still it is meaningful to summarize our experimental observations on the multicomponent (electrolyte) diffusion system in the presence of the electric current in the statement: the excess mass fluxes are proportional to the electric current density. This means that we have:

$$\dot{j}'_{Na} = M_{Na}\tau_{Na}\dot{j}_{\ell} \tag{83}$$

$$\dot{j}'_{K} = M_{K}\tau_{K}\dot{j}_{\ell} \tag{84}$$

$$\dot{j}'_{Cl} = -M_{Cl}\tau_{Cl}\dot{j}_{\ell} \tag{85}$$

$$\dot{j}'_{I} = -M_{I}\tau_{I}\dot{j}_{\ell} \tag{86}$$

Here the factors of proportionality, τ_i (i = Na,K,Cl,I) are chosen such that all $\tau_i \geqslant 0$. The coefficients τ_i must be functions of the composition, otherwise no observable effect of the electric current on the analytical field would exist.

The excess constituent mass fluxes $\dot{j}_i'$ are markedly different in their behaviour from the constituent mass fluxes $\dot{j}_i$ which were introduced in the preceding chapter. If we add a certain amount of component XY (e.g. NaCl) to the system at a given position, then, when diffusion occurs, in the neighbourhood of this position the two constituent fluxes $\dot{j}_{Na}$ and $\dot{j}_{Cl}$ will have the same direction and similar or even equal magnitudes, depending on the nature of the system. This is a natural consequence of the constraint of pairwise addition of constituents, i.e. the two constituents can only be added in a close local neighbourhood, the neighbourhood being of "microscopic" dimensions. In the presence of an electric current, the two excess constituent fluxes belonging to a given component have opposite directions which means that now we have a verification of the constraint of pairwise addition of the constituents such that the locations of addition are separated by a distance of macroscopic dimension.

The set of eqs.(83)-(86) supply us with the definition of the + and - signs which characterize the two classes in which all constituents in electrolyte solutions are to be grouped. If the constituent belongs to the "positive class" then the excess constituent flux has the same direction as the electric current vector, on the other hand, if the constituent belongs to the "negative class", then the excess constituent flux and the electric current density have opposite directions. Of course, this is the same classification as the distinction between positively and negatively charged ions in the conventional treatment. However, it is instructive to show that an electrochemistry can be formulated which does not use the concept of the electric charge. In order to remain as close as possible to the conventional treatment we shall also use the expressions, cationic constituents (cations, sign +) and anionic constituents (anions sign -); the justification will become clear later when the fluxes at the electrodes are described.

We are going to discuss some further aspects of eqs.(83)-(86). First we should realize that the four coefficients τ_i in these equations are not all independent. This is due to the fact that the divergences of the excess constituent fluxes are interrelated.

$$\frac{1}{M_{Na}} \operatorname{div} j'_{Na} + \frac{1}{M_K} \operatorname{div} j'_K = \frac{1}{M_{Cl}} \operatorname{div} j'_{Cl} + \frac{1}{M_I} \operatorname{div} j'_I \tag{88}$$

which follows from eqs.(75a) and (78), (79). If we insert eqs.(83)-(86) in eq.(88) we obtain:

$$\operatorname{div}\left(\tau_{Na} j_2 + \tau_K j_2 + \tau_{Cl} j_2 + \tau_I j_2\right) = 0$$

Now we have also the validity of eq.(71) and it follows:

$$\frac{d\tau_{Na}}{dx} + \frac{d\tau_K}{dx} + \frac{d\tau_{Cl}}{dx} + \frac{d\tau_I}{dx} = 0 \tag{89}$$

We may also express the fact that the four τ_i 's are interrelated by writing

$$\tau_{Na} + \tau_K + \tau_{Cl} + \tau_I = \mathcal{L} \tag{90}$$

where $\mathcal{L}$ is any arbitrary constant. As long as we have not specified the boundary conditions, apart from the validity of eq.(90), the quantities τ_{Na}, τ_K, τ_{Cl}, τ_I may involve any freely chosen additive constants. Thus we may write

$$\begin{aligned} \tau_{Na} &= A + \tilde{\tau}_{Na} \\ \tau_K &= B + \tilde{\tau}_K \\ \tau_{Cl} &= C + \tilde{\tau}_{Cl} \\ \tau_I &= \mathcal{L} - A - B - C - \tilde{\tau}_{Na} - \tilde{\tau}_K - \tilde{\tau}_{Cl} \end{aligned} \tag{91}$$

with A, B, C being any arbitrary constants, thus only the $\tilde{\tau}_i$'s depend on the position because intrinsically they depend on the composition of the solution. Thus we have

$$\frac{\partial \tau_i}{\partial x} = \frac{\partial \tau_i}{\partial \rho_{Na}} \frac{\partial \rho_{Na}}{\partial x} + \frac{\partial \tau_i}{\partial \rho_K} \frac{\partial \rho_K}{\partial x} + \frac{\partial \tau_i}{\partial \rho_{Cl}} \frac{\partial \rho_{Cl}}{\partial x} \qquad i = Na, K, Cl \tag{92}$$

It is clear from inspection of eqs.(83)-(86) that the τ_i 's are generalized transport numbers - although they are not numbers by dimension.

Now we introduce eq.(83)-(86) together with eq.(89) in eqs.(80)-(82) to get

$$\left(\frac{\partial \rho_s}{\partial t}\right)_{total} = -\operatorname{div} j_s - j_2\left[(M_{Na}+M_I)\frac{d\tau_{Na}}{dx} + (M_K+M_I)\frac{d\tau_K}{dx} - (M_{Cl}-M_I)\frac{d\tau_{Cl}}{dx}\right] \tag{93}$$

$$\left(\frac{\partial \rho_{Na}}{\partial t}\right)_{total} = -\operatorname{div} j_{Na} - M_{Na}\frac{d\tau_{Na}}{dx} j_2 \tag{94}$$

$$\left(\frac{\partial \rho_{Cl}}{\partial t}\right)_{total} = -\operatorname{div} j_{Cl} + M_{Cl}\frac{d\tau_{Cl}}{dx} j_2 \tag{95}$$

We can also describe the change with time of the analytical field in the component property space. Then we would have (from eq.(21a)-(23a)

$$\left(\frac{\partial \rho^*_{KCl}}{\partial t}\right)_{total} = -\operatorname{div} \tilde{j}_{KCl} - M_{KCl}\frac{d\tau_K}{dx} j_2 \tag{93a}$$

$$\left(\frac{\partial \rho^*_{NaCl}}{\partial t}\right)_{total} = -\operatorname{div} \tilde{j}_{NaCl} + M_{NaCl}\left(\frac{d\tau_{Cl}}{dx} + \frac{d\tau_K}{dx}\right) j_2 \tag{94a}$$

$$\left(\frac{\partial \rho^*_{NaI}}{\partial t}\right)_{total} = -\operatorname{div} \tilde{j}_{NaI} - M_{NaI}\left(\frac{d\tau_{Na}}{dx} + \frac{d\tau_K}{dx} + \frac{d\tau_{Cl}}{dx}\right) j_2 \tag{95a}$$

In all these formulas the $d\tau_i/dx$ are expressions like eq.(92), i.e. the second terms on the right-hand side of eqs.(93-105a) are linear functions of the partial mass density gradients.

3.3 The delocalized conservation of mass.

The general situation we have described in the preceding sections is illustrated in Fig.4. Let $\rho_i(x,t)$ be the partial mass density of any one of the constituents. In Fig.4b the distribution of $\partial\rho_i/\partial t$ which is connected with the concentration profile according to Fig.4a, is shown. We have

$$A\int_{x_1}^{x_2}\frac{\partial\rho_i}{\partial t}dx = \frac{dm_i}{dt} = 0 \tag{96}$$

The definition of the two coordinates x_1 and x_2 is as follows: for $x < x_1$, $\partial\rho_i/\partial x = 0$ and for $x > x_2$, $\partial\rho_i/\partial x = 0$. Eq.(96) expresses the fact that the total mass of constituent i in the system is constant.

Fig.4c corresponds to the situation described by eqs.(46a-d). We may rewrite these equations in the form

$$\frac{\partial\rho_\ell}{\partial t} = -\,\mathrm{div}\,\bar{J}_\ell + \left(\frac{\partial\rho_\ell}{\partial t}\right)_{reaction}$$

$$\dots\dots\dots\dots\dots$$

$$\frac{\partial\rho_{\ell'}}{\partial t} = -\,\mathrm{div}\,\bar{J}_{\ell'} + \left(\frac{\partial\rho_{\ell'}}{\partial t}\right)_{reaction}$$

$$\dots\dots\dots\dots\dots$$

where $\bar{J}_\ell, \dots \bar{J}_{\ell'} \dots$ are molecular mass fluxes and $(\partial\rho_\ell/\partial t)_{reaction}$ is the increase of density of species ℓ per unit time due to the chemical reaction producing this species. From Fig. 4c one sees that

$$\int_{x_1}^{x_2}\left(\frac{\partial\rho_\ell}{\partial t}\right)_{reaction} dx \neq 0$$

However if this integral > 0, then there is always an integral

$$\int_{x}^{x_2}\left(\frac{\partial\rho_{\ell'}}{\partial t}\right)_{reaction} dx < 0$$

for another reaction partner, so that - together with the divergence of the diffusion fluxes - eq.(96) holds for all the constituents involved. This is the principle of local mass conservation.

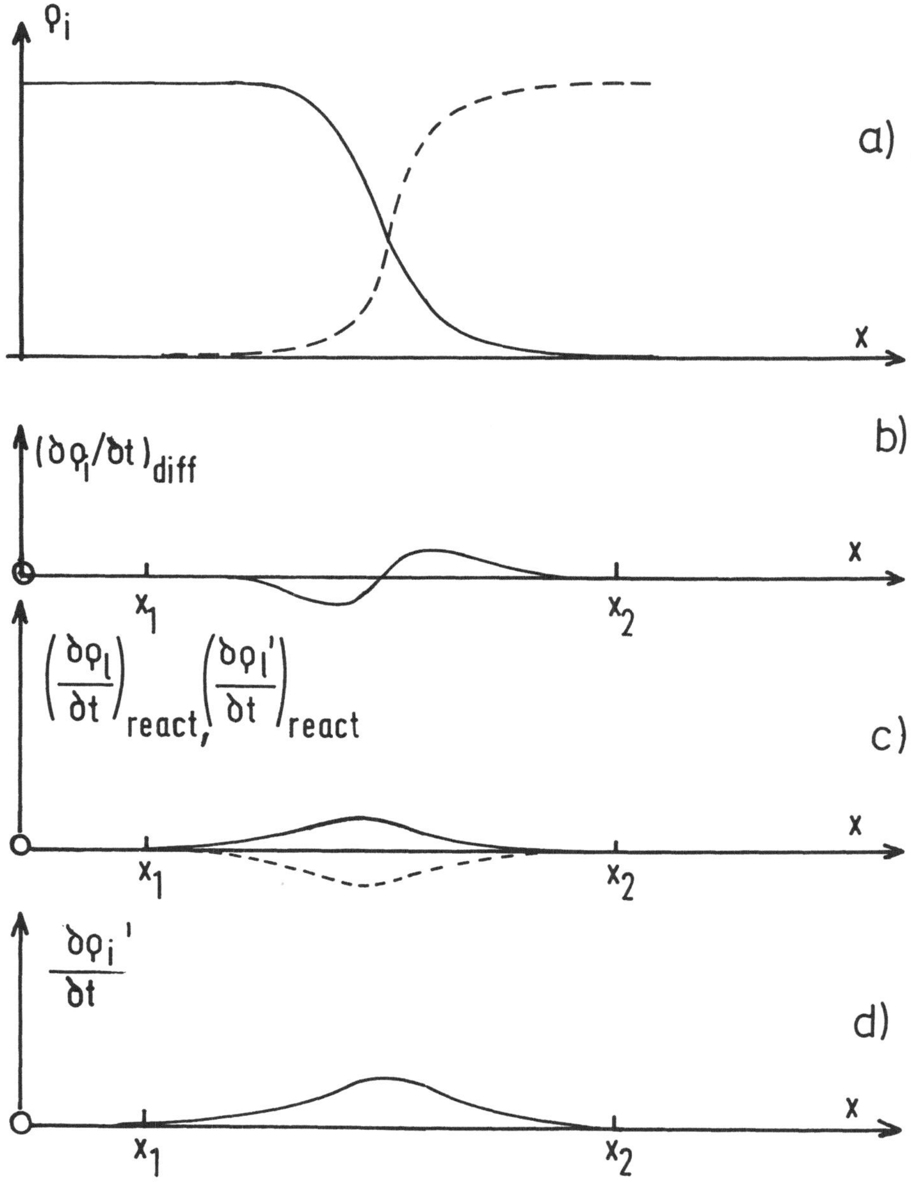

Fig. 4 : a) Schematic representation of partial mass density profile of the constituent i (any other constituent j given as dashed curve), b) rate of change of ρ_i as a consequence of ordinary diffusion, c) rate of change of components ℓ and ℓ' (dashed line) due to chemical reaction between i, j . . . , d) excess rate of change of constituent i in the presence of an electric current.

However in Fig. 4d we show the rate of increase of the excess constituent mass density. One sees that we have

$$\int_{x_1}^{x_2} \frac{\partial \rho_i}{\partial t} dx = \int_{x_1}^{x_2} \frac{\partial \rho_i'}{\partial t} dx > 0$$

and consequently, there is no conservation of the constituent mass when an electric current exists in the system. Of course, for the total physical system considered, the constituent mass must be preserved. We conclude that the mass created in the region $x_1 < x < x_2$ must have been removed somewhere else, however the location of the range $x_1' < x < x_2'$ where

$$\int_{x_1'}^{x_2'} \frac{\partial \rho_i'}{\partial t} dx < 0$$

may be as far away as we wish. For instance, if the system is a closed loop, then there will be a region $x \gg x_2$ in which ρ_i decreases so that finally we have again the density of the left-hand part of the figure, i.e. that shown for $x < x_1$. And since in that region $\partial \rho_i / \partial x$ has the opposite sign, $\partial \rho_i' / \partial t$ has the opposite sign as well. This is the effect of delocalized mass conservation which we mentioned before on page 24.

One important consequence of eqs.(93)-(96) is the fact that in Fig. 4d, $\partial \rho_i' / \partial t$ changes sign if we reverse the direction of the electric current. Then we have

$$\int_{x_1}^{x_2} \frac{\partial \rho_i}{\partial t} dx = \int_{x_1}^{x_2} \frac{\partial \rho_i'}{\partial t} dx < 0$$

however, the magnitude of this integral remains the same.

We could also be interested in the excess rate of increase of the component mass NaCl, for example. Then eq.(94a) can be consulted and for which we may write:

$$\frac{\partial \rho_{NaCl}^*}{\partial t} = - \operatorname{div} \tilde{j}_{NaCl} + \frac{\partial \rho_{NaCl}^{*\prime}}{\partial t}$$

where $\partial \rho_{NaCl}^{*\prime} / \partial t$ is the excess rate of increase of the NaCl component partial density. We have again:

$$A \int_{x_1}^{x_2} \frac{\partial \rho_{NaCl}^*}{\partial t} dx = A \int_{x_1}^{x_2} \frac{\partial \rho_{NaCl}^{*\prime}}{\partial t} dx = \frac{dm_{NaCl}}{dt} > 0 \qquad (97)$$

Eq.(97) implies the statement that in the inhomogeneous electrolyte solution in the presence of an electric current we have a local real component mass production. Of course, in reality we have only a transport of mass from remote regions to the one considered here, $x_1 < x < x_2$. But if we consider only this region, the amount of NaCl which is needed to construct the identical local system, increases.

The physical facts described in Fig.4 need some further comment. In Fig. 4a the decrease of $\rho_i(x)$ (for instance ρ_{Na}) with increasing coordinate x is assumed to be due to the fact that our electrolyte solution is a multicomponent system and that the concentration of constituent i (Na) decreases to a low value in the range $x_1 < x < x_2$, whereas the total solute concentration is essentially constant. Now we may also have a binary electrolyte solution, a NaCl solution, say. Let the constituent i be Na, then Fig.4a implies that the NaCl concentration decreases in the range $x_1 < x < x_2$. In such a situation experimental experience tells us that $\partial \rho'_{NaCl} / \partial t < 0$ if $j_2 > 0$, contrary to the behaviour sketched in Fig.4d. Thus, as is well known, the transport number of Na in a NaCl solution decreases with increasing salt concentration. It follows from these considerations that the reason for a change in the transport number has to be carefully checked (here the generalized transport number, however, the situation is not different for the conventional transport number t_i, see below). τ_i may vary because, at constant total solute concentration, the concentration of the constituent i varies strongly, or it may change because in a binary solution, τ_i is dependent on the salt concentration.

We turn now to some remarks concerning the partial constituent density distribution which are of fundamental significance. It has been shown that in the situation depicted in Fig.4d, the Na constituent mass (i = Na) must be taken from some remote region $x_1' < x < x_1$ in order to fulfil the requirement of mass conservation. In the region $x_1' < x < x_1$ the constituent Na may be present, however, in any arbitrary form. This fact is depicted schematically in Fig.5a. For $x_2' < x < x_1$ Na is present in the form of NaI, whereas in the remaining extension of the loop we have a NaCl solution. Now the important point is that the constituent I may be replaced by any arbitrary constituent X, with all arguments given below remaining valid. For simplicity let us continue to use the example of NaI as the solute in contact with the solute NaCl. Along the whole extension of the loop, the solution can be constructed from the components NaCl and NaI. The transformation relations between the constituent and construction space are

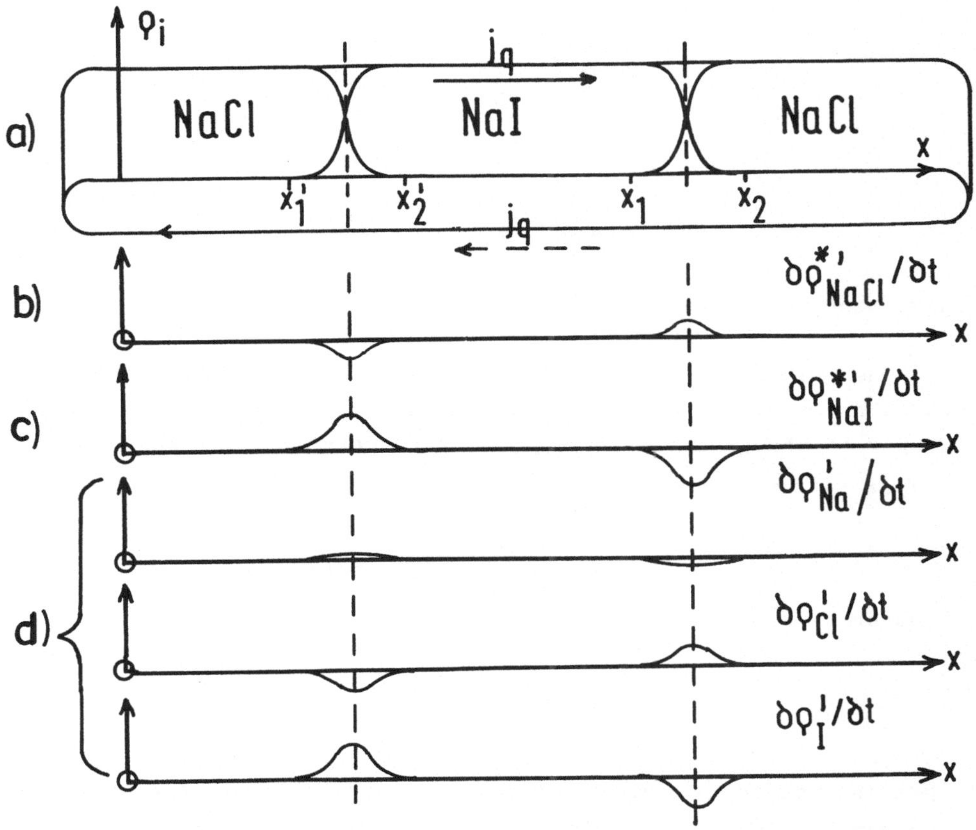

Fig.5 : A schematic represnetation of the non-localized mass conservation in the presence of the electric current.
a) Partial mass density profiles around the two boundary regions $x_1' \leqslant x \leqslant x_2'$, $x_1 \leqslant x \leqslant x_2$.
b) Excess rate of change of mass density of component NaCl,
c) Excess rate of change of mass density of component NaI,
d) Excess rates of changes of partial mass densities of constituents Na, Cl, and I.

$$\rho^*_{NaCl} = M_{NaCl} \frac{\rho_{Cl}}{M_{Cl}} \tag{98}$$

$$\rho^*_{NaI} = M_{NaI} \left(\frac{\rho_{Na}}{M_{Na}} - \frac{\rho_{Cl}}{M_{Cl}} \right) \tag{99}$$

Thus the excess rates of increase of the construction components are

$$\frac{\partial \rho^{*'}_{NaCl}}{\partial t} = M_{NaCl} \frac{d\tau_{Cl}}{dx} j_q \tag{100}$$

$$\frac{\partial \rho^{*'}_{NaI}}{\partial t} = -M_{NaI} \left(\frac{d\tau_{Na}}{dx} + \frac{d\tau_{Cl}}{dx} \right) j_q \tag{101}$$

where the relations

$$\frac{\partial \rho'_{Na}}{\partial t} = - M_{Na} \frac{d\tau_{Na}}{dx} \dot{J}_2 \qquad (94b)$$

$$\frac{\partial \rho'_{Cl}}{\partial t} = M_{Cl} \frac{d\tau_{Cl}}{dx} \dot{J}_2 \qquad (95b)$$

have been used (see eqs.(94),(95)).

The quantities $\partial \rho^{*'}_{NaCl} / \partial t$, $\partial \rho^{*'}_{NaI} / \partial t$ according to eqs.(100) and (101) are shown schematically in Fig.5b. They are derived from the slopes of the ρ_i given in Fig.5a, which determine the gradients of τ_i , i = Na, Cl. It may be seen from Fig.5b that

$$\int_{x_1}^{x_2} \frac{\partial \rho^{*'}_{NaCl}}{\partial t} dx \neq 0$$

however

$$\oint \frac{\partial \rho^{*'}_{NaCl}}{\partial t} dx = \oint \frac{\partial \rho^{*'}_{NaI}}{\partial t} dx = 0$$

that is, the total amounts of components in the loop are preserved, whereas the local amounts are not preserved. In Fig.5c the excess rates of increase of constituents Na, Cl, and I according to eqs.(94b) and (95b) are also schematically shown. It may be seen that we have

$$\frac{\partial \rho'_{Na}}{\partial t} = 0$$

$$\oint \frac{\partial \rho'_{Cl}}{\partial t} dx = \oint \frac{\partial \rho'_{I}}{\partial t} dx = 0$$

which is a statement to the effect that the constituent masses are conserved in the system.

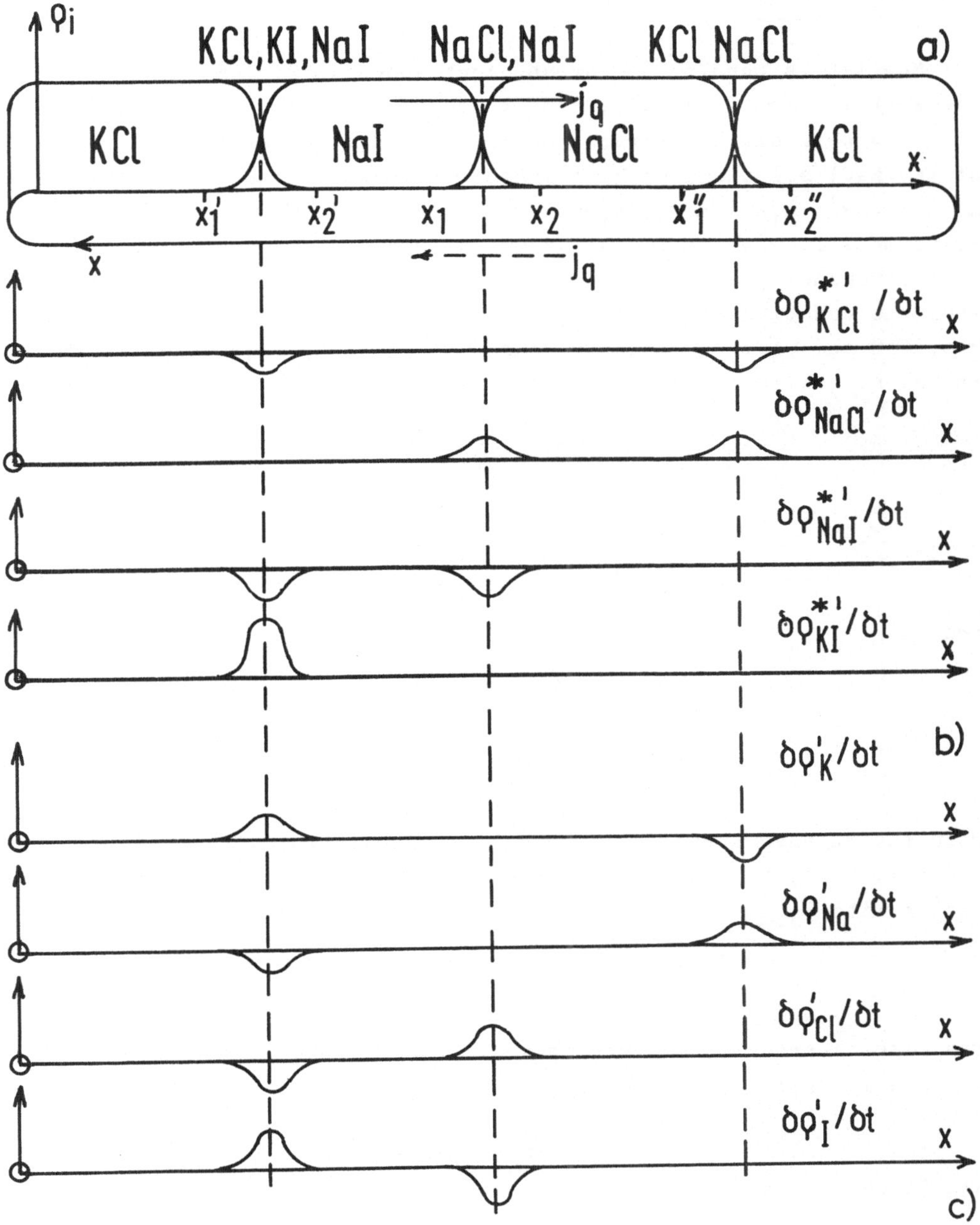

Fig.6. Schematic representation of an arrangement of three different electrolyte solutions forming a closed loop. a) Partial mass density distribution given in the construction space representation. b) The excess rates of changes of the various components given as a function of the cell coordinate x. c) The excess rates of changes of the various constituent partial mass densities given as a function of x. For other details see text.

3.4 A prototype of a galvanic cell

Now let us turn to the arrangement shown in Fig.6a. In addition to the system shown in Fig.5a, here a compartment containing a KCl solution has been placed between the NaCl and NaI compartments on the right of the NaCl solution. We have inhomogeneities in the three ranges, $x_1' < x < x_2'$; $x_1 < x < x_2$; $x_1'' < x < x_2''$. At the respective boundaries the components from which the solution is to be constructed in the ranges of inhomogeneity are indicated. We need the transformation relations from the constituent space Na, Cl, K to the respective construction spaces. For the range $x_1 < x < x_2$ eqs.(98) and (99) are valid. Furthermore, it may be easily derived that we have (see section 1,2)

$$\left.\begin{aligned} \rho^*_{NaI} &= M_{NaI}\frac{\rho_{Na}}{M_{Na}} \\ \rho^*_{KCl} &= M_{KCl}\frac{\rho_{Cl}}{M_{Cl}} \\ \rho^*_{KI} &= M_{KI}\left(\frac{\rho_K}{M_K} - \frac{\rho_{Cl}}{M_{Cl}}\right) \end{aligned}\right\} \text{ for } x_1' < x < x_2'$$

and

$$\left.\begin{aligned} \rho^*_{KCl} &= M_{KCl}\frac{\rho_K}{M_K} \\ \rho^*_{NaCl} &= M_{NaCl}\frac{\rho_{Na}}{M_{Na}} \end{aligned}\right\} \text{ for } x_1'' < x < x_2''$$

Then the excess rates of increase of the component densities, $\partial\rho_i^{*\prime}/\partial t$, i = KCl, NaCl, NaI, KI, are obtained if we use eqs.(94b) and (95b) together with

$$\frac{\partial\rho_K'}{\partial t} = -M_K\frac{d\tau_K}{dx}j_2$$

The results are sketched schematically in Fig.6b. It may be seen that we have for all four components

$$\oint\frac{\partial\rho_i^{*\prime}}{\partial t}dx \neq 0$$

and in particular

$$\oint\frac{\partial\rho^{*\prime}_{NaCl}}{\partial t}dx > 0, \qquad \oint\frac{\partial\rho^{*\prime}_{KI}}{\partial t}dx > 0 \tag{102}$$

$$\oint \frac{\partial \varrho^{*\prime}_{NaI}}{\partial t} dx < 0, \qquad \oint \frac{\partial \varrho^{*\prime}_{KCl}}{\partial t} dx < 0 \tag{103}$$

which indicates that there is a production of the components NaCl and KI and a depletion of the components KCl and NaI. Consequently in the diffusion arrangement shown in Fig.6a the existence of an electric current in the system is conncected with the chemical reaction

$$NaI + KCl \longrightarrow NaCl + KI \tag{I}$$

Finally, we consider Fig.6c, where the excess rates of increase of the four constituents are given. We see from this part of the figure that we have full mass conservation for each individual constituent, i.e.

$$\oint \frac{\partial \varrho_i'}{\partial t} dx = 0 \qquad i = Na, K, Cl, I$$

A diffusion arrangement as shown in Fig.6a, in which a chemical reaction can proceed such that all the excess rates of component production are proportional to the electric current density, is called a galvanic cell. It may be emphazised that in the region of the KCl-NaI boundary, i.e. $x_1' < x < x_2'$, of course there is also a chemical reaction producing NaCl and KI from NaI and KCl. This reaction satisfies what we have called the law of local mass con ervation. To the author's knowledge a galvanic cell consisting of a ring-shaped vessel which is filled with three different electrolyte solutions has only been described in the literature once before[1].

We recall that in the reaction equation (I) the constituents I and K may be replaced by any other constituents, that is, we may also write

$$NaX + YCl \longrightarrow NaCl + XY \tag{Ia}$$

A special case of reaction (Ia) is given by the situation X = Na, Y = Cl, then we have the reaction

$$Na + 1/2\ Cl_2 \longrightarrow NaCl(aq) \tag{Ib}$$

The fact that Na(metal) and Cl_2(gas) are no longer aqueous mixtures should not disturb us; we have never introduced the presupposition that the multicomponent character of the parts of the loop is constant along the extension of the loop. Of course, the reaction

$$Na + 1/2\ Cl_2 \longrightarrow NaCl(solid) \tag{Ic}$$

which occurs at the boundary $x_1' < x < x_2'$ and which is governed by the requirement of local mass conservation is easily eliminated by placing a system ZZ between NaNa and ClCl where the substance Z does not react chemically with either of the substances Na and Cl_2. From eq.(89) it follows that for the system XX, YY, ZZ we must have

$$\frac{d\tau_i}{dx} = 0$$

as a consequence of symmetry when both constituents occurring in the two constituent form of this equation are equal. It is perhaps unnecessary to mention that the system ZZ in very many cases of ordinary galvanic cells is a copper wire.

3.5 The conventional transport numbers

When defining our generalized transport numbers τ_i we have stated that these quantities may contain any arbitrary additive constant (see eq.(91)). For the conventional transport numbers normally used in electrochemistry all these constants are set equal to zero:

$$A = B = C = 0.$$

In order to specify the numerical value of the constant $\mathcal{L}$ occurring in eq.(90) to get agreement with the conventional scheme, we consider a boundary $0 < x < \delta x$ between the "pure" component X^*X^* (at $x \leq 0$, for instance a metal, Na, say) and an electrolyte solution which contains the constituent X^* (at $x \geq \delta x$). We construct the solution containing an supporting electrolyte such that the X^* constituent concentration is very small, $\rho_{X^*} \approx 0$. Then according to eqs.(78) and (83) the differential equation for τ_{X^*} is

$$\frac{\partial \rho'_{X^*}}{\partial t} = - M_{X^*} \frac{d\tau_{X^*}}{dx} j_2 \equiv - M_{X^*} \chi_{X^*} \frac{dt_{X^*}}{dx} j_2 \qquad (104)$$

where χ_{X^*} is a constant to be determined experimentally and t_{X^*} is the conventional transport number as will be seen shortly. We integrate this equation over the range of the boundary $0 \leq x \leq \delta x$

$$\int_0^{\delta x} \frac{\partial \rho'_{X^*}}{\partial t} dx = - M_{X^*} \chi_{X^*} \int_{t_{X^*X^*}}^{0} dt_{X^*} \cdot j_2 = M_{X^*} \chi_{X^*} t_{X^*X^*} j_2$$

$t_{X^*}(+\delta x)$ is necessarily zero because in the solution the X^* constituent concentration is assumed to be vanishingly small (i.e. for all $x \geq \delta x$). Now the convention is generally accepted that

$$t_{X^*X^*} \equiv 1$$

which gives

$$\int_0^{\delta x} \frac{\partial \rho_{X^*}}{\partial t} dx = -\frac{1}{A} \frac{dm_{X^*X^*}}{dt} = +M_{X^*} \chi_{X^*} j_2$$

where $dm_{X^*X^*}/dt$ is the rate of change of the mass of X^*X^*. $dm_{X^*X^*}/dt$ can be measured: the result for χ_{X^*} is

$$\chi_{X^*} = 1/\mathcal{F} \tag{105}$$

if the constituents present are subject to the constraint of pairwise addition, i.e. the solutes are NaCl, NaI, ... XY.... $\mathcal{F}$ is the Faraday constant. Thus we arrive at the result (see eq.(104))

$$\tau_{X^*} = \frac{t_{X^*}}{\mathcal{F}}, \quad \tau_{X^*X^*} = \frac{1}{\mathcal{F}}$$

e.g.

$$\tau_{Na} = \frac{1}{\mathcal{F}} t_{Na}, \quad \tau_{NaNa} = \frac{1}{\mathcal{F}} \tag{106}$$

at $x < 0$ all other constituents occurring in the solution are absent, then, in an equation of type (90) all other generalized transport numbers must be zero, here $\tau_{Cl} = \tau_K = \tau_I = 0$. So, we get, for instance

$$(\tau_{Na})_{x \leq 0} = \tau_{NaNa} = \mathcal{L} = \frac{1}{\mathcal{F}} \tag{107}$$

Since under the condition stated, the result eq.(105) is independent of the special choice of the constitutent, we have also

$$\tau_K = \frac{1}{\mathcal{F}} t_K ; \quad \tau_{Cl} = \frac{1}{\mathcal{F}} t_{Cl} ; \quad \tau_I = \frac{1}{\mathcal{F}} t_I \tag{106a}$$

and if we insert the results of eqs.(106),(106a) and (107) in eq.(90) we obtain

$$t_{Na} + t_K + t_{Cl} + t_I = 1 \tag{108}$$

The generalization to components of more complicated composition is straightforward. Consider the component $A_{\nu^+}B_{\nu^-}$, where the stoichiometric numbers ν^+, ν^- in our language represent the addition constraint for the two constituents A and B. Then, the set of numbers ν^+, ν^- may be used to define another set of numbers z^+, z^- by the relation

$$\frac{\nu^+}{\nu^-} = \frac{z^-}{z^+}$$

choosing one z properly, for instance $z^-(Cl) = 1$. The connection with the conventional charge numbers is obvious. Under these conditions, instead of eq.(90) one derives from the analog of eq.(88)

$$z^{+}\tau_A + z^{-}\tau_B = \mathcal{L}$$

and instead of eq.(106) the experimental result is

$$\begin{aligned} \chi_A &= \frac{1}{z^{+}\mathcal{F}} \; ; \quad \tau_A = \frac{1}{z^{+}\mathcal{F}} t_A \\ \chi_B &= \frac{1}{z^{-}\mathcal{F}} \; ; \quad \tau_B = \frac{1}{z^{-}\mathcal{F}} t_B \end{aligned} \tag{109}$$

so that we have again

$$t_A + t_B = 1 \tag{108a}$$

It is clear that the relations (108) and (108a) are well known from the conventional definition of transport numbers. It will be realized that the definition of the transport numbers given in our treatment is different from the conventional one. The latter introduces the t_i 's as the fraction of the electric current carried by the respective ions. This may be called an intrinsic definition of the t_i 's whereas our definition has an operational basis (see, however the more rigorous definition given by SPIRO[2)]).

Let us now return to our eqs.(83)-(86) defining the excess constituent mass fluxes. With eqs.(106) these quantities take the form:

$$\begin{aligned} \dot{j}'_{Na} &= M_{Na}\tau_{Na}\dot{j}_{\Sigma} = \frac{M_{Na}}{\mathcal{F}} t_{Na}\dot{j}_{\Sigma} \equiv M_{Na}\mathcal{J}'_{Na} \\ &\cdots\cdots\cdots\cdots \\ \dot{j}'_{Cl} &= -M_{Cl}\tau_{Cl}\dot{j}_{\Sigma} = -\frac{M_{Cl}}{\mathcal{F}} t_{Cl}\dot{j}_{\Sigma} \equiv M_{Cl}\mathcal{J}'_{Cl} \\ &\cdots\cdots\cdots\cdots \end{aligned} \tag{110}$$

However we may also write, using the formulation of eq.(110) after the identity sign:

$$\begin{aligned} \mathcal{J}'_{Na} &= \mathcal{J}'_{Na^+} = \frac{1}{\mathcal{F}} t_{Na^+}\dot{j}_{\Sigma} = \frac{1}{\mathcal{F}}\dot{j}_{\Sigma}^{Na^+} \\ \mathcal{J}'_{K} &= \mathcal{J}'_{K^+} = \frac{1}{\mathcal{F}} t_{K^+}\dot{j}_{\Sigma} = \frac{1}{\mathcal{F}}\dot{j}_{\Sigma}^{K^+} \\ \mathcal{J}'_{Cl} &= \mathcal{J}'_{Cl^-} = -\frac{1}{\mathcal{F}} t_{Cl^-}\dot{j}_{\Sigma} = \frac{1}{\mathcal{F}}\dot{j}_{\Sigma}^{Cl^-} \\ \mathcal{J}'_{I} &= \mathcal{J}'_{I^-} = -\frac{1}{\mathcal{F}} t_{I^-}\dot{j}_{\Sigma} = \frac{1}{\mathcal{F}}\dot{j}_{\Sigma}^{I^-} \end{aligned} \tag{111}$$

$\mathcal{J}_i'$, i = Na, K, ... are the excess molecular or atomic constituent fluxes. In contrast to this, the quantities $\mathcal{J}_{Na^+}'$, $\mathcal{J}_{K^+}'$, $\mathcal{J}_{Cl^-}'$, and $\mathcal{J}_{I^-}'$ are understood to be the respective ionic fluxes. Thus the designation "atomic" (or molecular) constituent is replaced by "ionic". The excess constituent fluxes are not observable quantities;only n-1 divergences of these fluxes are observable when the multicomponent electrolyte solution contains n constituents. Partly there may also be combined divergences, and in particular, in a binary electrolyte solution the combined divergences of both constituents are observable. This illuminates the "intrinsic" character of the concept of ionic fluxes. The third column of eqs.(111) expresses the fact that the transport numbers are understood to give the fraction of the electric current which is transported by the respective ionic flux, as already mentioned. Finally, the quantities $j_2^{Na^+}$, $j_2^{K^+}$, $j_2^{Cl^-}$, ... are the "ionic" electric current densities.

On page 7 we have discussed the definition of"free ion concentrations" (see eqs.(6a)-(8a)). Now we see that the set of eqs.(111) suggests a special form of definition of (free) ionic species, namely

$$c_{Na^+} = c_{Na} ; \quad c_{K^+} = c_K ; \qquad c_{Cl^-} = c_{Cl} ; \quad c_{I^-} = c_I$$

that is, the ion concentrations are equal to the constituent concentrations. This correlates to the statement that the excess atomic (or molecular) constitutent flux is equal to the ionic flux. In fact, according to our scheme outlined on page 7 in general the concentrations of the ionic species are defined by relations such as $c_{Na^+} = c_{Na^+}(c_{Na}, c_K, c_{Cl}), \ldots$ and the changes with time of these quantities are not equal to the divergences of the ionic fluxes according to eqs.(111). The number of ions, according to the defining relations, may even be practically zero. In such a situation one has to be careful that the treatment does not involve two different definitions of ionic species which are not consistent.

In our further treatment we shall continue to use the denotation $t_{Na} \ldots t_{Cl}$ in order to achieve a higher degree of generality. The reader who has difficulties following this path may with safety replace all our transport numbers t_{Na}..... by the ionic transport numbers t_{Na^+}.....

$$t_{Na} = t_{Na^+}, \ldots \quad t_{Cl} = t_{Cl^-}, \ldots$$

3.6 The fundamental equations of electrochemistry

Using the special choice given by eq. (106), eqs. (93)-(95) take the form

$$\frac{\partial \rho_s}{\partial t} = -\operatorname{div} j_s - \frac{j_2}{\mathcal{F}}\Big[(M_{Na}+M_I)\frac{dt_{Na}}{dx} + (M_K+M_I)\frac{dt_K}{dx} - (M_{Cl}-M_I)\frac{dt_{Cl}}{dx}\Big] \qquad (112)$$

$$\frac{\partial \rho_{Na}}{\partial t} = -\operatorname{div} j_{Na} - \frac{M_{Na}}{\mathcal{F}} j_2 \frac{dt_{Na}}{dx} \qquad (113)$$

$$\frac{\partial \rho_{Cl}}{\partial t} = -\operatorname{div} j_{Cl} + \frac{M_{Cl}}{\mathcal{F}} j_2 \frac{dt_{Cl}}{dx} \qquad (114)$$

Very many of the results given below are derived from these equations. For this reason we call them the fundamental equations of electrochemistry. Here the fundamental equations are written down in the total solute density-constituents density coordinate system. Transformation to the construction coordinate system or the full constituents coordinate system is straightforward; such transformed descriptions will be occasionally used below.

In eqs.(112)-(114) the variation of the t_j's along the diffusion path x is determined by the variation of the composition of the system

$$\frac{dt_j}{dx} = \sum_\ell \frac{\partial t_j}{\partial \rho_\ell}\frac{\partial \rho_\ell}{\partial x} \qquad j = s, Na, Cl$$

then with eqs.(63) we obtain

$$\frac{\partial \rho_s}{\partial t} = \frac{\partial}{\partial x}\Big(\sum_\ell D_{s\ell}\frac{\partial \rho_\ell}{\partial x}\Big) - \frac{1}{\mathcal{F}} j_2 \sum\Big[(M_{Na}+M_I)\frac{\partial t_{Na}}{\partial \rho_\ell} + (M_K+M_I)\frac{\partial t_K}{\partial \rho_\ell} - (M_{Cl}-M_I)\frac{\partial t_{Cl}}{\partial \rho_\ell}\Big]\frac{\partial \rho_\ell}{\partial x} \qquad (115)$$

$$\frac{\partial \rho_{Na}}{\partial t} = \frac{\partial}{\partial x}\Big(\sum_\ell D_{Na\ell}\frac{\partial \rho_\ell}{\partial x}\Big) - \frac{M_{Na}}{\mathcal{F}} j_2 \sum_\ell \frac{\partial t_{Na}}{\partial \rho_\ell}\frac{\partial \rho_\ell}{\partial x} \qquad (116)$$

$$\frac{\partial \rho_{Cl}}{\partial t} = \frac{\partial}{\partial x}\left(\sum_{l} D_{Cl\,l}\frac{\partial \rho_l}{\partial x}\right) + \frac{M_{Cl}}{\mathcal{F}} j_{\xi} \sum_{l} \frac{\partial t_{Cl}}{\partial \rho_l}\frac{\partial \rho_l}{\partial x} \tag{117}$$

and finally, there is an equation which is not independent

$$\frac{\partial \rho_w}{\partial t} = \frac{\partial}{\partial x}\left(\sum_{l} D_{wl}\frac{\partial \rho_l}{\partial x}\right) \tag{118}$$

These are the diffusion equations in the presence of an electric current; their solution

$$\rho_s(x,t)\ ;\quad \rho_{Na}(x,t)\ ;\quad \rho_{Cl}(x,t)\ ;\quad \text{and}\ \rho_w(x,t)$$

represents the analytical field for $j_{\xi} \geq 0$.

It may be seen that in equations (115)-(118) we have omitted the subscript "total" for simplicity, the quantities on the left-hand side of these equations are the experimentally observed rates of composition change.

A remark concerning the definition of the mass fluxes has to be made here. As before, in the absence of the electric current, the mass fluxes are given by the derivatives of the field quantities ρ_s, ρ_{Na}, ρ_{Cl} with respect to the position:

$$-j_l(x,t) = D_{ls}\frac{\partial \rho_s(x,t)}{\partial x} + D_{lNa}\frac{\partial \rho_{Na}(x,t)}{\partial x} + D_{lCl}\frac{\partial \rho_{Cl}(x,t)}{\partial x}$$

$$l = \text{s, Na, Cl} \tag{119}$$

however, the conceptual situation has changed. Now eqs.(119) are the definitions of the mass fluxes. Previously, in the absence of an electric current, the mass fluxes were defined by their negative divergences and their vanishing values at the boundaries. In contrast to this, now eq.(112) implies the following statement. The change of local solute mass density per unit time is the sum of the negative divergence of the total solute mass flux (j_s) and the negative divergence of the excess solute mass flux (j_s'). As already mentioned, the latter contribution we may also call the rate of production of real, i.e. observable solute mass. Correspondingly, the statement of eq.(116) is. The change of the local Na constituent mass density per unit time equals the sum of the negative divergence of the constituent mass flux j_{Na} and the negative divergence of the excess constituent mass flux $\dot{j}_{Na}$, the latter negative divergence being the rate of production of real, i.e. analytically detectable Na constituent mass.

For completeness we write the fundamental equations of electrochemistry for the case that the constituents forming the electrolyte solution are Ca, K, Cl, and I.

$$\frac{\partial \rho_s}{\partial t} = - \operatorname{div} \dot{j}_s - \frac{j_I}{\mathcal{F}} \left[\left(\tfrac{1}{2} M_{Ca} + M_I \right) \frac{dt_{Ca}}{dx} + \left(M_K + M_I \right) \frac{dt_K}{dx} - \left(M_{Cl} - M_I \right) \frac{dt_{Cl}}{dx} \right] \qquad (112a)$$

$$\frac{\partial \rho_{Ca}}{\partial t} = - \operatorname{div} \dot{j}_{Ca} - \frac{M_{Ca}}{2\mathcal{F}} j_I \frac{dt_{Ca}}{dx} \qquad (113a)$$

$$\frac{\partial \rho_{Cl}}{\partial t} = - \operatorname{div} \dot{j}_{Cl} + \frac{M_{Cl}}{\mathcal{F}} j_I \frac{dt_{Cl}}{dx} \qquad (114a)$$

3.7 A comparison of the use of generalized and conventional transport numbers

Let us return to our diffusion system which so far has undefined boundary arrangements. Such a system may be a loop with an extended homogeneous part or the electrodes may be very far away. Suppose that ρ_s = const., ρ_{Na} = const., ρ_{Cl} = const., at all $x < x_{01}$ and all $x > x_{02}$, i.e. we have

$$\left.\begin{aligned} \frac{\partial \rho_s}{\partial x} &= \frac{\partial \rho_{Na}}{\partial x} = \frac{\partial \rho_{Cl}}{\partial x} = 0 \\ &\text{and} \\ \frac{\partial \rho_s}{\partial t} &= \frac{\partial \rho_{Na}}{\partial t} = \frac{\partial \rho_{Cl}}{\partial t} = 0 \end{aligned}\right\} \quad \begin{aligned} x &< x_{01} \\ x &> x_{02} \end{aligned}$$

In the range $x_{01} < x < x_{02}$ the system is non-uniform. Then, considering our definition eq.(119) we find:

at $x < x_{01}$ $\qquad \dot{j}_s = \dot{j}_{Na} = \dot{j}_{Cl} = 0$

at $x > x_{02}$ $\qquad \dot{j}_s = \dot{j}_{Na} = \dot{j}_{Cl} = 0$

i.e. the total solute mass flux and the two constituent mass fluxes vanish if we are sufficiently far away from the region of inhomogeneity. Likewise at $x < x_{01}$ $\quad div\, j'_{Na} = div\, j'_{Cl} = div\, j'_K = 0$

at $x > x_{02}$ $\quad div\, j'_{Na} = div\, j'_{Cl} = div\, j'_K = 0$

i.e. the <u>divergences</u> of the excess constituent fluxes vanish when we consider them sufficiently far away from the composition gradients. However, the excess constituent fluxes themselves are indefinite at $x < x_{01}$, or at $x > x_{02}$. At one of these positions the independent j'_i s may be chosen entirely arbitrarily. We give two examples. At any $x_0 < x_{01}$ where the composition is constant, we denote this composition by "input composition". Somewhere at $x > x_{01}$ there are concentration gradients for the various constituents. We apply eqs.(91) and obtain for the three independent excess mass fluxes:

$$j'_S = \{(M_{Na} + M_I)(A + \tilde{\tau}_{Na}(x_0)) + (M_K + M_I)(B + \tilde{\tau}_K(x_0)) - (M_{Cl} - M_I) \cdot (C + \tilde{\tau}_{Cl}(x_0)) - M_I \mathcal{L}\} j_2$$

$$j'_{Na} = M_{Na}(A + \tilde{\tau}_{Na}(x_0))\, j_2 \tag{120}$$

$$j'_{Cl} = -M_{Cl}(C + \tilde{\tau}_{Cl}(x_0))\, j_2$$

Now we chose the arbitrary constant $\mathcal{L}$ such that

$$\mathcal{L}_{x_0} = \frac{(M_{Na} + M_I)(A + \tilde{\tau}_{Na}(x_0)) + (M_K + M_I)(B + \tilde{\tau}_K(x_0)) - (M_{Cl} - M_I)(C + \tilde{\tau}_{Cl}(x_0))}{M_I} \tag{121}$$

Then we find from eq.(120)

$$j'_S(x_0) = 0$$

Thus we see that at the input position x_0, the total solute mass flux <u>and</u> the excess solute mass flux vanish if we set $\mathcal{L} = \mathcal{L}_{x_0}$ according to eq.(121). The constituent mass fluxes also vanish, but the excess constituent mass fluxes do not. It is not the objective of the present work to give a microscopic representation of diffusion processes in the presence of an electric current, however, in order to satisfy the reader's temptation to translate all statements to a molecular language we briefly indicate that the behaviour we have found is typical for a rotation. The salt molecule rotates about a certain point being at rest on the average. The movement of the sum of all these points is the total mass flux. In fact, it is zero at $x = x_0$. The excess constituent mass fluxes represent the rotating parts of the salt molecules, the excess total solute mass flux is a certain linear combination of these movements [3,4,5].

Now we consider the position range $x > x_{01}$, as already mentioned, at $x \geq x_{02}$ we are "beyond" the composition inhomogeneity. Then at $x_{01} < x$ the three excess mass fluxes are

$$j_s'(x) = \left\{ (M_{Na}+M_I) \int_{x_0}^{x} \frac{d\tilde{\tau}_{Na}}{dx} dx + (M_K+M_I) \int_{x_0}^{x} \frac{d\tilde{\tau}_K}{dx} dx - (M_{Cl}-M_I) \int_{x_0}^{x} \frac{d\tilde{\tau}_{Cl}}{dx} dx \right\} j_\Sigma$$

$$j_{Na}' = M_{Na} j_\Sigma \left\{ A + \tilde{\tau}_{Na}(x_0) + \int_{x_0}^{x} \frac{d\tilde{\tau}_{Na}}{dx} dx \right\}$$

$$j_{Cl}' = - M_{Cl} j_\Sigma \left\{ C + \tilde{\tau}_{Cl}(x_0) + \int_{x_0}^{x} \frac{d\tilde{\tau}_{Cl}}{dx} dx \right\}$$

In general $\partial\tilde{\tau}_j / \partial\rho_\ell \neq 0$, that is $d\tilde{\tau}_j / dx \neq 0$, and thus at the "output", $x > x_{02}$ where $\partial\rho_i / \partial x = 0$, the excess total solute mass flux is no longer zero. The gain or loss of the excess mass fluxes in the region $x_{01} < x < x_{02}$ has contributed to the change of the local solute mass distribution and this in turn causes a modification of the solute mass flux j_s, $j_s \neq 0$ in the range $x_{01} < x < x_{02}$, $j_s = 0$ for $x > x_{02}$. This is the coupling between the electric current and the mass flux. The same situation is also valid for the constituent fluxes. More details will be given in the next chapter.

Next we apply the choice of the constants $\mathcal{L}$, A, B, C as given by eqs.(106),(106a) and (107). Then we have at the input coordinate x_0

$$(j_s'(x_0))_t = \left\{ (M_{Na}+M_I) t_{Na} + (M_K+M_I) t_K - (M_{Cl}-M_I) t_{Cl} - M_I \right\} \frac{j_\Sigma}{\mathcal{F}} \tag{122}$$

$$j_{Na}' = \frac{M_{Na}}{\mathcal{F}} t_{Na} j_\Sigma \tag{123}$$

$$j_{Cl}' = - \frac{M_{Cl}}{\mathcal{F}} t_{Cl} j_\Sigma \tag{124}$$

For illustration let us assume for the moment that we have

$$t_{Na} = t_K = t_{Cl} = t_I = 1/4$$

then

$$(j_s'(x_0))_t = \frac{1}{4} \frac{1}{\mathcal{F}} j_\Sigma (M_{Na} + M_K - M_{Cl} - M_I) \neq 0$$

Thus now at the input the excess mass flux does not vanish.
At $x > x_{01}$ we have

$$(j_s'(x))_t = (j_s'(x_0))_t + \Big[(M_{Na}+M_I)\int_{x_0}^{x}\frac{dt_{Na}}{dx}dx + (M_K+M_I)\int_{x_0}^{x}\frac{dt_K}{dx}dx - (M_{Cl}-M_I)\int_{x_0}^{x}\frac{dt_{Cl}}{dx}dx\Big]\frac{j_q}{\mathcal{F}}$$

Now, considering eq.(91) and the definitions eqs.(104) and (105) we see that

$$(j_s'(x))_t - (j_s'(x_0))_t = j_s'(x) - j_s'(x_0)$$

for any $x > x_0$. In summary, we obtain the result that the excess total solute mass flux is undetermined if the boundary conditions are not specified, however the difference between the output and input quantities is a well-defined quantity and independent of the arbitrary choice of the four constants A,B,C, $\mathcal{L}$. The same holds true for the constituent mass fluxes. Other features of the process to be considered here have already been described in connection with Figs. 5 and 6.

References

1) F. Dolezalek and F. Krüger, Z.Elektrochem. 1906, 12, 669
2. M. Spiro, in:"Techniques of Chemistry" (A. Weissberger, ed.) Vol. I, Part IIA, Wiley-Interscience, New York 1971,p.209
3) H.G. Hertz, Ber.Bunsengesellschaft Phys.Chem., 1977, 81, 656
4) H.G. Hertz, in: "Protons and Ions Involved in Fast Dynamic Phenomena", 1978, Elsevier Scientific Publishing Co., Amsterdam
5) A. Geiger and H.G. Hertz, J.C.S. Faraday I, 1980, 76, 135

Chapter 4

The moving boundary method

4.1 The experimental determination of transport numbers

The equations of the preceding section - no explicit considerations of the electrodes - have their most important application in conjunction with the moving boundary method which serves to measure transport numbers.[1] In a typical experiment we have two different cations and only one anion. In order to remain within the general framework of the main treatment we consider the special constituents Na, Li, and Cl. The replacement of K by Li is necessary for technical reasons as will be seen later. Now the set of eqs.(112)-(114) only contains two equations: the third one, eq. (114) is no longer necessary. Thus we have

$$\frac{\partial \rho_s}{\partial t} = \sum_{l=s,Na} D_{sl} \frac{\partial^2 \rho_l}{\partial x^2} - \frac{j_2}{\mathcal{F}}\left[(M_{Na}+M_{Cl})\frac{dt_{Na}}{dx} + (M_{Li}+M_{Cl})\frac{dt_{Li}}{dx}\right] \tag{125}$$

and

$$\frac{\partial \rho_{Na}}{\partial t} = \sum_{l=s,Na} D_{Nal} \frac{\partial^2 \rho_l}{\partial x^2} - \frac{j_2}{\mathcal{F}} M_{Na} \frac{dt_{Na}}{dx} \tag{126}$$

$$\frac{\partial \rho_w}{\partial t} = \sum_{l=s,Na} D_{wl} \frac{\partial^2 \rho_w}{\partial x^2}$$

where the concentration dependence of the diffusion coefficients has been neglected. The concentration distributions are sketched in Fig.7. The purpose of this arrangement is the measurement of the transport number of Na. Of course this is the transport number in the binary homogeneous electrolyte NaCl + water which we may denote by t_{Na}. To begin with, let us consider the first term on the right-hand side of eq.(126) which is $-\mathrm{div}\, j_{Na}$.

There is a negative Na constituent mass flux in the range around x_b where c_{Na} varies as depicted schematically in Fig.8. As a consequence we have for $-\partial j_{Na}/\partial x$ the behaviour shown in Fig.9.

Now we integrate eq.(126) with respect to x, the result is

$$\int_0^\infty \frac{\partial c_{Na}}{\partial t}\, dx = -\frac{j_2}{\mathcal{F}}\int_0^\infty \frac{dt_{Na}}{dx}\, dx = -\frac{j_2}{\mathcal{F}} t_{Na} \tag{127}$$

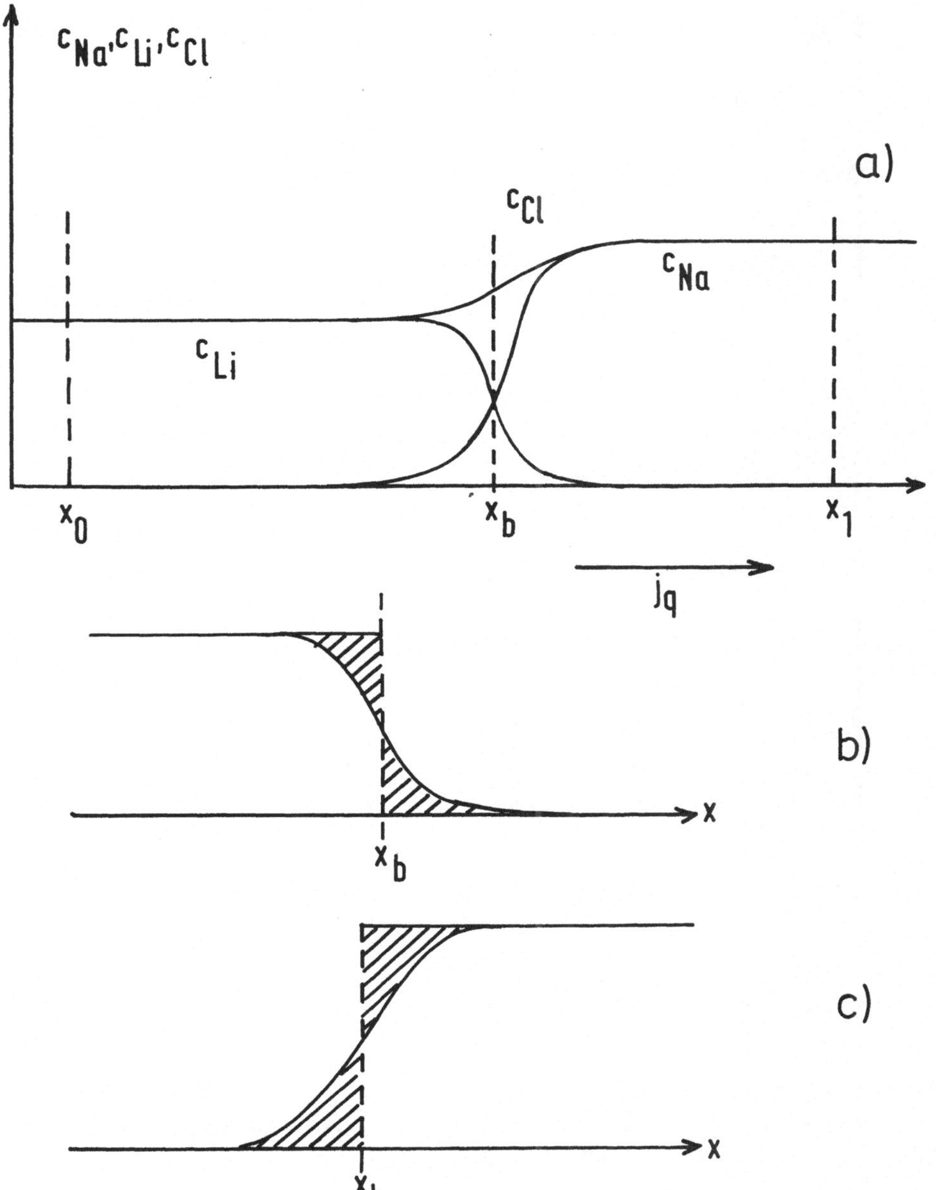

Fig. 7 a) Concentration distributions around the moving boundary (schematic);

b) and c) definition of coordinate x_b of moving boundary.

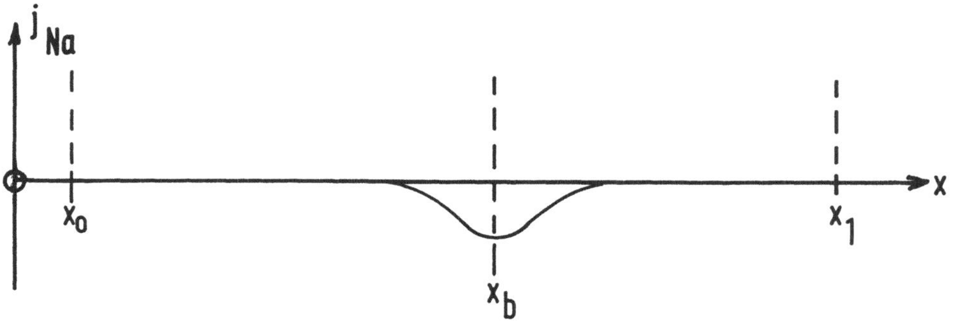

Fig.8 Constituent mass flux j_{Na} belonging to the concentration profile shown in Fig.7.

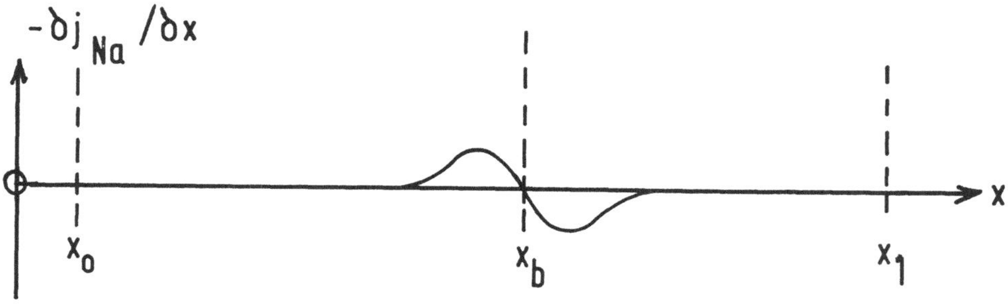

Fig. 9 Rate of change of Na constituent concentration due to diffusive flux belonging to the concentration profile shown in Fig.7

because $t_{Na} = 0$ where $c_{Na} = 0$ and $t_{Na}(x) = t_{Na}$ = const. in the region $x > x_b$ where c_{Na} = const. Due to the fact that the constituent diffusion flux j_{Na} cannot "produce" any mass of Na, the integral over the first term on the right of eq.(126) vanishes.

Next we write:

$$\frac{\partial}{\partial t}\int_0^\infty c_{Na}\,dx = -\frac{j_e}{\mathcal{F}}t_{Na}$$

where

$$\int_0^{\infty} c_{Na}\, dx$$

is the total amount of Na in the system between the two electrodes (when normalized to unit cross section). However we are not interested in the situation for $x > x_1$ as shown in Fig.7. Therefore, considering the dashed line at $x = x_b$ in Fig.7 we have

$$\frac{\partial}{\partial t}(x_1 - x_b)\, c_{Na} = -\frac{j_Q}{\mathcal{F}}\, t_{Na} \tag{128}$$

The dashed line at $x = x_b$ in Fig.7, is drawn such that the hatched areas are equal (see Fig.7b).

Integration of eq.(128) with respect to time gives

$$c_{Na}\,\Delta(x_1 - x_b) = c_{Na}\left[-x_b^{end} - (-x_b^{begin})\right]$$

$$= -\frac{1}{\mathcal{F}}\, t_{Na} \int_{t_{begin}}^{t_{end}} j_Q\, dt$$

or

$$c_{Na}(x_b^{end} - x_b^{begin}) = \frac{1}{\mathcal{F}}\, t_{Na} \int_{t_{begin}}^{t_{end}} j_Q\, dt$$

Multiplication by the cross section A gives the well-known formula

$$c_{Na} V_{Na} = \frac{1}{\mathcal{F}}\, Q\, t_{Na} \tag{129}$$

where

$$V_{Na} = A(x_b^{end} - x_b^{begin})$$

and

$$Q = \int_{t_{begin}}^{t_{end}} j_Q\, dt$$

Measurement of V_{Na} and Q allows the experimental determination of the transport number t_{Na}.

Next we calculate the concentration variation of the other cationic constituent i.e. Li. We stated previously that when the total solute density and the constituent density for Na are given, then the Li constituent partial density - or concentration - is also given. It follows that the time derivative of c_{Li} is also known:

$$\frac{\partial c_i}{\partial t} = \frac{\partial \rho_s/\partial t - \partial c_{Na}/\partial t\,(M_{Na} + M_{ce})}{M_{Li} + M_{ce}}$$

Introduction of eqs.(125) and (126) yields:

$$\frac{\partial c_{Li}}{\partial t} = \frac{\sum_e D_{se}\frac{\partial^2 \rho_e}{\partial x^2} - \frac{j_e}{\mathcal{F}}(M_{Li} + M_{ce})\frac{dt_{Li}}{dx} - \left(1 + \frac{M_{ce}}{M_{Na}}\right)\sum_e D_{Nae}\frac{\partial^2 \rho_e}{\partial x^2}}{M_{Li} + M_{ce}}$$

$$= \frac{\sum_e D_{se}\frac{\partial^2 \rho_e}{\partial x^2} - \left(1 + \frac{M_{ce}}{M_{Na}}\right)\sum_e D_{Nae}\frac{\partial^2 \rho_e}{\partial x^2}}{M_{Li} + M_{ce}} - \frac{j_e}{\mathcal{F}}\frac{dt_{Li}}{dx} \qquad (130)$$

After integration with respect to x one obtains again:

$$\frac{\partial}{\partial t}\int_0^\infty c_{Li}\,dx = \frac{j_e}{\mathcal{F}}\,t_{Li} \qquad (131)$$

because the diffusion flux terms give integrals which vanish and $t_{Li}(x) = t_{Li} = \text{const.}$ at all those $x < x_b$ for which c_{Li} = const. And of course, we have $t_{Li} = 0$ for all x where $c_{Li} = 0$.

$$\int_0^\infty c_{Li}\,dx$$

is the total amount of Li in the system referred to unit cross-section. As before, we define an arbitrary position x_o between the anode and $x = x_b$, located such that c_{Li} = const. at $x = x_o$ (see Fig.7). Now we assume that the hatched area construction holds also for the Li distribution at the beginning of the experiment, $t = t_{begin}$ (see Fig.7c). This gives for eq.(131)

$$c_{Li}\frac{\partial}{\partial t}(x_b - x_o) = \frac{j_e}{\mathcal{F}}\,t_{Li} \qquad (131a)$$

and after integration with respect to time one obtains

$$c_{Li}\,\Delta(x_b - x_o) = c_{Li}\left(x_b^{end} - x_b^{begin}\right)$$

$$= \frac{t_{Li}}{\mathcal{F}}\int_{t_{begin}}^{t_{end}} j_e\,dt$$

and finally, after multiplication by the cross-section A one finds

$$c_{Li} V_{Li} = \frac{t_{Li}}{\mathcal{F}} Q \tag{132}$$

with

$$V_{Li} = A(x_b^{end} - x_b^{begin})$$

Now, in order to obtain

$$V_{Li} = V_{Na}$$

we must have (see eqs.(129) and (132)

$$\frac{c_{Li}}{c_{Na}} = \frac{t_{Li}}{t_{Na}} \tag{133}$$

which is the Kohlrausch relation.

The requirement $V_{Li} = V_{Na}$

$$= (x_b^{end} - x_b^{begin})_{Li} = (x_b^{end} - x_b^{begin})_{Na}$$

means that the points of maximum slope for Na and Li concentration profiles always coincide and this in turn is one of the conditions for the conservation of a sharp boundary during its movement.

4.2 The self-regulating mechanism of the moving boundary

Even if at the beginning of the experiment, eq.(133) is not fulfilled, then, during the passage of the current the agreement with this relation is automatically approached. This is one aspect of the "self-regulating" mechanism which may be understood if we write eq.(130) in abbreviated form

$$\frac{\partial c_{Li}}{\partial t} = -\frac{j}{\mathcal{F}} \frac{dt_{Li}}{dx} \tag{130a}$$

neglecting the diffusion term - for simplicity -.

Let us first assume that in the respective homogeneous regions, $c_{Li} = c_{Na}$ at the beginning of the experiment (more generally $c_{Li} > c_{Na} \cdot t_{Li}/t_{Na}$). The situation is sketched in Fig.10a. Our main interest lies in the Li concentration distribution. In fact, t_{Li} is a function of the coordinate x, and thus there is an increase per unit time of the Li concentration at any point where $dt_{Li}/dx < 0$ (see eq.130a). But dt_{Li}/dx does not depart from zero because c_{Li} varies with the position x as shown in Fig.10a. The transport number of Li in LiCl solution is almost constant as the LiCl concentration varies. It is the presence of Na which causes the position dependence of t_{Li},

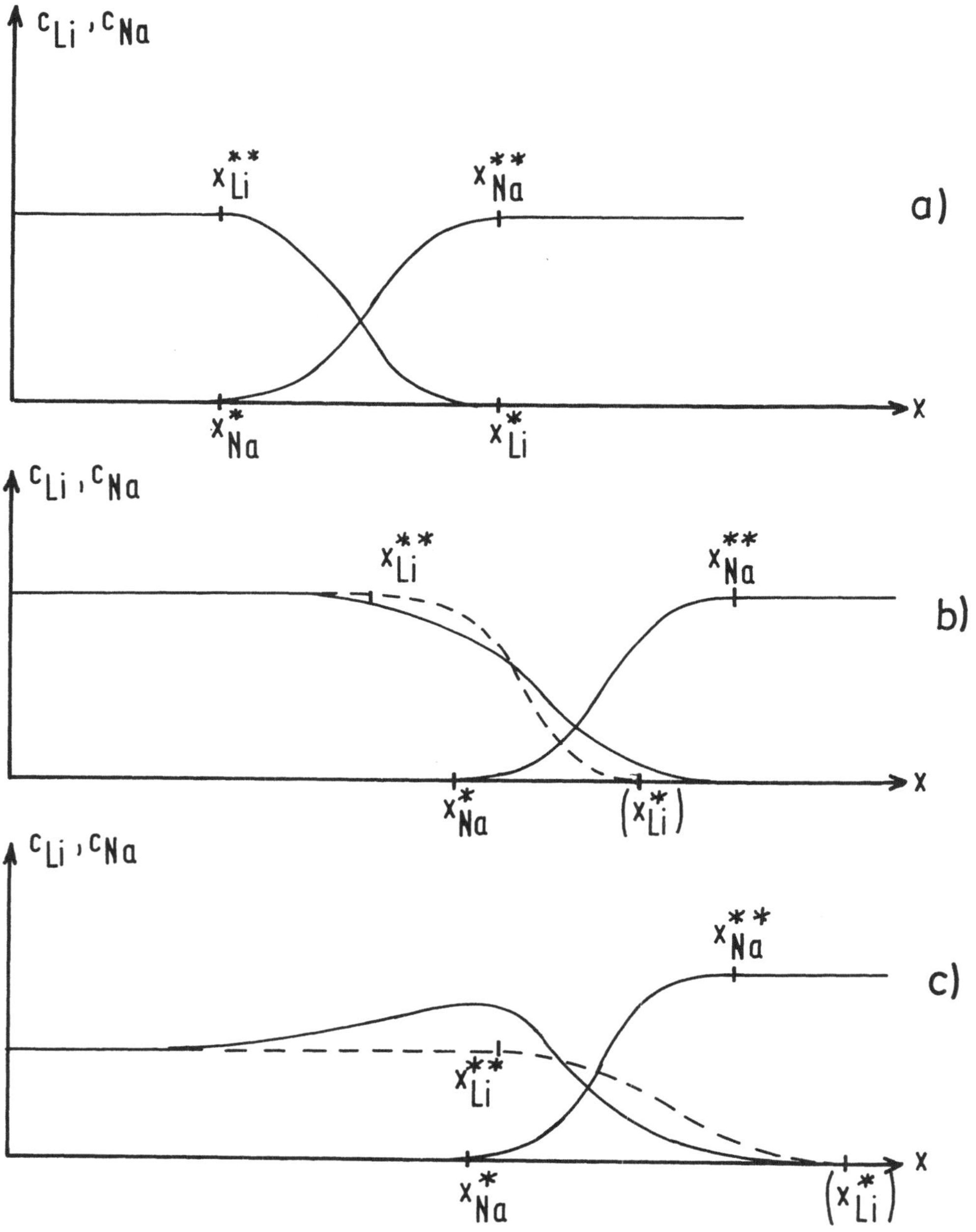

Fig.10 : Concentration profiles and self-regulation mechanism in the moving boundary between a LiCl and NaCl solution. For more details see text.

and indeed we have $t_{Li} \rightarrow 0$ at $x = x_{Li}^{*}$, where $c_{LiCl} \rightarrow 0$. This could never occur if there is not another cation present in the system. At x_{Na}^{*} we have $c_{Na} = 0$, and for all $x > x_{Na}^{**}$, $dc_{Na}/dx = 0$; likewise for all $x < x_{Li}^{**}$ $dc_{Li}/dx = 0$. $x_{Li}^{**} = x_{Na}^{*}$, and $x_{Li}^{*} = x_{Na}^{**}$. Now we consider Fig. 10b which describes the situation after the electric current has been switched on for a period. According to eqs. (128) and (131a) the two concentration profiles move with different speeds when $c_{Li} = c_{Na}$. Since $t_{Li} < t_{Na}$, the Li concentration profile moves more slowly. The dashed curve in Fig.10b shows what we would have if the Li profile only suffered a translation without a change of shape. For $x < x_{Na}^{*}$ $dc_{Li}/dx < 0$, however as we saw above, this does not influence the transport number. Since here $dc_{Na}/dx = 0$, $dt_{Li}/dx \approx 0$ and dc_{Li}/dt is virtually zero. This means that dc_{Li}/dt is already zero before the initial concentration value is fully established. Consequently, as indicated by the full curve in Fig.10b, during the movement of the profile, the Li concentration has decreased relative to the value given by the dashed line. For $x > x_{Na}^{*}$ the amount of Na admixed to the solution is greater than in a situation where the Na and Li distribution had the same point of inflection. Thus the transport number of Li, having a tendency to remain constant for $x < x_{Na}^{*}$, for $x > x_{Na}^{*}$ declines to zero more rapidly which causes $-dt_{Li}/dx$ to be large in this region. Consequently - as given by eq.(130a) - dc_{Li}/dt is also greater, causing an excess increase of the Li concentration as indicated in Fig.10b.

Next we consider the reverse situation, that is $c_{Li} < c_{Na} \cdot t_{Li}/t_{Na}$. Now, according to eq.(130a), the Li concentration profile travels faster than that of the Na distribution. In Fig.10c the dashed profile is again the one we would expect if only translation and no change of shape occurred. However, now in the region $x < x_{Li}^{**}$ there is already an appreciable amount of Na admixed to the solution because Na moves so slowly. The consequence is a depression of the Li transport number in the sense to give $dt_{Li}/dx < 0$ and this in turn produces an excess creation of Li constituent, where according to the original shape of the Li distribution no such increase should occur. The result for $x < x_{Li}^{**}$ is sketched in Fig.10c. Finally, since at $x = x_{Li}^{**}$, t_{Li} has already become close to zero, for the position $x > x_{Li}^{**}$ there is no longer much possibility for t_{Li} to change. Thus dt_{Li}/dx is small and we obtain incomplete restoration of the original distribution which is again demonstrated in Fig.10c. We see that the situation $c_{Li} = c_{Na} \cdot t_{Li}/t_{Na}$ is the only one in which the concentration distribution in the range $x < x_{Li}^{**}$ is preserved during the action of the electric current. Moreover, from experience one can derive that, apart from eq.(133), there is also the asymmetry condition

$$t_{Na} > t_{Li} \tag{133a}$$

which should hold, where here Na is defined as the constituent occupying the space with $x > x_b$ if the electric density j_2 has a positive direction, thus Na is the leading constituent or leading cation. The asymmetry rule is necessary to have another "self-regulating" mechanism working properly which causes the concentration profile to sharpen during the movement of the boundary.

In the following section we shall give a qualitative treatment also of this self-regulating or "adjusting" effect. Let us assume that the instantaneous concentration profile of Na is given by the equations

$$c_{Na}(x) = 0 \qquad \text{for} \qquad x \leq x^* \equiv x^*_{Na}$$

and

$$c_{Na}(x) = c_{Na} \frac{((x-x^*)/\lambda)^n}{1+((x-x^*)/\lambda)^n} \qquad \text{for} \qquad x > x^* \equiv x^*_{Na} \tag{134}$$

where λ is a constant scaling factor.
According to eq.(134) c_{Na} is the constant Na concentration for $x \gg x^*$. The exponent n describes the sharpness of the concentration profile; this sharpness increases with increasing n and for $n \rightarrow$ very large, eq.(134) describes a rectangular step function. The point of inflection occurs at

$$x_i = x^* + \lambda \left(\frac{n^2}{n(n+2)} \right)^{1/n}$$

For sufficiently large n values we can use the approximation

$$x_i = x^* + \lambda \tag{135}$$

Now the moving boundary implies the fact that x_i varies with time:

$$x_i = ut \tag{136}$$

where u is the speed of the boundary. Since λ is a constant, x^* also changes with time. Insertion of eqs.(135) and (136) into eq.(134) yields

$$c_{Na}(x,t) = 0 \qquad \text{for} \qquad x \leq ut - \lambda$$

$$c_{Na}(x,t) = c_{Na}\frac{[(x-ut+\lambda)/\lambda]^{n(t)}}{1+[(x-ut+\lambda)/\lambda]^{n(t)}} \quad (137)$$

for $\qquad x \geq ut-\lambda$

We must understand in this formula that n is a function of the time, thus the change of n with time incorporates the change of the shape of the profile during the application of the current. Next we write for the transport number

$$t_{Na}(x) = t_{Na}\frac{c_{Na}(x)}{\alpha c_{Na}} \quad (138)$$

where, as before t_{Na} is the transport number in the binary system $NaCl + H_2O$. $c_{Na}(x)$ is the Na concentration distribution along x at a given instant. The quantity α incorporates the inequality (133a), it has the properties

$\alpha < 1 \qquad$ for $\qquad x \approx ut-\lambda$

$\alpha \to 1 \qquad$ for $\qquad x > ut$

Thus, since $t_{Li} < t_{Na}$, in those regions where both constituents Na and Li are present, i.e. $ut-\lambda < x < ut+\lambda$, the excess constituent flux of Na is larger than that given by the relative amounts of constituents present. Obviously $u = dx_b/dt$, so that eq.(128) yields:

$$u = \frac{j_q}{\mathcal{F}}\frac{t_{Na}}{c_{Na}} \quad (139)$$

Now we introduce eqs.(137),(138), and (139) in eq.(126) and we neglect the diffusion term. The result is

$$\frac{1}{n}\frac{dn}{dt}[(x-ut+\lambda)/\lambda]\ln(x-ut+\lambda)/\lambda$$

$$= \frac{j_q}{\mathcal{F}}\frac{t_{Na}}{c_{Na}\lambda}\left(\frac{\alpha-1}{\alpha}\right) + \frac{j_q}{\mathcal{F}}\frac{t_{Na}}{n\alpha^2 c_{Na}}\frac{d\alpha}{dx}\cdot\left\{1+[(x-ut+\lambda)/\lambda]^n\right\}(x-ut+\lambda)/\lambda \quad (140)$$

For $x < ut$, the right-hand side of this equation is negative, because $\alpha < 1$ and $(x-ut+\lambda)/\lambda \approx 0$, $d\alpha/dx \approx 0$, so that the magnitude of the second term is small. Furthermore, on the left-hand side:

$$\ln(x-ut+\lambda)/\lambda < 0$$

$$(x-ut+\lambda)/\lambda > 0$$

it follows that in the range $ut - \lambda < x < ut$ we have

$$dn/dt > 0$$

At $x = ut$ both sides of eq.(140) are zero, and in the range $ut < x < ut + \lambda$ the right-hand side of eq.(140) becomes positive. Since $\alpha \longrightarrow 1$, now the second term on the right dominates. Also on the left

$$[(x - ut + \lambda)/\lambda] \ln (x - ut + \lambda)/\lambda > 0$$

so that we get the result

$$dn/dt > 0$$

in the range $ut < x < ut + \lambda$. Thus we have demonstrated that n increases in the course of time, the concentration profile becomes steeper; in the opposite case, that is $\alpha > 1$ and $d\alpha/dx < 0$, we would have the reverse behaviour.

The Li concentration may also be described by the analog of eq.(137)

$$C_{Li}(x,t) = 0 \qquad \text{for} \qquad x > ut + \lambda$$

$$C_{Li}(x,t) = C_{Li} \frac{[(ut + \lambda - x)/\lambda]^{n(t)}}{1 + [(ut + \lambda - x)/\lambda]^{n(t)}}$$

$$\text{for} \qquad x \leq ut + \lambda$$

Now we have

$$t_{Li}(x) = t_{Li} \frac{C_{Li}(x)}{\alpha^* C_{Li}}$$

with

$$\alpha^* > 1 \qquad \text{for} \qquad x \approx ut + \lambda$$

$$\alpha^* \longrightarrow 1 \qquad \text{for} \qquad x < ut$$

thus $d\alpha^*/dx > 0$. Then, using eq.(130a) we obtain an expression, which apart from the interchange of the signs of x and ut, has the same form as eq.(140), however the factor $(\alpha - 1)/\alpha$ has to be replaced by $(1 - \alpha^*)/\alpha^*$. Then by the same arguments as applied above it can be shown that for the Li distribution we have also $dn/dt > 0$, that is, a sharpening of the boundary during the movement. If the concentration profile becomes steeper and steeper, then the diffusion flux increases and finally a stationary state is formed in which the shape no longer varies with time.

The moving boundary between two solutions containing the same electrolyte of different concentrations is also given by our eq.(129):

$$V(c_{Na'} - c_{Na''}) = \frac{1}{\mathcal{F}} Q(t_{Na'} - t_{Na''})$$

With respect to this problem, equations have been reported in the literature which are of similar appearance to the ones given in our treatment[1]. However the diffusion coefficients in these formulas are directly related to the ionic mobilities and this is not a requirement of our theory. Also, the total conceptual situation is different (see also the report of NEWMAN[2] where the equations given are similar to ours, and see also the analysis of the work of MILLER, of HAASE and RICHTER, SCHÖNERT and others to be given in chapter 7).

References

1) D.A. MacInnes and L.G. Longsworth, Chem.Rev. 1932, 11, 171

2) L.G. Longsworth, J.Amer.Chem.Soc., 1943,65, 1755.

3) J. Newman, in: Advances in Electrochemistry and Electrochemical Engineering, Vol.5, ed.C.W. Tobias, Interscience London, 1967.

Chapter 5

The diffusion process at the electrodes in the presence of an electric current.

5.1 Mass fluxes and excess mass fluxes at a metal electrode

In the previous chapter we described the changes of the concentration profiles in a liquid boundary in the presence of an electric current, in particular, the movement of the boundary was the main object of interest. When the electrodes are of solid material - or in other cases are not electrolytes and not miscible with the electrolyte - then it is more suitable to give the main emphasis to the description of the mass fluxes onto and away from the electrode material and to devote only a fairly brief discussion to the concentration variation in the electrolyte solution. It should be mentioned here that in this book the word electrode is reserved for one single metallic phase - sometimes in combination with a non-metallic solid phase - it is not understood to refer to the combined system metal-electrolyte, for instance to the "half-cell" Cu(metal)/$CuSO_4$ solution. Still, as has been demonstrated in Chapter 3, the boundary between two different liquid electrolyte phases in certain respects acts like an electrode although not an electrode in the sense of the definition just given. Therefore, in the last section of this chapter we shall also give an outline of the effect of the mass fluxes and the excess mass fluxes at a liquid junction between two bulk electrolytes.

As before, we consider an electrolyte solution containing the four constituents Na, K, Cl, and I. For simplicity we confine ourselves to a "one dimensional" system with constant cross-section, x being the coordinate along which the various fluxes occur. Let us assume that a metal electrode is at x = 0 and that our diffusion system extends from x = 0 to $x = \ell$ where $x = \ell$ is the coordinate of the other electrode, the "end electrode". Thus the two electrodes represent the boundaries of our diffusion problem. In contrast to the boundaries of an ordinary diffusion cell, these boundaries are such that the various mass fluxes do not necessarily vanish at the corresponding coordinates x = 0, $x = \ell$. As we shall see $j_k \neq 0$, k = s,Na,Cl if the electric current density $j_Z \neq 0$. Let us choose the positive direction of the vector j_Z such that it coincides with the positive x direction $[\rho_{Na}]_{el}$, the partial constituent density of Na in the metal electrode is different from zero, but for the three other constituents we have $[\rho_k]_{el} = 0$, k = K, Cl, I. Therefore we call our electrode a sodium electrode. In the end electrode at least one of the constituents has a non-zero partial density, but the details are not of importance here.

Let the partial densities have the following simple distributions around x = 0 and at t = 0. Thus the electric current J_z has just been switched on, the system had attained its equilibrium distributions for $t < 0$.

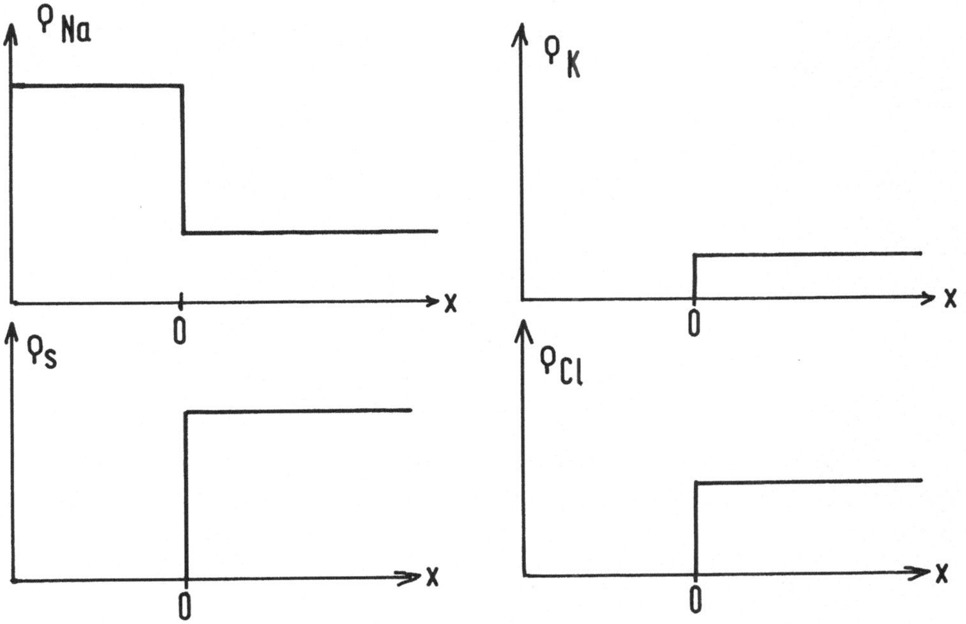

Fig. 11 Total solute partial density and constituent partial mass densities at the boundary Na electrode/electrolyte in an equilibrium situation.

As has already been mentioned, we wish to calculate the various mass fluxes at x = 0, i.e. at the boundary of the Na electrode/electrolyte solution. We start from our fundamental equations of electrochemistry, eqs.(112)-(114), which, for convenience, are written once again.

$$\frac{\partial \rho_s}{\partial t} = - \mathrm{div}\, j_s - \left[(M_{Na} + M_I) \frac{dt_{Na}}{dx} + (M_K + M_I) \frac{dt_K}{dx} - (M_{Cl} - M_I) \frac{dt_{Cl}}{dx} \right] \frac{j_z}{\mathcal{F}} \qquad (112)$$

$$\frac{\partial \rho_{Na}}{\partial t} = -\operatorname{div} j_{Na} - \frac{M_{Na}}{\mathcal{F}} j_{\Sigma} \frac{dt_{Na}}{dx} \qquad (113)$$

$$\frac{\partial \rho_{Cl}}{\partial t} = -\operatorname{div} j_{Cl} + \frac{M_{Cl}}{\mathcal{F}} j_{\Sigma} \frac{dt_{Cl}}{dx} \qquad (114)$$

When the coordinate x is taken with respect to the electrode surface and at the same time x = 0 is assumed to be constant relative to the cell coordinate system, this is an approximation. Actually the electrode process represents a moving boundary problem and the fluxes with respect to the cell walls contain an additive contribution which however does not need to be written down explicitly. As we have outlined in Chapter 3 the absolute value of the Na transport number at the boundary is already fixed:

$$t_{Na}(x) = 1 \quad \text{at} \quad x = 0$$

furthermore, since $\rho_{Cl} = \rho_K = \rho_I = 0$ for $x \leq 0$ we must have $t_{Cl}(x) = t_K(x) = t_I(x) = 0$ at x = 0. Next we introduce linear x dependences for the various t's

$$t_{Na}(x) = 1 - \frac{1 - t_{Na}}{\delta'_x} x \qquad (141)$$

$$t_K(x) = t_K \frac{x}{\delta'_x} \quad , \quad t_{Cl}(x) = t_{Cl} \frac{x}{\delta'_x} \qquad (142)$$

where the t_i's are the respective transport numbers in the region $x > \delta'_x$ where the composition of the solution is constant. This functional form of the transport numbers is sketched in Fig.12.

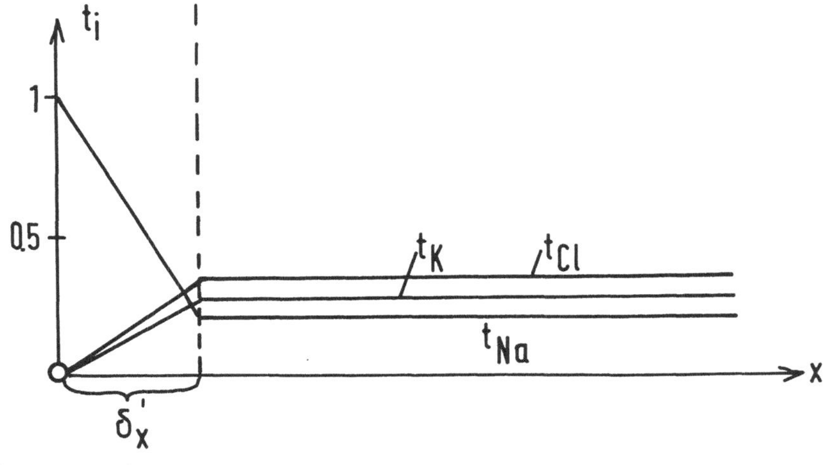

Fig. 12 Assumed linear position dependences of the various transport numbers at the Na electrode.

Now, considering eqs.(122)-(124) the excess total solute mass flux and the Na and Cl excess constituent mass fluxes at x $\geqslant \delta' x$ can be immediately written down.

$$j_s' = \{(M_{Na}+M_I)t_{Na} + (M_K+M_I)t_K - (M_{Cl}-M_I)t_{Cl} - M_I\}\frac{j_\varepsilon}{\mathcal{F}} \tag{143}$$

$$j_{Na}' = \frac{M_{Na}}{\mathcal{F}} t_{Na} j_\varepsilon \tag{144}$$

$$j_{Cl}' = -\frac{M_{Cl}}{\mathcal{F}} t_{Cl} j_\varepsilon \tag{145}$$

and we have also

$$j_I' = -\frac{M_I}{\mathcal{F}}(1 - t_{Na} - t_K - t_{Cl}) j_\varepsilon \tag{146}$$

These are the <u>excess</u> mass fluxes; now we compute the mass fluxes (see the definitions on page 59). If we use the expressions eqs.(141) and (142) in eqs. (112)-(114) and perform the integration with respect to x, we obtain:

$$\int_0^{\delta' x} \frac{\partial \rho_s}{\partial t} dx = - j_s - [(M_{Na}+M_I)(t_{Na}-1) + (M_K+M_I)t_K - (M_{Cl}-M_I)t_{Cl}]\frac{j_\varepsilon}{\mathcal{F}} \tag{147}$$

$$\int_0^{\delta' x} \frac{\partial \rho_{Na}}{\partial t} dx = - j_{Na} + (1-t_{Na})\frac{j_\varepsilon}{\mathcal{F}} M_{Na} \tag{148}$$

$$\int_0^{\delta' x} \frac{\partial \rho_{Cl}}{\partial t} dx = - j_{Cl} + t_{Cl}\frac{j_\varepsilon}{\mathcal{F}} M_{Cl} \tag{149}$$

Of course these results are independent of the special choice of the functional form of $t_i(x)$ according to eqs.(141) and (142). Now we consider the limit $\delta' x \to 0$. Then the left-hand sides of eqs.(147)-(149) vanish and our result is:

$$(j_s)_{x=0} = \{(M_{Na}+M_I)(1-t_{Na}) - (M_K+M_I)t_K + (M_{Cl}-M_I)t_{Cl}\}\frac{j_\varepsilon}{\mathcal{F}} \tag{150}$$

$$(j_{Na})_{x=0} = (1-t_{Na})\frac{j_\varepsilon}{\mathcal{F}} M_{Na} \tag{151}$$

$$(j_{Cl})_{x=0} = t_{Cl}\frac{j_\varepsilon}{\mathcal{F}} M_{Cl} \tag{152}$$

Having these fluxes, the two other constituent mass fluxes are also known, by virtue of eqs.(61) and (62). One derives

$$(j_K)_{x=0} = -t_K \frac{j_\varrho}{\mathcal{F}} M_K$$

$$(j_I)_{x=0} = (1 - t_{Na} - t_K - t_{Cl}) \frac{j_\varrho}{\mathcal{F}} M_I = t_I \frac{j_\varrho}{\mathcal{F}} M_I$$

These are the total solute mass flux and the constituent mass fluxes at the electrode surface, $\delta x \to 0$ at $t = 0$.

Let us now discuss our results. Assume that the current density is positive, $j_\varrho > 0$. Then we have a positive total solute mass flux j_s. For, if we assume for a moment $t_{Na} = t_K = t_{Cl} = t_I = 1/4$, we get:

$$(j_s)_{x=0} = (M_{Na} + M_{Cl} + M_{Na} + M_I + M_{Na} - M_K) \frac{j_\varrho}{4\mathcal{F}}$$

which is > 0, that is, the mass flux vector is directed into the electrolyte solution. This is not only due to the special choice of the system. Had we taken the constituents Li, Cl, Br, and Cs, then, considering that $t_{Li} < 1/4$, we would get $j_s \approx 0$, however > 0.

With all the $t_j = 1/4$, the excess total solute mass flux in the Na, K, Cl, I system is (see eq.(143))

$$j_s' = \frac{1}{4}(M_{Na} + M_K - M_{Cl} - M_I) \frac{j_\varrho}{\mathcal{F}}$$

which is < 0 if $j_\varrho > 0$.

The constituent mass flux j_{Na} and the excess constituent mass flux j_{Na}' are also positive, the electrode acts as anode. We recall that the constituent mass flux is connected with the gradients of the partial mass densities, whereas the excess constituent mass is directly connected with the electric current density and in both cases the negative divergence of the flux is the experimentally accessible quantity. It is the result of a suitable analytical operation with or without the extraction of a sample. As already mentioned, on a molecular scale, the excess constituent fluxes represent the rotations of salt molecules.

The Cl and I constituent mass fluxes have positive values, also. The balancing of Cl and I masses to fulfil the conservation of mass is effected by the two excess constituent fluxes j_{Cl}' and j_I' which are negative. The K constituent mass flux is negative; it is the only negative contribution to the total solute mass flux, however the excess constituent mass flux j_K' is positive in order to warrant fulfilment of the conservation of mass. In Fig.13 we give a representation of the various fluxes at the metal electrode. Constituent and solute mass fluxes are drawn in full, the dashed arrows represent excess constituent mass fluxes. It has been assumed that all transport numbers are equal to 1/4.

The three mass fluxes eq.(150)-(152) play an important role in the computation of the analytical field $\rho_S(x,t)$, $\rho_{Na}(x,t)$, $\rho_{Cl}(x,t)$, namely they represent the boundary conditions at x = 0. So far they are the boundary conditions only at time t = 0, however this will be generalized shortly. Thus the analytical field has to satisfy the following boundary conditions:

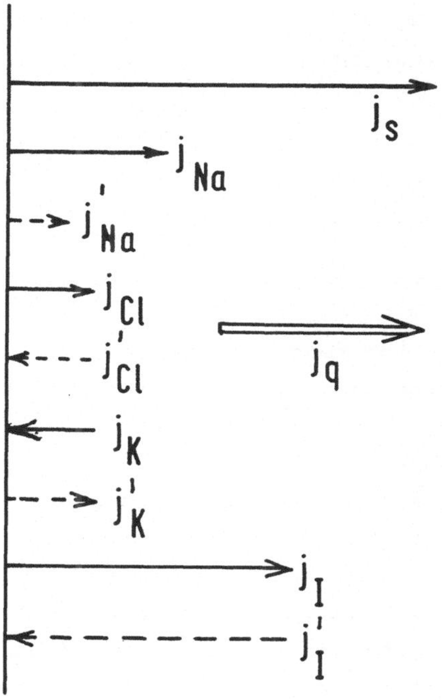

Fig. 13 The total solute mass flux and the various constituent mass fluxes at the Na electrode surface. All transport numbers are taken to be 1/4. Dashed arrows are the corresponding excess mass fluxes.

$$-(j_S)_{x=0} = -\left\{(M_{Na}+M_I)(1-t_{Na}) - (M_K+M_I)t_K + (M_{Cl}-M_I)t_{Cl}\right\}\frac{j_q}{\mathcal{F}}$$

$$= D_{SS}\frac{\partial \rho_S}{\partial x} + D_{SNa}\frac{\partial \rho_{Na}}{\partial x} + D_{SCl}\frac{\partial \rho_{Cl}}{\partial x} \tag{153}$$

$$-(j_{Na})_{x=0} = -(1-t_{Na})\frac{j_q}{\mathcal{F}}M_{Na}$$

$$= D_{NaS}\frac{\partial \rho_S}{\partial x} + D_{NaNa}\frac{\partial \rho_{Na}}{\partial x} + D_{NaCl}\frac{\partial \rho_{Cl}}{\partial x} \tag{154}$$

$$-(j_{Cl})_{x=0} = -t_{Cl} \frac{j_2}{\mathcal{F}} M_{Cl}$$

$$= D_{Cl\,s} \frac{\partial \rho_s}{\partial x} + D_{Cl\,Na} \frac{\partial \rho_{Na}}{\partial x} + D_{Cl\,Cl} \frac{\partial \rho_{Cl}}{\partial x} \tag{155}$$

These are three equations for the three quantities

$$\left(\frac{\partial \rho_s}{\partial x}\right)_{x=0}, \quad \left(\frac{\partial \rho_{Na}}{\partial x}\right)_{x=0}, \quad \left(\frac{\partial \rho_{Cl}}{\partial x}\right)_{x=0}$$

In the present case, these gradients are negative because $j_\ell > 0$, ℓ = s,Na,Cl. As a consequence the slopes of the rectangular density distributions shown in Fig. 11 have the following properties:

$$\left[\lim_{x\to 0} \frac{\partial \rho_s}{\partial x}\right]_{x<0} = \infty \quad , \quad \left(\frac{\partial \rho_s}{\partial x}\right)_{x=0} < 0$$

$$\left(\frac{\partial \rho_s}{\partial x}\right)_{x>0} = 0$$

and similarly for the two constituents Na and Cl.

In the next step we consider the situation for t > 0. We assume that the experimental conditions are such that j_2 = const. In the derivation of the boundary fluxes eqs.(147)-(152) the transport numbers are no longer those of the uniform solution, rather they are the local ones at the electrode which result when integration is extended only to the coordinate $\delta' x$, $\delta' x \to 0$. The variation with time of the analytical field is given by, or has to satisfy, the equations (153)-(155). $t_j = t_j(\rho_s, \rho_{Na}, \rho_{Cl})$ (j = s, Na,Cl) have to be known functions of the composition. It is to be expected that the t_j depend only weakly on the total salt concentration if the relative amounts of the constituents show little change. However, the boundary conditions given by eqs.(153),(155) should slowly change even if the current density is kept constant. This is a consequence of the weak concentration dependence of the transport numbers.

In practical cases the solution of eqs.(115)-(117) for a four constituent mixture in general will be very difficult and has probably not yet been done. However, if KCl and KI have comparatively high concentrations and NaCl is present only in a very small amount, then we have $t_{Na} \to 0$ and it follows from eq.(151) that

$$-\frac{j_2}{\mathcal{F}} M_{Na} = D_{Na\,Na} \frac{\partial \rho_{Na}}{\partial x} \tag{156}$$

and since $\partial \rho_{Na}/\partial x$ is small, we get $j_s \approx 0$, $j_{Cl} \approx 0$ and $\partial \rho_s/\partial x =$

$= \partial \varrho_{Cl}/\partial x = 0$ throughout the total system.

In this situation one usually speaks of a supporting electrolyte, which is here given by the two components KI and KCl. However, in most practical cases the supporting electrolyte contains only one component. This leads us to a brief discussion of the mass fluxes at the Na electrode if the electrolyte solution is constructed from the two components NaCl and KCl. Now the set of three equations (150)-(152) has to be replaced by the set of two equations:

$$(j_S)_{x=0} = \left\{ M_{NaCl}(1-t_{Na}) - M_{KCl}\, t_K \right\} \frac{j}{\mathcal{F}} \tag{150a}$$

$$(j_{Na})_{x=0} = (1-t_{Na}) \frac{j}{\mathcal{F}} M_{Na} \tag{151a}$$

and in the limiting situation $\varrho_{NaCl} \to 0$, eq.(156) is also valid with $j_S \approx 0$.

Finally, it is obvious that in the case of the binary electrolyte NaCl + H_2O, eq.(151) disappears and we are left with the boundary condition

$$(j_S)_{x=0} = M_{NaCl}(1-t_{Na}) \frac{j}{\mathcal{F}} = -D_{NaCl} \frac{\partial \varrho_{NaCl}}{\partial x} \tag{150b}$$

5.2 The Hittorf method to measure transport numbers.

As the reader will have realized, the formulas governing the diffusion processes at the electrode are fairly complicated. Still, the general equations (112)-(114) are of great importance for practical use and this can be seen if they are integrated with respect to x and the integration is extended into a region $x \geq x_{unif}$ where the gradients of all the partial mass densities vanish. Here we have also

$$(j_S)_{x_{unif}} = (j_{Na})_{x_{unif}} = (j_{Cl})_{x_{unif}} = 0 \tag{157}$$

The result of the integration is

$$\int_0^{x_{unif}} \frac{\partial \varrho_s}{\partial t} dx = -\left[(j_s)_{x_{unif}} - (j_s)_{x=0}\right]$$

$$-\frac{j_\varrho}{\mathcal{F}}\Big\{(M_{Na}+M_I)\left[(t_{Na})_{x_{unif}} - (t_{Na})_{x=0}\right] + (M_K+M_I)\left[(t_k)_{x_{unif}} - (t_K)_{x=0}\right]$$

$$-(M_{Cl}-M_I)\left[(t_{Cl})_{x_{unif}} - (t_{Cl})_{x=0}\right]\Big\}$$

$$\int_0^{x_{unif}} \frac{\partial \varrho_{Na}}{\partial t} dx = -\left[(j_{Na})_{x_{unif}} - (j_{Na})_{x=0}\right] - \frac{M_{Na}}{\mathcal{F}} j_\varrho \left[(t_{Na})_{x_{unif}} - (t_{Na})_{x=0}\right]$$

$$\int_0^{x_{unif}} \frac{\partial \varrho_{Cl}}{\partial t} dx = -\left[(j_{Cl})_{x_{unif}} - (j_{Cl})_{x=0}\right] + \frac{M_{Cl}}{\mathcal{F}} j_\varrho \left[(t_{Cl})_{x_{unif}} - (t_{Cl})_{x=0}\right]$$

Now we substitute the left-hand parts of eqs.(153)-(155) and eqs.(157) in the last relations to obtain

$$A\frac{\partial}{\partial t}\int_0^{x_{unif}} \varrho_s dx = \frac{dm_s}{dt} = \Big\{(M_{Na}+M_I)\left[1-(t_{Na})_{x_{unif}}\right] - (M_K+M_I)(t_K)_{x_{unif}}$$

$$+(M_{Cl}-M_I)(t_{Cl})_{x_{unif}}\Big\}\frac{j_\varrho A}{\mathcal{F}}$$

$$A\frac{\partial}{\partial t}\int_0^{x_{unif}} \varrho_{Na} dx = \frac{dm_{Na}}{dt} = \left[1-(t_{Na})_{x_{unif}}\right]\frac{M_{Na}A}{\mathcal{F}} j_\varrho$$

$$A\frac{\partial}{\partial t}\int_0^{x_{unif}} \varrho_{Cl} dx = \frac{dm_{Cl}}{dt} = (t_{Cl})_{x_{unif}}\frac{M_{Cl}A}{\mathcal{F}} j_\varrho$$

As before, A is the cross-section of the cell and m_j is the total mass of solute or constituent contained in the volume $x_{unif}\cdot A$. So far the cross-section has been assumed to be constant. If this restriction is taken away, then on the left-hand side, A has to be placed inside the integral. On the right-hand side A is the cross-section at $x = x_{unif}$. Finally one gets after integration with respect to time:

$$\Delta m_s = \Big\{(M_{Na}+M_I)\left[1-(t_{Na})_{x_{unif}}\right] - (M_K+M_I)(t_K)_{x_{unif}}$$

$$+(M_{Cl}-M_I)(t_{Cl})_{x_{unif}}\Big\}\cdot\frac{1}{\mathcal{F}}\int_0^t J_\varrho dt \qquad (158)$$

$$\Delta m_{Na} = \frac{M_{Na}}{\mathcal{F}}\left[1-(t_{Na})_{x_{unif}}\right]\cdot\int_0^t J_\xi\,dt \qquad (159)$$

$$\Delta m_{Cl} = \frac{M_{Cl}}{\mathcal{F}}(t_{Cl})_{x_{unif}}\cdot\int_0^t J_\xi\,dt \qquad (160)$$

These are the famous Hittorf relations which allow experimental determination of the three independent transport numbers t_{Na}, t_K, t_{Cl} at any desired concentration. In the best known case we have a binary electrolyte and eq.(158) reduces to the relation

$$\Delta m_s = (M_{Na}+M_{Cl})\left[1-(t_{Na})_{x_{unif}}\right]\cdot\frac{1}{\mathcal{F}}\int_0^t J_\xi\,dt \qquad (161)$$

where $m_s = m_{NaCl}$.

In our sketch of the fluxes (Fig.12) at the electrode boundary the Na excess constituent mass flux j_{Na}' was positive. This is the usual situation. However in a limited number of systems one observes that the cationic excess constituent mass flux is negative when the electric current density $j_\xi > 0$. This behaviour is particularly well-known for concentrated zinc and cadmium halide solutions. Now the cationic transport number is negative. This effect is due to strong complex formation between the metal constituent and the halide. In this situation the negative halide excess constituent mass flux flowing towards the anode is so strongly coupled to the cationic excess constituent mass flux that the latter attains a small negative value, its direction being reversed. Correspondingly, at the anode, when $j_\xi > 0$ we have a positive metal constituent mass flux and a negative excess constituent mass flux.

The fluxes at the gas electrode have the same features as those at the metal electrode and need no detailed treatment.

5.3 Electrodes of the second kind

We consider, as an example, the silver-silver chloride electrode. The constituent distribution curves are given in Fig.14.

At $x = x_1$ we have a drop of t_{Ag} from $t_{Ag} = 1$ to $t_{Ag}^{AgCl} < 1$ in AgCl. As usual we write down the relation

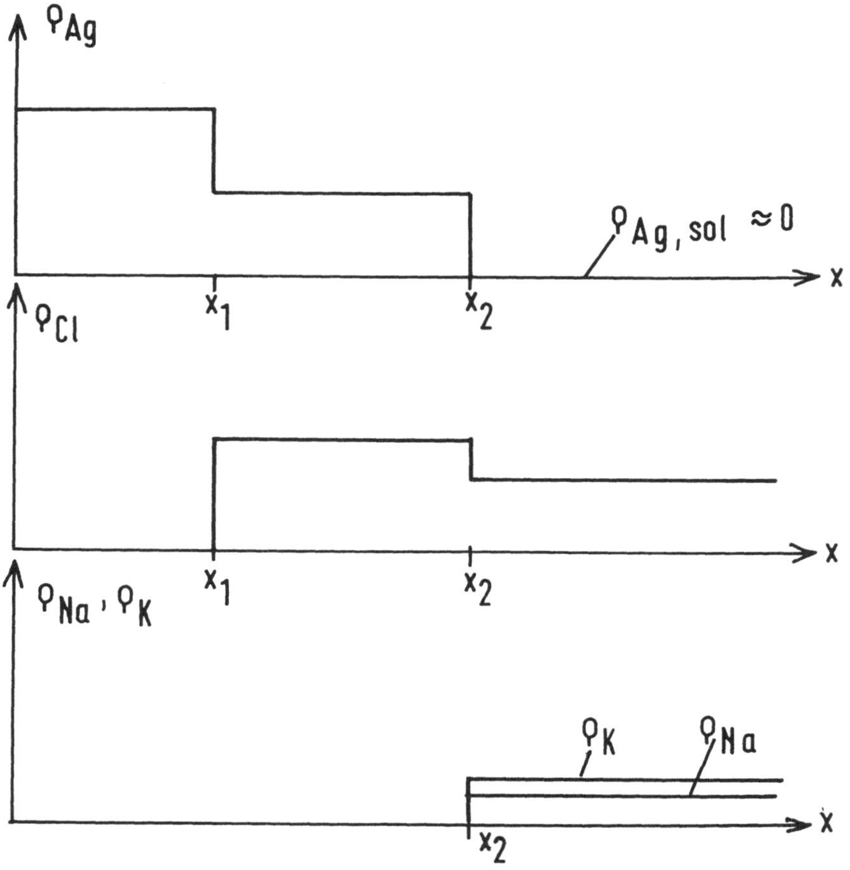

Fig. 14 The constituent partial mass density distributions for a Ag/AgCl electrode in contact with an electrolyte solution containing NaCl and KCl.

$$\frac{\partial \rho_{Ag}^{(AgCl)}}{\partial t} = - \frac{j_e}{F} \frac{dt_{Ag}}{dx} \qquad x \geqslant x_1$$

which is an application of the fundamental law of electrochemistry to the region of inhomogeneity at $x \geqslant x_1$. Integration over an interval $x_1 < x < x_1 + \delta x$ with $\delta_x \to 0$ yields

$$A \int_{x_1}^{x_1+\delta x} \frac{\partial \rho_{Ag}^{(AgCl)}}{\partial t} dx = M_{Ag} \left(1 - t_{Ag}^{(AgCl)}\right) \frac{j_e}{F} A$$

The excess Ag constituent mass flux into AgCl is

$$j_{Ag}' = M_{Ag} t_{Ag}^{(AgCl)} \frac{j_e}{F}$$

and thus the rate of change of the total mass of metallic silver is

$$\frac{dm_{Ag}}{dt} = -\frac{M_{Ag}}{\mathcal{F}} j_2 A$$

Now, since $\rho_{Ag}^{(AgCl)}$ is fully determined by T and p we must have $\partial\rho_{Ag}^{(AgCl)}/\partial t = 0$ and consequently the boundary between Ag and AgCl moves to the left if at x = 0 the silver is at rest. At x_2, ρ_{Ag} drops to practically zero due to the low solubility of AgCl. Thus we obtain

$$A\int_{x_1}^{\infty}\frac{\partial\rho_{Ag}}{\partial t}dx = M_{Ag}\, t_{Ag}^{(AgCl)}\frac{j_2}{\mathcal{F}}A$$

This is a positive displacement of the surface AgCl-liquid electrolyte relative to the boundary Ag/AgCl.

According to Fig.14 we have assumed that the aqueous electrolyte solution contains the constituents Na, K, and Cl. Then in the solution at the boundary x = x_2 the changes with time of the total solute mass density and the Na constituent density are:

$$\frac{\partial\rho_s}{\partial t} = -\operatorname{div} j_s - \left[(M_{Na}+M_{Cl})\frac{dt_{Na}}{dx} + (M_K+M_{Cl})\frac{dt_K}{dx}\right]\frac{j_2}{\mathcal{F}}$$

$$\frac{\partial\rho_{Na}}{\partial t} = -\operatorname{div} j_{Na} - M_{Na}\frac{dt_{Na}}{dx}\frac{j_2}{\mathcal{F}}$$

Since no Na can penetrate into the solid electrolyte AgCl, we have t_{Na} = 0 at the boundary x_2. Thus, after integration we find:

$$(j_s)_{x=x_2} = -\left[(M_{Na}+M_{Cl})t_{Na} + (M_K+M_{Cl})t_K\right]\frac{j_2}{\mathcal{F}}$$

$$(j_{Na})_{x=x_2} = -\frac{M_{Na}}{\mathcal{F}} j_2 t_{Na}$$

and, taking into account that

$$j_s = j_{Na} + j_K + j_{Cl}$$

and

$$\frac{j_{Na}}{M_{Na}} + \frac{j_K}{M_K} = \frac{j_{Cl}}{M_{Cl}}$$

the other two constituent mass fluxes are found to be:

$$(j_K)_{x=x_2} = -\frac{M_K}{\mathcal{F}} j_2 t_K$$

$$(j_{Cl})_{x=x_2} = -(1-t_{Cl})\frac{M_{Cl}}{\mathcal{F}} j_2$$

Thus all the fluxes, the total solute mass flux and the constituent mass fluxes, are negative if $j_\xi > 0$. The electrode acts as an anode. The excess constituent mass fluxes are

$$j_{Na}' = t_{Na} \frac{M_{Na}}{\mathcal{F}} j_\xi$$

$$j_K' = t_K \frac{M_K}{\mathcal{F}} j_\xi$$

$$j_{Cl}' = - t_{Cl} \frac{M_{Cl}}{\mathcal{F}} j_\xi$$

The various fluxes are shown in Fig. 15

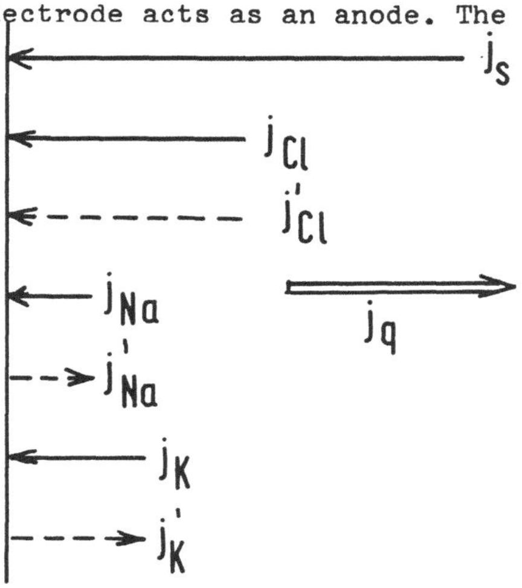

Fig. 15 Mass fluxes and excess mass fluxes at the Ag/AgCl electrode. Both cation transport numbers are set equal to 1/4. Excess mass fluxes are drawn as broken lines.

5.4 The fluxes at the redox electrode

We begin with the characterization of the electrolyte solution. Let it contain the four constituents:

Fe(II), Fe(III), Na, and Cl.

The respective partial mass densities are

$$\rho_{Fe(II)} \equiv \rho_2 \; ; \qquad \rho_{Fe(III)} \equiv \rho_3 \; ; \qquad \rho_{Cl} \; ; \qquad \rho_{Na}$$

The two constituent densities ρ_2 and ρ_3 may be considered to be defined by the relations

$$\rho_S = \rho_2 + \rho_3 + \rho_{Na} + \rho_{Cl}$$

and

$$\frac{1}{M_{Fe}}(2\rho_2 + 3\rho_3) + \frac{\rho_{Na}}{M_{Na}} = \frac{\rho_{Cl}}{M_{Cl}}$$

Assume the system is fully determined in that it has been constructed from the three components NaCl, $FeCl_2$, and $FeCl_3$. Then transformation to the three variables ρ_S, ρ_{Na}, ρ_{Cl}, is straightforward and we have

$$\rho_2 = 3\left[\rho_S - \rho_{Cl}\left(1 + \frac{1}{3}\frac{M_{Fe}}{M_{Cl}}\right) - \rho_{Na}\left(1 - \frac{1}{3}\frac{M_{Fe}}{M_{Na}}\right)\right] \qquad (162)$$

$$\rho_3 = 2\left[\rho_{Cl}\left(1+\frac{1}{2}\frac{M_{Fe}}{M_{Cl}}\right)+\rho_{Na}\left(1-\frac{1}{2}\frac{M_{Fe}}{M_{Na}}\right)-\rho_s\right] \tag{163}$$

which may be rearranged to give:

$$\rho_s = \rho_{Cl}\left(1+\frac{1}{3}\frac{M_{Fe}}{M_{Cl}}\right)+\rho_{Na}\left(1-\frac{1}{3}\frac{M_{Fe}}{M_{Na}}\right)+\frac{\rho_2}{3} \tag{162a}$$

$$\rho_s = \rho_{Na}\left(1-\frac{1}{2}\frac{M_{Fe}}{M_{Na}}\right)+\rho_{Cl}\left(1+\frac{1}{2}\frac{M_{Fe}}{M_{Cl}}\right)-\frac{\rho_3}{2} \tag{163a}$$

Now we can write down the diffusion equations in the absence of the electric current.

$$\frac{\partial \rho_s}{\partial t} = -\,\mathrm{div}\, j_s \quad ; \quad \frac{\partial \rho_{Na}}{\partial t} = -\,\mathrm{div}\, j_{Na} \quad ; \quad \frac{\partial \rho_{Cl}}{\partial t} = -\,\mathrm{div}\, j_{Cl}. \tag{164}$$

with

$$\begin{aligned}
-j_s &= D_{ss}\frac{\partial \rho_s}{\partial x} + D_{sNa}\frac{\partial \rho_{Na}}{\partial x} + D_{sCl}\frac{\partial \rho_{Cl}}{\partial x}\\
-j_{Na} &= D_{Nas}\frac{\partial \rho_s}{\partial x} + D_{NaNa}\frac{\partial \rho_{Na}}{\partial x} + D_{NaCl}\frac{\partial \rho_{Cl}}{\partial x}\\
-j_{Cl} &= D_{Cls}\frac{\partial \rho_s}{\partial x} + D_{ClNa}\frac{\partial \rho_{Na}}{\partial x} + D_{ClCl}\frac{\partial \rho_{Cl}}{\partial x}
\end{aligned} \tag{165}$$

Then the densities ρ_2 and ρ_3 are well-defined at any point although the constituents Fe(ν), ν = II,III, do not appear explicitly in the analytical field.

Now we place the electrolyte solution between two boundaries. At x = 0 we place a platinum electrode; the electrode at x = ℓ is a chlorine electrode. Let the electric current density j_{ξ} have a positive direction when it points towards the right, i.e. towards the chlorine electrode. Then the electric current is connected with the chemical reaction $2\,FeCl_2 + Cl_2 \longrightarrow 2\,FeCl_3$. Of course, when $j_{\xi} \neq 0$ the diffusion equations in the form eq.(164) are no longer valid. As has been described previously , one has to add the negative divergences of the excess total solute and constituent mass fluxes, $-\mathrm{div}\, j_s'$, $-\mathrm{div}\, j_{Na}'$, and $-\mathrm{div}\, j_{Cl}'$ on the right-hand side of eqs.(164). The eqs.(165) remain valid. However, considering eqs.(162a) or (163a) we see that dt_2/dx or dt_3/dx is also needed. And in the case of the redox electrode it is not clear at the present stage what values the integrals

$$\int_0^{\delta x} \frac{dt_2}{dx}\,dx \quad ; \quad \int_0^{\delta x} \frac{dt_3}{dx}\,dx$$

have at the boundary x = 0. Therefore, in order to analyze the fluxes in detail, it is convenient to change the variables to the set ρ_s, ρ_2, ρ_3, where ρ_2 and ρ_3 may be considered to be defined by the two eqs. (162) and (163). For these variables the general diffusion equations are:

$$\frac{\partial \rho_s}{\partial t} = -\operatorname{div} j_s - \operatorname{div} j_s'$$
$$\frac{\partial \rho_2}{\partial t} = -\operatorname{div} j_2 - \operatorname{div} j_2' \qquad (166)$$
$$\frac{\partial \rho_3}{\partial t} = -\operatorname{div} j_3 - \operatorname{div} j_3'$$

and

$$-j_k = \sum D_{ki} \frac{\partial \rho_i}{\partial x} \qquad i,k = s,2,3$$

The excess mass fluxes are

$$j_s' = \frac{\dot{j}_2}{\mathcal{F}}\left[\frac{1}{2}(2M_{Cl}+M_{Fe})t_2 + \frac{1}{3}(3M_{Cl}+M_{Fe})t_3 + (M_{Cl}+M_{Na})t_{Na} - M_{Cl}\right] \qquad (167)$$

$$j_2' = \frac{M_{Fe}\, t_2}{2\mathcal{F}}\, \dot{j}_2 \qquad (168)$$

$$\dot{j}_3' = \frac{M_{Fe}\, t_3}{3\mathcal{F}}\, \dot{j}_2 \qquad (169)$$

Combination of eqs.(166)-(169) yields the system of diffusion equations describing the analytical field $\rho_s(x,t)$, $\rho_2(x,t)$, and $\rho_3(x,t)$. The only additional information we need is the boundary conditions at the platinum and the Cl_2 electrode. We begin with the platinum electrode at x = 0. As before we consider a thin layer at the electrode surface, $0 \leq x \leq \delta x$. The iron does not enter into the Pt metal, thus we must have $t_2 = t_3 = 0$ at x = 0.

As a consequence we write

$$t_i(x) = \frac{t_i}{\delta x}\, x \qquad i = 2,3 \qquad (170)$$

$$\text{for} \quad 0 \leq x \leq \delta x$$

where the t_i's are the transport numbers in the solution at $x = \delta x$. We see immediately, with eq.(170) and $\dot{j}_2 > 0$, the excess constituent mass fluxes $\dot{j}_2'$ and $\dot{j}_3'$ are positive. What is the mechanism that supplies the mass for these positive excess constituent mass fluxes? The mass cannot come from the metallic phase as it does not contain Fe or Cl. Of course, the mass comes from the electrolyte solution by the aid of the mass fluxes. The balancing of the mass and excess mass fluxes is determined by the chemical reaction

$$FeCl_2 + 1/2\ Cl_2 \longrightarrow FeCl_3$$

within the layer δx. This chemical reaction is coupled to the electric current density and for the local mass production rate due to this chemical reaction, we write:

local Fe(III) constitutent mass production rate per unit length: $\frac{j_e}{\mathcal{F}} M_{Fe} \frac{1}{\delta x}$

local Fe(II) constituent mass depletion rate per unit length: $-\frac{j_e}{\mathcal{F}} M_{Fe} \frac{1}{\delta x}$

local Cl constituent mass production rate per unit length: $\frac{j_e}{\mathcal{F}} M_{Cl} \frac{1}{\delta x}$

For Cl the requirement of mass conservation is guaranteed through the Cl mass delivery or mass removal at the other electrode at $x = \ell$ (see below). Taking into account this chemical reaction we have the following mass balancing equations in the interval $0 \leq x \leq \delta x$: (see eqs.(166)-(169)

$$\frac{\partial \rho_s}{\partial t} = -\operatorname{div} j_s - \frac{j_e}{\mathcal{F}}\left[\frac{1}{2}(2M_{Cl} + M_{Fe})\frac{dt_2}{dx} + \frac{1}{3}(3M_{Cl} + M_{Fe})\frac{dt_3}{dx} + (M_{Cl} + M_{Na})\frac{dt_{Na}}{dx}\right] + \frac{j_e}{\mathcal{F}} M_{Cl} \frac{1}{\delta x} \tag{171}$$

$$\frac{\partial \rho_2}{\partial t} = -\operatorname{div} j_2 - \frac{j_e}{2} M_{Fe} \frac{dt_2}{dx} - \frac{j_e}{\mathcal{F}} M_{Fe} \frac{1}{\delta x} \tag{172}$$

$$\frac{\partial \rho_3}{\partial t} = -\operatorname{div} j_3 - \frac{j_e}{3} M_{Fe} \frac{dt_3}{dx} + \frac{j_e}{\mathcal{F}} M_{Fe} \frac{1}{\delta x} \tag{173}$$

After inserting eqs.(170) in these equations we integrate them with respect to x. This gives after some rearrangement:

$$\int_0^{\delta x} \frac{\partial \rho_s}{\partial t} dx = -j_s - \frac{j_e}{\mathcal{F}}\left[M_{Fe}\left(\frac{1}{2}t_2 + \frac{1}{3}t_3\right) + M_{Na} t_{Na} - M_{Cl} t_{Cl}\right]$$

$$\int_0^{\delta x} \frac{\partial \rho_2}{\partial t} dx = -j_2 - \frac{j_e}{\mathcal{F}}\left(1 + \frac{1}{2}t_2\right)$$

$$\int_0^{\delta x} \frac{\partial \rho_3}{\partial t} dx = -j_3 + \frac{j_e}{\mathcal{F}}\left(1 - \frac{1}{3}t_3\right)$$

where the relation

$$t_2 + t_3 + t_{Na} + t_{Cl} = 1 \tag{174}$$

has been used. Now we may go to the limit $\delta x \to 0$ and it follows:

$$(\dot{j}_s)_{x=0} = -\frac{\dot{j}_2}{\mathcal{F}}\left[M_{Fe}\left(\frac{1}{2}t_2 + \frac{1}{3}t_3\right) + M_{Na}t_{Na} - M_{Cl}t_{Cl}\right] \tag{175}$$

$$(\dot{j}_2)_{x=0} = -\frac{\dot{j}_2}{\mathcal{F}}\left(1 + \frac{1}{2}t_2\right)M_{Fe} \tag{176}$$

$$(\dot{j}_3)_{x=0} = +\frac{\dot{j}_2}{\mathcal{F}}\left(1 - \frac{1}{3}t_3\right)M_{Fe} \tag{177}$$

Applying equations analogous to eqs.(61) and (62) we obtain the two remaining constituent mass fluxes:

$$(\dot{j}_{Na})_{x=0} = -\frac{M_{Na}}{\mathcal{F}}t_{Na}\dot{j}_2 \tag{178}$$

$$(\dot{j}_{Cl})_{x=0} = \frac{M_{Cl}}{\mathcal{F}}t_{Cl}\dot{j}_2 \tag{179}$$

and of course, the two corresponding excess constituent mass fluxes are:

$$\dot{j}_{Na}' = \frac{M_{Na}}{\mathcal{F}}t_{Na}\dot{j}_2 \tag{180}$$

$$\dot{j}_{Cl}' = -\frac{M_{Cl}}{\mathcal{F}}t_{Cl}\dot{j}_2 \tag{181}$$

Finally, if we apply once again, eq.(174) in eq.(175), we find

$$(\dot{j}_s)_{x=0} = -\frac{\dot{j}_2}{\mathcal{F}}\left[\frac{1}{3}M_{Fe}\left(1 + \frac{1}{2}t_2\right) + \left(M_{Na} - \frac{1}{3}M_{Fe}\right)t_{Na} - \left(M_{Cl} + \frac{1}{3}M_{Fe}\right)t_{Cl}\right] \tag{175a}$$

which, together with eqs.(176) and (177) gives the mass fluxes in the ρ_s, ρ_{Na}, ρ_{Cl} representation of the system from which we originally started (see eq.(162)-(165), in particular eq.(162a)).

When we now proceed to the discussion of the results just derived, we see from a comparison of eqs.(167) and (175) that at x = 0 we have $\dot{j}_s = -\dot{j}_s'$, i.e. the magnitudes of the total solute mass flux and the excess mass flux are equal. The total solute mass flux at x = 0 in the present system is negative. All the fluxes are sketched schematically in Fig. 16. Again the excess mass fluxes are drawn as broken lines. For demonstration, let us set all four transport numbers equal to 1/4.

Then

$$j_s = -\frac{j_2}{4\mathcal{F}}\left[\frac{5}{6}M_{Fe} + M_{Na} - M_{Cl}\right] < 0$$

if

$$j_2 > 0$$

If we had LiBr instead of NaCl as the"supporting electrolyte", then $j_s > 0$, if $j_2 > 0$. Yet, the total diffusive molar flux at x = 0 is always < 0 if $j_2 > 0$. The material for the two Fe excess con stituent fluxes has to be supplied. So, considering the analytical field at the redox electrode, the gradient of the total solute density may be positive ($j_s < 0$) or negative ($j_s > 0$) if $j_2 > 0$, depending on the nature of the supporting electrolyte.

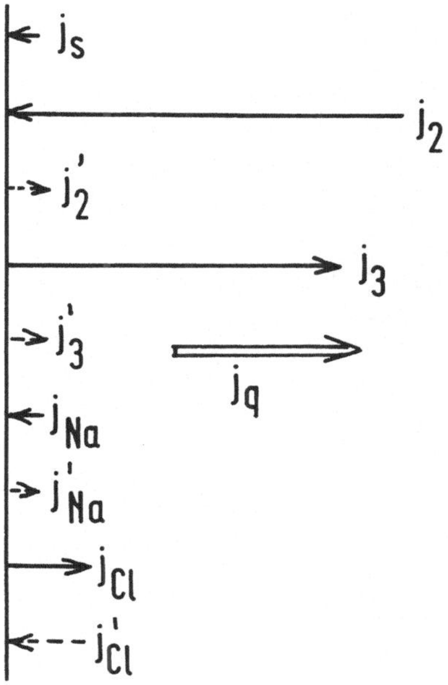

Fig.16: Schematic representation of mass and excess mass fluxes at redox electrode. Transport numbers are estimated.

With these boundary conditions the diffusion equations in the half-cell containing the redox electrode can be solved at least in principle. We mention again that the boundary conditions are themselves time dependent even if j_2 = const. This is due to the slight concentration dependence of the transport numbers. On the other hand, it is also possible to solve the diffusion equations in the ρ_s, ρ_{Na}, ρ_{Cl} representation. These are eqs.(165) to which terms $dt_{Na}/dx\,(j_2 M_{Na}/\mathcal{F})$ etc. have to be added. The boundary conditions are given by eqs.(175), (176), and (177).

It is straightforward to write down the boundary fluxes at the chlorine electrode which is located at x = ℓ . Again we first consider a thin layer $\ell - \delta x < x < \ell$. Then the transport numbers as a function of x in the range δx are

$$t_2(x) = t_2 - \frac{t_2}{\delta x}(x - \ell + \delta x) \tag{182}$$

$$t_3(x) = t_3 - \frac{t_3}{\delta x}(x - \ell + \delta x) \tag{183}$$

$$t_{Na}(x) = t_{Na} - \frac{t_{Na}}{\delta x}(x - \ell + \delta x) \tag{184}$$

$$t_{Cl}(x) = 1 - \frac{1 - t_{Cl}}{\delta x}(l - x) \tag{185}$$

where the last equation expresses the fact that the electrode is electrochemically active with respect to Cl, i.e. $t_{Cl}(l) = 1$.

In eq.(171) we replace $\rho_s = \rho_s(\rho_2, \rho_3, \rho_{Na})$ by $\rho_s = \rho_s(\rho_2, \rho_3, \rho_{Cl})$ and then we have

$$\frac{\partial \rho_s}{\partial t} = -\operatorname{div} j_s - \frac{j_2}{\mathcal{F}}\left[\frac{1}{2}(M_{Fe} - 2M_{Na})\frac{dt_2}{dx} + \frac{1}{3}(M_{Fe} - 3M_{Na})\frac{dt_3}{dx} - (M_{Cl} + M_{Na})\frac{dt_{Cl}}{dx}\right]$$

$$\frac{\partial \rho_2}{\partial t} = -\operatorname{div} j_2 + \frac{j_2}{2\mathcal{F}} M_{Fe}\frac{dt_2}{dx}$$

$$\frac{\partial \rho_3}{\partial t} = -\operatorname{div} j_3 + \frac{j_2}{3\mathcal{F}} M_{Fe}\frac{dt_3}{dx}$$

Application of eqs.(182)-(185) in these relations, integration with respect to x, and going to the limit $\delta x \to 0$ yields the three boundary conditions (after some rearrangement):

$$(j_s)_{x=l} = -\frac{j_2}{\mathcal{F}}\left[M_{Fe}\left(\frac{1}{2}t_2 + \frac{1}{3}t_3\right) + M_{Na}t_{Na} + M_{Cl}(1 - t_{Cl})\right] \tag{186}$$

$$(j_2)_{x=l} = -\frac{M_{Fe}}{2\mathcal{F}}t_2 j_2 \tag{187}$$

$$(j_3)_{x=l} = -\frac{M_{Fe}}{3\mathcal{F}}t_3 j_2 \tag{188}$$

and furthermore we have

$$(j_{Na})_{x=l} = -\frac{M_{Na}}{\mathcal{F}}t_{Na} j_2 \tag{189}$$

$$(j_{Cl})_{x=l} = -\frac{M_{Cl}}{\mathcal{F}}(1 - t_{Cl}) j_2 \tag{190}$$

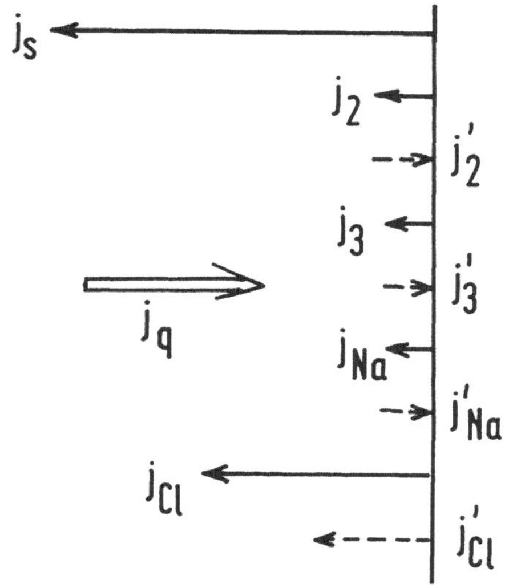

Fig.17: Schematic representation of mass fluxes and excess mass fluxes at the surface between the Cl_2 electrode and an electrolyte solution containing NaCl, $FeCl_2$, and $FeCl_3$.

It will be seen that we have

$$(j_s)_{x=\ell} = (j_s)_{x=0} - \frac{j_2}{\mathcal{F}} M_{Cl}$$

$$(j_2)_{x=\ell} = (j_2)_{x=0} + \frac{j_2}{\mathcal{F}} M_{Fe}$$

$$(j_3)_{x=\ell} = (j_3)_{x=0} - \frac{j_2}{\mathcal{F}} M_{Fe}$$

Thus, the total increase of mass of the electrolyte solution per unit time is $A j_2 M_{Cl} / \mathcal{F}$ (A= cross-section of the system). Since all mass fluxes are negative at $x = \ell$ the total solute mass density gradient and all the partial constituent mass density gradients are positive at the surface of the chlorine electrode. The diffusion equations may now be solved also in the half cell containing the Cl_2 electrode, it may be solved in the ρ_s, ρ_2, ρ_3 representation, according to eq.(166) or in the $\rho_s, \rho_{Na}, \rho_{Cl}$ representation according to the scheme eq.(165) with added local excess total solute and constituent mass production rates.

Of course, the case of greatest practical interest is the one in which we have $\rho_2, \rho_3 \ll \rho_s$. Now we have $\partial\rho_s/\partial x \approx 0$ and $t_2, t_3 \ll t_{Na}, t_{Cl}$ consequently, at the boundary x = 0

$$(j_s)_{x=0} \approx 0$$

$$-(j_2)_{x=0} = D_{22}\frac{\partial\rho_2}{\partial x} + D_{23}\frac{\partial\rho_3}{\partial x}$$

$$-(j_3)_{x=0} = D_{32}\frac{\partial\rho_3}{\partial x} + D_{33}\frac{\partial\rho_3}{\partial x}$$

However, as $\rho_i \to 0$, i = 1,2, the cross diffusion coefficients must vanish, thus

$$\frac{j_2}{\mathcal{F}}\left(1 + \frac{1}{2}t_2\right)M_{Fe} = D_{22}\frac{\partial\rho_2}{\partial x} \qquad \text{at } x = 0$$

$$-\frac{j_2}{\mathcal{F}}\left(1 - \frac{1}{3}t_3\right)M_{Fe} = D_{33}\frac{\partial\rho_3}{\partial x} \qquad \text{at } x = 0$$

These are the boundary conditions for the set of diffusion equations:

$$\rho_s \approx \text{const.}$$

$$\frac{\partial\rho_2}{\partial t} = \frac{\partial}{\partial x} D_{22}\frac{\partial\rho_2}{\partial x} - \frac{j_2}{2\mathcal{F}} M_{Fe}\frac{dt_2}{dx}$$

$$\frac{\partial\rho_3}{\partial t} = \frac{\partial}{\partial x} D_{33}\frac{\partial\rho_3}{\partial x} - \frac{j_2}{3\mathcal{F}} M_{Fe}\frac{dt_3}{dx}$$

for $x \geqslant 0$, and finally neglecting the concentration dependence of D_{ii} and t_i , we get

$$\frac{\partial \rho_2}{\partial t} = D_{22} \frac{\partial^2 \rho_2}{\partial x^2}$$
$$\frac{\partial \rho_3}{\partial t} = D_{33} \frac{\partial^2 \rho_3}{\partial x^2}$$

subject to the above boundary conditions.

For completeness we give some formulas for the treatment of the system Pt/$FeCl_3$, $FeCl_2/Cl_2$. Now the analytical field may be given in the representations ρ_s, ρ_{cl}, or ρ_s, ρ_2, or ρ_2, ρ_3. Let us first choose the last possibility. Then, in the bulk of the liquid, $\rho_s(x,t)$ and $\rho_3(x,t)$ have to satisfy the differential equations:

$$\frac{\partial \rho_s}{\partial t} = -\operatorname{div} j_s - \frac{j_2}{\mathcal{F}}\left[\frac{1}{2}(2M_{cl}+M_{Fe})\frac{dt_2}{dx} + \frac{1}{3}(3M_{cl}+M_{Fe})\frac{dt_3}{dx}\right]$$

$$\frac{\partial \rho_3}{\partial t} = -\operatorname{div} j_3 - \frac{j_2}{3\mathcal{F}} M_{Fe} \frac{dt_3}{dx}$$

with

$$-j_s = D_{ss}\frac{\partial \rho_s}{\partial x} + D_{s3}\frac{\partial \rho_3}{\partial x}$$
$$-j_3 = D_{3s}\frac{\partial \rho_s}{\partial x} + D_{33}\frac{\partial \rho_3}{\partial x}$$

At the electrode surface the boundary conditions are:
(the redox electrode is at x = 0)

$$(j_s)_{x=0} = -\frac{j_2}{\mathcal{F}}\left[M_{Fe}\left(\frac{1}{2}t_2 + \frac{1}{3}t_3\right) - M_{cl}t_{cl}\right]$$

$$(j_3)_{x=0} = +\frac{j_2}{\mathcal{F}}\left(1-\frac{1}{3}t_3\right)M_{Fe}$$

$$(j_s)_{x=\ell} = -\frac{j_2}{\mathcal{F}}\left[M_{Fe}\left(\frac{1}{2}t_2 + \frac{1}{3}t_3\right) + M_{cl}(1-t_{cl})\right]$$

$$(j_3)_{x=\ell} = -\frac{j_2}{3\mathcal{F}} t_3 M_{Fe}$$

with

$$t_2 + t_3 + t_{cl} = 1$$

Furthermore, we have the excess mass fluxes at the boundaries:

$$(j_2')_{x=0} = -\frac{j_2}{\mathcal{F}}\left(1+\frac{1}{2}t_2\right)M_{Fe}$$
$$(j_{cl}')_{x=0} = \frac{j_2}{\mathcal{F}} M_{cl} t_{cl}$$
$$(j_2')_{x=\ell} = -\frac{j_2}{2\mathcal{F}} M_{Fe} t_2$$
$$(j_{cl}')_{x=\ell} = \frac{j_2}{\mathcal{F}} M_{cl} t_{cl}$$

These boundary conditions have also to be used, if one works with the ρ_s, ρ_2, or ρ_s, ρ_{Cl} representation of the analytical field. For instance in the ρ_s, ρ_{Cl} representation we have at x = 0

$$\frac{j_q}{F}\left[M_{Fe}\left(\frac{1}{2}t_2 + \frac{1}{3}t_3\right) - M_{Cl}t_{Cl}\right] = D_{ss}\frac{\partial \rho_s}{\partial x} + D_{sCl}\frac{\partial \rho_{Cl}}{\partial x}$$

$$-\frac{j_q}{F}M_{Cl}t_{Cl} = D_{Cls}\frac{\partial \rho_s}{\partial x} + D_{ClCl}\frac{\partial \rho_{Cl}}{\partial x}$$

In a stationary state, let $\partial\rho_i/\partial x$ be equal to $\Delta\rho_i/\Delta x$, then the partial mass densities show distributions as depicted in Fig.18

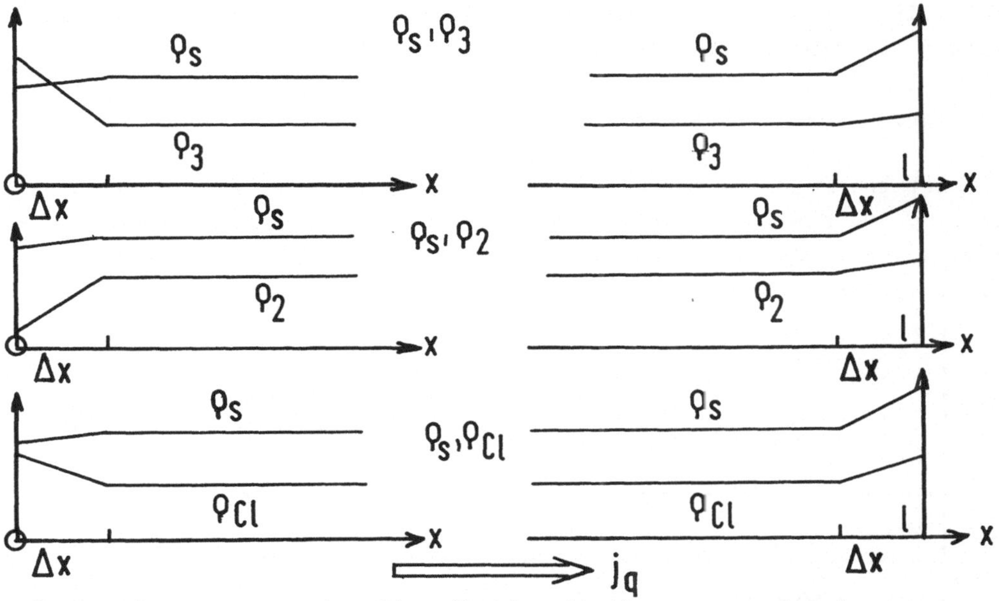

Fig.18: Partial mass density distribution in the same system however, represented in different sets of independent variables. At the left we have the redox electrode at the right the chlorine electrode.

5.5 Mass production at the liquid junction

In Fig. 13 it will be seen that at the Na electrode there is a positive Na excess constituent mass flux. This means that j_{Na}' is directed into the electrolyte solution. Then, for example, at the chlorine electrode, which is the end electrode at $x = \ell$, we again find the Na excess mass flux. Here j_{Na}' points towards the electrode and at $x = \ell$ it is reversed to give a Na constituent mass flux j_{Na} being directed into the electrolyte solution (see Fig. 17). This is the typical behaviour for any other excess constituent mass flux which starts from an electrode being electrochemically active with respect to the constituent considered. We have now to answer the question: what happens if the end electrode ($x = \ell$) is in contact with an electrolyte solution which does not contain the constituent whose excess mass flux starts at the active electrode ($x = 0$)? For instance, the Na constituent concentration may go to zero along the diffusion path $0 < x < \ell$. In this situation we have a junction between two different electrolyte solutions. Indeed, the treatment of this problem brings us back to the moving boundary we have already discussed in chapter 4. However now we wish to consider this region of the electrolyte solution with the special aim of obtaining a clear understanding of the properties of the excess mass fluxes starting at the electrodes.

a) Liquid junction between a NaI and a KCl solution.

Let the boundary between the two solutions be at $x = x_0$ as shown in Fig. 19. For $x < x_0$, apart from a thin boundary layer around x_0 we have a "pure" NaI solution, for $x > x_0$ the electrolyte is a "pure" KCl solution.

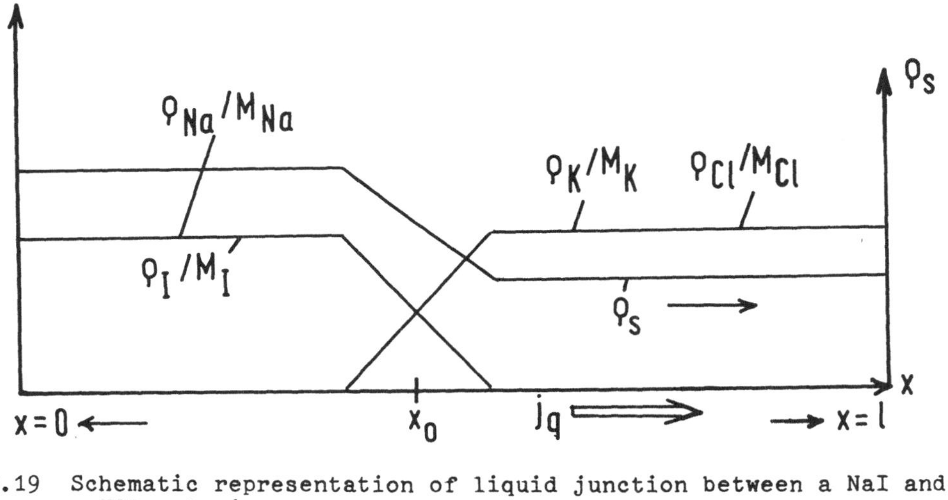

Fig.19 Schematic representation of liquid junction between a NaI and a KCl solution.

Then, as long as we remain in the bulk of the electrolyte solutions, $0 < x < \ell$, we have

$$\int_0^\ell \operatorname{div} j_s \, dx = \int_0^\ell \operatorname{div} j_{Na} \, dx = \int_0^\ell \operatorname{div} j_{Cl} \, dx = 0 \qquad (191)$$

because the mass fluxes can only transport mass from one part of the diffusion region to the other. Then, when we write down the fundamental equations of electrochemistry (eqs.(112)-(114)), integrate them over the bulk of the liquid phases, $0 < x < \ell$, and take into account eq.(191), we obtain:

$$\int_0^\ell \frac{\partial \rho_s}{\partial t} dx = -\int_0^\ell \frac{\partial j_s}{\partial x} dx - \Big[(M_{Na} + M_I) \int_0^\ell \frac{dt_{Na}}{dx} dx + (M_K + M_I) \int_0^\ell \frac{dt_K}{dx} dx - (M_{Cl} - M_I) \int_0^\ell \frac{dt_{Cl}}{dx} dx \Big] \frac{j_q}{\mathcal{F}}$$

$$= \Big[(M_{Na} + M_I)(t_{Na}^{(\ell)} - t_{Na}^{(0)}) + (M_K + M_I)(t_K^{(\ell)} - t_K^{(0)}) - (M_{Cl} - M_I)(t_{Cl}^{(\ell)} - t_{Cl}^{(0)}) \Big] \frac{j_q}{\mathcal{F}}$$

Likewise

$$\int_0^\ell \frac{\partial \rho_{Na}}{\partial t} dx = -\frac{M_{Na}}{\mathcal{F}} j_q (t_{Na}^{(\ell)} - t_{Na}^{(0)})$$

$$\int_0^\ell \frac{\partial \rho_{Cl}}{\partial t} dx = \frac{M_{Cl}}{\mathcal{F}} j_q (t_{Cl}^{(\ell)} - t_{Cl}^{(0)})$$

In the special case that $t_{Na}^{(\ell)} = 0$, $t_K^{(0)} = t_{Cl}^{(0)} = 0$, we have:

$$\frac{\partial \Delta m_s}{\partial t} = \int_0^\ell \frac{\partial \rho_s}{\partial t} dx = \Big[(M_{Na} + M_I) t_{Na}^{(0)} - (M_K + M_I) t_K^{(\ell)} + (M_{Cl} - M_I) t_{Cl}^{(\ell)} \Big] \frac{j_q}{\mathcal{F}} \tag{192}$$

$$\frac{\partial \Delta m_{Na}}{\partial t} = \frac{M_{Na}}{\mathcal{F}} j_q t_{Na}^{(0)} \tag{193}$$

$$\frac{\partial \Delta m_{Cl}}{\partial t} = \frac{M_{Cl}}{\mathcal{F}} j_q t_{Cl}^{(\ell)} \tag{194}$$

Furthermore, for the constituents K and I one finds:

$$\frac{\partial \Delta m_K}{\partial t} = -\frac{M_K}{\mathcal{F}} j_q t_K^{(\ell)} \tag{195}$$

$$\frac{\partial \Delta m_I}{\partial t} = \frac{M_I}{\mathcal{F}}(t_{Na}^{(0)} - t_K^{(\ell)} - t_{Cl}^{(\ell)})\dot{j}_q = -\frac{M_I}{\mathcal{F}}(1 - t_{Na}^{(0)})\dot{j}_q \tag{196}$$

Now we compare eq.(193) with eq.(144). We see that the rate of Na constituent mass production integrated over the bulk of the liquid exactly equals the excess constituent mass flux at the electrolyte boundary. And also, comparison of eq.(196) with eq.(150), where we have set $t_K^{(0)} = t_{Cl}^{(0)} = 0$, leads us to the result that the rate of removal of I constituent mass integrated over the whole liquid phase exactly equals M_I/M_{NaI} times the total solute mass flux into the electrolyte at x = 0. Locally the production and depletion of constituent masses is distributed over the range of inhomogeneity of the system, i.e. the liquid junction. Partly the diffusive flux acts against the production (or depletion), partly both effects add their contribution to the local change of partial constituent mass density. If the end electrode at $x = \ell$ is a Cl_2 electrode (or Ag/AgCl electrode) then, mutatis mutandis, everything stated for Na and I holds true also for K and Cl^-. We have a removal of K from the region around x_0, and a production of the constituent Cl.

b) The liquid junction between a NaCl and a KCl solution.

The distributions of the mass densities around x_0 are sketched in Fig.20.

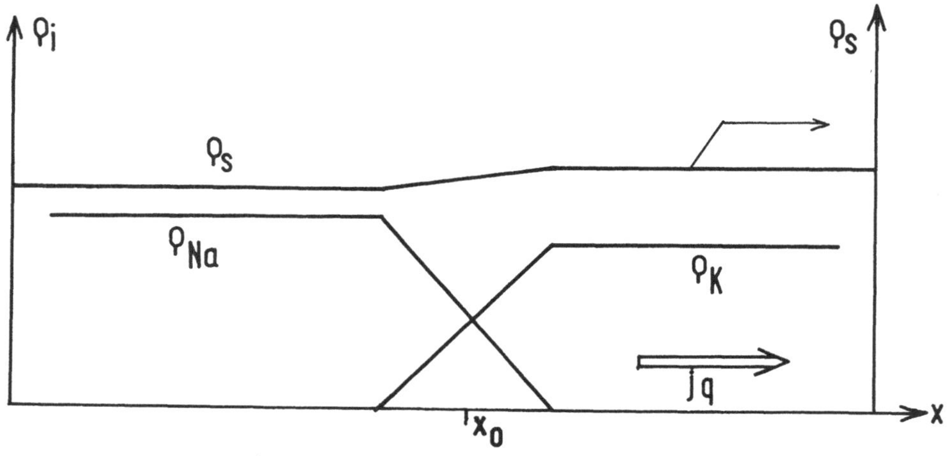

Fig. 20 Schematic representation of liquid junction between a NaCl and a KCl solution

Now the mass productions are only determined by the two integrals

$$\int_0^\ell \frac{\partial \rho_s}{\partial t} dx = - \frac{j_2}{F}\left[(M_{Na}+M_{Cl}) \int_0^\ell \frac{dt_{Na}}{dx} dx + (M_K + M_{Cl}) \int_0^\ell \frac{dt_K}{dx} dx \right]$$

and

$$\int_0^\ell \frac{\partial \rho_{Na}}{\partial t} dx = - \frac{j_2}{F} M_{Na} \int_0^\ell \frac{dt_{Na}}{dx} dx = - \frac{j_2}{F} M_{Na} (t_{Na}^{(\ell)} - t_{Na}^{(0)})$$

We have to consider the special case $t_{Na}^{(\ell)} = 0 = t_K^{(0)}$ which gives

$$\frac{\partial \Delta m_s}{\partial t} = \frac{j_2}{F}\left[(M_{Na}+M_{Cl}) t_{Na}^{(0)} - (M_K + M_{Cl}) t_K^{(\ell)} \right]$$

$$\frac{\partial \Delta m_{Na}}{\partial t} = \frac{j_2}{F} M_{Na} t_{Na}^{(0)}$$

Then it follows from the fact that the system is fully determined by these two equations:

$$\frac{\partial \Delta m_K}{\partial t} = - \frac{j_2}{F} M_K t_K^{(\ell)}$$

$$\frac{\partial \Delta m_{Cl}}{\partial t} = \frac{j_2}{F} M_{Cl} (t_{Na}^{(0)} - t_K^{(\ell)})$$

These formulas give us the following information: (i) The Na constituent mass production around the liquid junction region is positive. NaCl is formed at the electrode and at the liquid boundary. (ii) There is a depletion of K around the boundary x_0. KCl is formed at the electrode at $x = \ell$ (or K is removed from the electrolyte solution if the end electrode is a K metal electrode).(iii) $\partial \Delta m_{Cl} / \partial t$ is negative because $t_{Na} < t_K$. More NaCl is formed at the anode (x = 0) than KCl at the cathode $x = \ell$ (or, we can also say, because the Cl constituent flux is smaller at $x = \ell$ than at x = 0).

c) The liquid junction between a redox half-cell containing $FeCl_2$, $FeCl_3$, and NaCl, and a half-cell containing an HNO_3 solution.

At x = 0 there is a platinum electrode and the end electrode at $x = \ell$ is a hydrogen electrode. Now the system is to be described by the set of 5 quantities

$$\rho_s, \rho_2, \rho_3, \rho_H, \rho_{NO_3}$$

The total solute density ρ_s is given by the expression:

$$\begin{aligned}\rho_s &= \rho_{Na} + \rho_H + \rho_2 + \rho_3 + \rho_{NO_3} \\ &= (M_{Na}+M_{Cl})c_{Na} + (M_H+M_{Cl})c_H + (M_{Fe}+3M_{Cl})c_3 \\ &\quad + (M_{Fe}+2M_{Cl})c_2 + (M_{NO_3}-M_{Cl})c_{NO_3}\end{aligned}$$

and the corresponding variations with time are:

$$\begin{aligned}\frac{\partial \rho_s}{\partial t} &= -\operatorname{div} j_s - \frac{j_q}{\mathcal{F}}\Big[(M_{Na}+M_{Cl})\frac{dt_{Na}}{dx} + (M_H+M_{Cl})\frac{dt_H}{dx} + \frac{1}{3}(M_{Fe}+3M_{Cl})\frac{dt_3}{dx} \\ &\quad + \frac{1}{2}(M_{Fe}+2M_{Cl})\frac{dt_2}{dx} - (M_{NO_3}-M_{Cl})\frac{dt_{NO_3}}{dx}\Big]\end{aligned}$$

$$\frac{\partial \rho_2}{\partial t} = -\operatorname{div} j_2 - \frac{j_q}{2\mathcal{F}} M_{Fe}\frac{dt_2}{dx}$$

$$\frac{\partial \rho_3}{\partial t} = -\operatorname{div} j_3 - \frac{j_q}{3\mathcal{F}} M_{Fe}\frac{dt_3}{dx}$$

$$\frac{\partial \rho_H}{\partial t} = -\operatorname{div} j_H - \frac{j_q}{\mathcal{F}} M_H\frac{dt_H}{dx}$$

$$\frac{\partial \rho_{NO_3}}{\partial t} = -\operatorname{div} j_{NO_3} + \frac{j_q}{\mathcal{F}} M_{NO_3}\frac{dt_{NO_3}}{dx}$$

Then the total productions of masses in the electrolyte solution are given by the integrals over these equations:

$$\begin{aligned}\frac{\partial \Delta m_s}{\partial t} &= -\frac{j_q}{\mathcal{F}}\Big[(M_{Na}+M_{Cl})(t_{Na}^{(\ell)}-t_{Na}^{(0)}) + (M_H+M_{Cl})(t_H^{(\ell)}-t_H^{(0)}) \\ &\quad + \frac{1}{3}(M_{Fe}+3M_{Cl})(t_3^{(\ell)}-t_3^{(0)}) + \frac{1}{2}(M_{Fe}+2M_{Cl})(t_2^{(\ell)}-t_2^{(0)}) \\ &\quad - (M_{NO_3}-M_{Cl})(t_{NO_3}^{(\ell)}-t_{NO_3}^{(0)})\Big]\end{aligned}$$

$$\frac{\partial \Delta m_2}{\partial t} = -\frac{j_q}{2\mathcal{F}} M_{Fe}(t_2^{(\ell)}-t_2^{(0)})$$

. . . .

$$\frac{\partial \Delta m_{NO_3}}{\partial t} = \frac{j_q}{\mathcal{F}} M_{NO_3}(t_{NO_3}^{(\ell)}-t_{NO_3}^{(0)})$$

where, as before, the integrals over $\operatorname{div} j_i$ vanish.

We consider the special case which is depicted in Fig. 21:

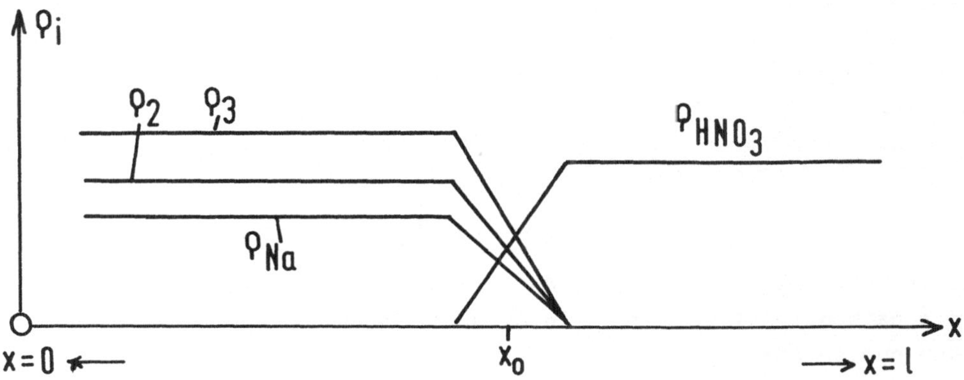

Fig. 21 Partial density distribution (schematic) in the liquid boundary between a redox half-cell containing $FeCl_2$, $FeCl_3$, and NaCl and a half-cell containing HNO_3

This gives

$$\frac{\partial \Delta m_s}{\partial t} = \frac{j_2}{\mathcal{F}}\Big[(M_{Na}+M_{Cl})t_{Na}^{(0)} - (M_H + M_{Cl})t_H^{(\ell)} + \tfrac{1}{3}(M_{Fe}+3M_{Cl})t_3^{(0)} + \tfrac{1}{2}(M_{Fe}+2M_{Cl})t_2^{(0)} + (M_{NO_3} - M_{Cl})t_{NO_3}^{(\ell)}\Big]$$

$$\left.\begin{aligned}
\frac{\partial \Delta m_2}{\partial t} &= \frac{j_2}{2\mathcal{F}} M_{Fe} t_2^{(0)} && \text{production of Fe(II)}\\
\frac{\partial \Delta m_3}{\partial t} &= \frac{j_2}{3\mathcal{F}} M_{Fe} t_3^{(0)} && \text{production of Fe(III)}\\
\frac{\partial \Delta m_H}{\partial t} &= -\frac{j_2}{\mathcal{F}} M_H t_H^{(\ell)} && \text{removal of H}\\
\frac{\partial \Delta m_{NO_3}}{\partial t} &= \frac{j_2}{\mathcal{F}} M_{NO_3} t_{NO_3}^{(\ell)} && \text{production of } NO_3
\end{aligned}\right\} \text{ for } j_2 > 0$$

It may be seen that the rates of production of Fe(II) and Fe(III), integrated over the bulk, equal exactly the excess constituent mass fluxes at the Pt electrode (x = 0) as given by eqs.(168) and (169). At the cathode ($x = \ell$) there is a constituent mass flux of hydrogen, $(1-t_H) j_2 M_H/\mathcal{F}$ and an excess hydrogen constituent flux $t_H j_2 M_H/\mathcal{F}$, the latter one being equal to the total rate of hydrogen removal from the electrolyte solution. There is a depletion of the constituent NO_3 at the cathode which now appears in the electrolyte solution around x_o. Furthermore, there is also a production of Na and a removal of Cl in the region around x_o which, however, do not appear explicitly in the above formulas.

We may remove the HNO_3 component and have a finite NaCl concentration along the total cell volume from x = 0 to x = ℓ . Then the second and third equation remain the same; however the first one should be rewritten as

$$\frac{\partial \Delta m_s}{\partial t} = \frac{j_2}{\mathcal{F}}\Big[(M_{Na}+M_{Cl})(t_{Na}^{(0)}-t_{Na}^{(\ell)}) + \frac{1}{3}(M_{Fe}+3M_{Cl})t_3^{(0)} + \frac{1}{2}(M_{Fe}+2M_{Cl})t_2^{(0)}\Big]$$

Finally we refrain from adding NaCl to any of the electrolytes, but have a cathode half-cell containing HNO_3. Thus the system is given by the components:

$FeCl_2$, $FeCl_3$, HNO_3

and it is fully defined by the total solute density ρ_s and the three constituent densities ρ_3, ρ_H, ρ_{NO_3}.
Now we have

$$\rho_s = (M_H+M_{Cl})c_H + (M_{Fe}+3M_{Cl})c_3 + (M_{Fe}+2M_{Cl})c_2 + (M_{NO_3}-M_{Cl})c_{NO_3}$$

and consequently, the rates of mass change in the bulk electrolyte are given as:

$$\frac{\partial \Delta m_s}{\partial t} = -\frac{j_2}{\mathcal{F}}\Big[(M_H+M_{Cl})(t_H^{(\ell)}-t_H^{(0)}) + \frac{1}{3}(M_{Fe}+3M_{Cl})(t_3^{(\ell)}-t_3^{(0)}) + \frac{1}{2}(M_{Fe}+2M_{Cl})(t_2^{(\ell)}-t_2^{(0)}) - (M_{NO_3}-M_{Cl})(t_{NO_3}^{(\ell)}-t_{NO_3}^{(0)})\Big]$$

. . . .

$$\frac{\partial \Delta m_{NO_3}}{\partial t} = \frac{j_2}{\mathcal{F}} M_{NO_3}(t_{NO_3}^{(\ell)}-t_{NO_3}^{(0)})$$

In the particular situation that the liquid junction connects a mixture of $FeCl_2$ + $FeCl_3$ dissolved in water and an aqueous solution of HNO_3, we obtain

$$\frac{\partial \Delta m_s}{\partial t} = \frac{j_2}{\mathcal{F}}\Big[\frac{1}{3}(M_{Fe}+3M_{Cl})t_3^{(0)} + \frac{1}{2}(M_{Fe}+2M_{Cl})t_2^{(0)} - (M_H+M_{Cl})t_H^{(\ell)} + (M_{NO_3}-M_{Cl})t_{NO_3}^{(\ell)}\Big]$$

$$\frac{\partial \Delta m_3}{\partial t} = \frac{j_2}{3\mathcal{F}} M_{Fe} t_3^{(o)}$$

$$\frac{\partial \Delta m_H}{\partial t} = -\frac{j_2}{\mathcal{F}} M_H t_H^{(\ell)}$$

$$\frac{\partial \Delta m_{NO_3}}{\partial t} = \frac{j_2}{\mathcal{F}} M_{NO_3} t_{NO_3}^{(\ell)}$$

It may be seen that all statements regarding mass production and depletion made above remain valid.

Chapter 6

Energy changes in electrochemical systems

6.1 Rate of internal energy change of a homogeneous system

In the preceding chapters we have given a detailed analysis of the local rates of change of the partial mass densities characterizing the analytical field. In the next step it is necessary to present a similar analysis for the local rate of change of the internal energy.

It is instructive to begin with a homogeneous system. Let the internal energy of such a system be $U = U(T, p, m_1, m_2, \ldots)$, i.e. U is a unique function of the set of variables T, p, m_1, m_2 . . . . . It is defined relative to a reference state T_0, p_0 and via the unidirectional delivery of adiabatic work

$$U = U(T, p, T_0, p_0, m_1, m_2, \ldots) = \left(-\int \sum_i \mathfrak{K}_i \, ds_i\right)_{\Delta T = 0}$$

$$= (W)_{\Delta T = 0} \tag{197}$$

ds_i are displacements of macroscopic bodies in ordinary space which are coupled to the system, the $\mathfrak{K}_i$ are the corresponding forces. "Adiabatic work" means that the temperature difference between the system and a reference envelope is kept equal to zero for all changes of the state of the system. The change of internal energy connected with an internal change of the system (chemical reaction, mixing) in the state T_0, p_0 is given by the relation

$$\Delta U = \left[U(T, p, m_1, m_2, \ldots)\right]_{\text{constructed at } T_0, p_0} - U(T, p, T_0, p_0, m_1, m_2, \ldots) \tag{197a}$$

with

$$\left[U(T, p, m_1, m_2, \ldots)\right]_{\text{constructed at } T_0, p_0} = (W)^*_{\Delta T = 0} \tag{197b}$$

where $(W)^{*}_{\Delta T=0}$ is the adiabatic work performed on the system whilst the internal change is proceeding. One sees that ΔU is the difference between two amounts of adiabatic unidirectional work, however ΔU itself in general cannot be represented as adiabatic work. Then at T_0, p_0 we have

$$U(T_0, p_0) = U(T_0, p_0)_{\text{before construction}} + \Delta U$$

Details cannot be given here, the most important feature of U is that it assigns to each transition in the abstract substance property space integrals or differences of integrals over the ordinary space involving a non-equivalence parameter, the force.

The rate of change of the function U is

$$\frac{dU}{dt} = -p\frac{dV}{dt} + \frac{dQ}{dt} \tag{198}$$

which is the first law of thermodynamics.
Since U is a well-defined function, eq.(198) defines the amount of heat Q delivered to the system per unit time (our formulation of the first law is the one given by Born[1] in a somewhat modified form). Now we write for the rate of change of the internal energy including the presence of a chemical reaction:

$$\frac{dU}{dt} = \left(\frac{\partial U}{\partial T}\right)_{p,m_1,m_2,..}\frac{dT}{dt} + \left(\frac{\partial U}{\partial p}\right)_{T,m_1,m_2..}\frac{dp}{dt} + \sum_i\left(\frac{\partial U}{\partial m_i}\right)_{T,p,m_j..}\frac{dm_i}{dt} \tag{199}$$

Let us assume that the change of the parameters T, p, m_i are caused by a sufficiently slow volume change so that the equilibrium situation is essentially retained. The choice of the rate of heat exchange dQ/dt, determines the variation of the quantities T and p, the m_i are functions of the construction masses and of T and p. Then, if we proceed from eq.(198) to the second law of thermodynamics we have

$$\left\{\frac{\partial U}{\partial T}\left(\frac{dT}{dV}\right)_{\frac{dQ}{dt}} + \frac{\partial U}{\partial p}\left(\frac{dp}{dV}\right)_{\frac{dQ}{dt}} + \sum_i\frac{\partial U}{\partial m_i}\left(\frac{dm_i}{dt}\right)_{\frac{dQ}{dt}}\right\}\frac{dV}{dt} - T\frac{dS}{dt} = -p\frac{dV}{dt} \tag{200}$$

where S is the entropy of the system. In eq.(200) the right-hand side represents the term

$$-\int \mathfrak{K}\, d\underline{s}$$

mentioned above, and the reversibility implies that eq.(200) is valid for both directions of $d\underline{s}$ whereas the definition of the internal energy essentially requires $-\int \mathfrak{K} d\underline{s} > 0$. For completeness we write eq. (200) for the case of an isothermal process:

$$\frac{dF}{dt} = \left\{ \left(\frac{\partial U}{\partial p}\right)_{T,m_i} \left(\frac{\partial p}{\partial V}\right)_T + \sum_i \left(\frac{\partial U}{\partial m_i}\right)\left(\frac{\partial m_i}{\partial p}\right)_T \left(\frac{\partial p}{\partial V}\right)_T \right\} \frac{dV}{dt} - T\frac{dS}{dt} = -p\frac{dV}{dt} \tag{200a}$$

where F is the Helmholtz free energy.

6.2 Rate of internal energy change in a normal non-uniform system.

After this preparation we turn to the non-equilibrium system. Let us begin with the diffusion problem treated in chapter 2. Now eqs.(199) and (200) have to be transcribed to a local formulation:

$$\begin{aligned} \rho\frac{d\mathcal{U}}{dt} &= \rho\left\{ \left(\frac{\partial \mathcal{U}}{\partial T}\right)_{p,m_i..} \frac{dT}{dt} + \left(\frac{\partial \mathcal{U}}{\partial p}\right)_{T,m_i..} \frac{dp}{dt} \right. \\ &\qquad \left. + \sum_i \left(\frac{\partial \mathcal{U}}{\partial \rho_i}\right)_{T,p,m_i} \frac{d\rho_i}{dt} \right\} \\ &= \rho\left(\frac{\partial \mathcal{U}}{\partial T}\right)_{p,m_i} \frac{dT}{dt} + \rho\left(\frac{\partial \mathcal{U}}{\partial p}\right)_{T,m_i} \frac{dp}{dt} + \sum_i \left(\frac{\partial U}{\partial m_i}\right)_{T,p,m_j} \frac{d\rho_i}{dt} \end{aligned} \tag{201}$$

where $\mathcal{U}$ is the internal energy per unit mass. In each volume element we have local energy conservation (see eq.(198))

$$\rho\frac{\partial \mathcal{U}}{\partial t} = -\operatorname{div} j_Q - p\frac{dv}{dt} \tag{202}$$

where j_Q is the heat flux

$$j_Q = -\lambda\frac{dT}{dx} \tag{203}$$

and dv/dt is the rate of volume change per unit volume. Kinetic and potential energy contributions have been neglected. The rates of change of the partial mass densities, $\partial\rho_i/\partial t$ are given by the diffusion equations described in chapter 2. Of course, in the most general case the diffusion process may also involve a chemical reaction. The term containing dp/dt usually is without practical importance, the stress tensor, introduced in eq.(70) contains the pressure, and since

σ varies with time, p undergoes a very slight change, also. The local volume changes, when integrated lead to the total rate of volume change

$$\frac{dV}{dt} = \int_0^\ell \frac{dv}{dt}\, dx \cdot A$$

however, this effect is usually small when we are dealing with liquid or solid systems and thus the volume work can be neglected. So the total change of internal nergy per unit time is given by the approximate relation

$$A\int_0^\ell \rho \frac{\partial u}{\partial t}\, dx = -A\int_0^\ell \mathrm{div}\, j_Q\, dx = -\int_f j_Q\, df$$
$$= \frac{dQ}{dt}$$

where the last integral is taken over the surface of the system.

6.3 Energy and momentum changes in the presence of an electric current

Now let us consider a non-uniform system in the presence of an electrical current. We observe the variation with time of the local quantities T(x,t), p(x,T), ρ_i(x,t). We find that the rates of change, $\partial T/\partial t$, $\partial p/\partial t$, $\partial \rho_i/\partial t$ are different from those we had in the situation $j_2 = 0$. This has already been explained for the case of $\partial \rho_i/\partial t$. In order to incorporate these additional effects in the local energy conservation relation we write, as we did for the analytical field, $\rho_1, \rho_2, \ldots$

$$\rho \frac{\partial u}{\partial t} + p\frac{dv}{dt} = -\mathrm{div}\, j_Q - \mathrm{div}\, j_\varepsilon^{\cdot\,\prime} \qquad (204)$$

where $j_\varepsilon^{\cdot\,\prime}$ is the excess energy flux. It will be seen below that it would be more correct to say "excess work flux", however, it is felt that "excess energy flux" is a more convenient notation. We note that, as before, j_Q is given by eq.(203).
For the excess energy flux we write in analogy to eqs.(83)-(86)

$$j_\varepsilon^{\cdot\,\prime} = \tau_\varepsilon \dot{j}_2 \qquad (205)$$

and we define τ_ε by eq.(205) to be the excess energy transport factor. Combination of eqs.(204) and (205) yields ($\mathrm{div}\, j_2 = 0$)

$$\rho \frac{\partial u}{\partial t} + p\frac{dv}{dt} = -\mathrm{div}\, j_Q - \frac{d\tau_\varepsilon}{dx} j_2 \qquad (206)$$

The excess energy flux and the excess constituent mass fluxes have one common property. Their absolute values are undetermined unless they are fixed either at the boundaries of the system or somewhere else. Thus we have

$$\tau_\varepsilon = \tau_\varepsilon^0 + \tau_\varepsilon^* = E_i + \tau_\varepsilon^* \qquad (207)$$

and

$$\frac{d\tau_\varepsilon}{dx} = \frac{d\tau_\varepsilon^*}{dx}$$

Anticipating some facts which will be described shortly, we call the constant E_i, the electrical potential at the input. Now let us consider a homogeneous part of the system, where $\partial\rho_i/\partial x = 0$, i = 1, 2 . . . Then experimental observation tells us that even here the divergence of the excess energy flux in general does not vanish. The divergence of j_ε' vanishes only in a superconducting material, i.e. it behaves like that of the excess constitutent mass fluxes. The experimental result in the general case is

$$-\frac{d\tau_\varepsilon}{dx} = r \cdot j_z \qquad (208)$$

where r is the specific resistance. One sees from eq.(208) that the gradient of the energy transport factor is proportional to the electric current density. With eq.(208), eq.(207) takes the form:

$$\rho\frac{\partial u}{\partial t} + p\frac{dv}{dt} = -\operatorname{div} j_Q + r j_z^2 \qquad (209)$$

If we now consider eq.(202) to be the expression of the local energy conservation, because the change of internal energy is either balanced by the volume work or by a heat flux connected with a gradient of temperature, then we are able to state that in the presence of the electric current there is a rate of production of internal energy. The question of energy conservation is still open; the energy must come from some remote part of the system, thus, just as we had a delocalized mass conservation in the presence of an electric current so we have a delocalized energy conservation when $j_z \neq 0$. From eq.(208) one sees that the excess energy transport factor in the homogeneous part of the system has the form

$$\tau_\varepsilon = \tau_\varepsilon^0 - \int_0^x r\,dx' \cdot j_z = E_i - R(x) j_z \cdot A \qquad (210)$$

where R is the resistance and $R(x) \cdot J_z$ is the electric potential connected with Ohm's law.

We turn now to the description of the non-uniform parts of the system. Now additional terms appear in the expression for the gradient of the excess energy transport factor. In fact we have

$$-\frac{d\tau_\varepsilon}{dx} = r j_2 - \rho\left(\frac{\partial u}{\partial T}\right)_{p,..}\frac{d\tau_T}{dx} - \rho\left(\frac{\partial u}{\partial p}\right)_{T,..}\frac{d\tau_p}{dx}$$

$$-\sum\left(\frac{\partial U}{\partial m_i}\right)_{T,p}\frac{d\tau_{\rho_i}}{dx} - \frac{d\tau_{\varepsilon,hp}}{dx} \tag{211}$$

The new terms which occur in eq.(211) as compared with eq.(208) are all explicitly connected with the inhomogeneity of the system, that is, the gradients $d\tau_i/dx$ are to be written in the form

$$\frac{d\tau_i}{dx} = \sum_i \frac{\partial\tau_i}{\partial\rho_i}\frac{\partial\rho_i}{\partial x}$$

thus, in the homogeneous parts of the system where $\partial\rho_i/\partial x = 0$ these contributions vanish. Apart from the last one all the new terms represent the local productions of energy which are connected with the rates of change of the parameters T, p, ρ_i determining the internal energy. The term with $d\tau_T/dx$ is connected with the Peltier effect and the set of terms containing the $d\tau_{\rho_i}/dx$ are connected with the local prodcution of real mass as described in chapter 3. The τ_{ρ_i} are our generalized transport numbers, but now transformed to the construction space since the energy equation is formulated in this mode of representation. Of course, we could also have written the energy equation involving another form of composition description.

The only term we wish to discuss here briefly is the one containing $d\tau_p/dx$. This term has no practical importance for our present treatment, however, for reasons of systematic completeness we may devote some words to the corresponding physical facts.

The equation of motion as the equation governing the change with time of the local linear momentum, has been briefly mentioned on page 34 (eq.(70)). As in the case of the two other conservation laws - that of the mass and that of the energy - this conservation law will also be modified in the presence of an electrical current. In full analogy to our previous approaches we write:

$$\rho\frac{\partial u}{\partial t} = -\mathrm{Div}\,\sigma - \mathrm{Div}\,\sigma'$$

where σ' is the excess stress tensor, it is the effect which is "caused" by the electric current over and above the effects to be found in the ordinary system. We have

$$\sigma' = \begin{pmatrix} \tau_x^* j_{\varrho x} & \tau_x^* j_{\varrho y} & \tau_x^* j_{\varrho z} \\ \tau_y^* j_{\varrho x} & \tau_y^* j_{\varrho y} & \tau_y^* j_{\varrho z} \\ \tau_z^* j_{\varrho x} & \tau_z^* j_{\varrho y} & \tau_z^* j_{\varrho z} \end{pmatrix}$$

and the divergence of this tensor is

$$\mathrm{Div}\,\sigma' = \begin{pmatrix} \frac{\partial}{\partial x}(\tau_x^* j_{\varrho x}) + \frac{\partial}{\partial y}(\tau_y^* j_{\varrho x}) + \frac{\partial}{\partial z}(\tau_z^* j_{\varrho x}) \\ \frac{\partial}{\partial x}(\tau_x^* j_{\varrho y}) + \frac{\partial}{\partial y}(\tau_y^* j_{\varrho y}) + \frac{\partial}{\partial z}(\tau_z^* j_{\varrho y}) \\ \frac{\partial}{\partial x}(\tau_x^* j_{\varrho z}) + \frac{\partial}{\partial y}(\tau_y^* j_{\varrho z}) + \frac{\partial}{\partial z}(\tau_z^* j_{\varrho z}) \end{pmatrix}$$

In a simple system with constant cross-section A we have $j_{\varrho z} = j_{\varrho y} = 0$, however $\partial\tau_x^*/\partial x \neq 0$ and $\partial\tau_z^*/\partial x \neq 0$ which gives

$$\mathrm{Div}\,\sigma' = \left\{ \frac{\partial\tau_x^*}{\partial x} j_\varrho, 0, 0 \right\} , \quad \mathrm{Div}\,\sigma' = \left\{ \frac{\partial\tau_z^*}{\partial z} j_\varrho, 0, 0 \right\}$$

so there is a linear momentum production at the walls and at all other inhomogeneities of the system - the electrokinetic effects. This modifies the pressure distribution according to eq.(70) - the stress tensor contains the pressure - which then in principle leads to a modification of the internal energy via the pressure dependence of U.

We have still to discuss the last term on the right-hand side of eq.(211). Let us assume for simplicity that there is only one dominant composition dependence. Then we have

$$\frac{d\tau_{\varepsilon,hp}}{dx} = \frac{d\tau_{\varepsilon,hp}}{d\rho_1} \frac{d\rho_1}{dx} \tag{212}$$

and it is experimentally observed that

$$\frac{d\tau_{\varepsilon,hp}}{d\rho_1} = -f(j_\varrho, \rho_1) \tag{213}$$

$f(j_2, \rho_1)$ is a function of the composition and the electric current density. The functional form is

$$f(j_2, \rho_1) = a_0 + a_1 j_2 + a_2 j_2^2 + a_3 j_2^3 + a_4 j_2^4 + \dots \quad (213a)$$

therefore we have chosen the denotation $\frac{d\tau_{\varepsilon,hp}}{dx}$: there is an energy production which depends on higher powers (hp) of the current density than the second. $f(j_2, \rho_1)$ is defined for positive and negative values of j_2, the positive direction of j_2 is the direction in which $d\rho_1/dx<0$.

Now we can introduce eqs.(211) and (213) in eq.(206) to obtain

$$\rho \frac{\partial u}{\partial t} + p \frac{dv}{dt}$$
$$= - \operatorname{div} j_Q + r j_2^2 + f(\rho_1, j_2) \frac{\partial \rho_1}{\partial x} j_2$$
$$- \rho \left(\frac{\partial u}{\partial t}\right)_{p,..} \frac{d\tau_T}{dx} j_2 - \rho \left(\frac{\partial u}{\partial p}\right)_{T,..} \frac{d\tau_p}{dx} j_2 - \sum \left(\frac{\partial U}{\partial m_i}\right)_{T,p..} \frac{d\tau_{\rho_i}}{dx} j_2 \quad (214)$$

or

$$\rho \frac{\partial u}{\partial t} + p \frac{dv}{dt} + \operatorname{div} j_Q$$
$$= r j_2^2 + f(\rho_1, j_2) \frac{\partial \rho_1}{\partial x} j_2$$
$$- \rho \left(\frac{\partial u}{\partial T}\right)_{p,..} \frac{d\tau_T}{dx} j_2 - \rho \left(\frac{\partial u}{\partial p}\right)_{T,..} \frac{d\tau_p}{dx} j_2 - \sum \left(\frac{\partial U}{\partial m_i}\right) \frac{d\tau_{\rho_i}}{dx} j_2 \quad (214a)$$

In Fig.22 we give a schematic representation of the physical situation which is described by eq.(214). It is assumed that the system has two regions of inhomogeneity: $x \approx x_{01}$ and $x \approx x_{02}$. The ranges of x for which we have position independent values of $\rho \partial u/\partial t$ are described by the first two terms of eq.(214). If the heat flux j_Q through the walls of the cell were absent then we would have the narrow lines instead of the heavily drawn lines. Around the two inhomogeneities we have a strong decrease of internal energy per unit time. Note that this is one possible example; there might also be an increase of the internal energy, depending on the nature of the system. In this range, apart from the two already mentioned, there are three further contributions to the rate of change of the internal energy. These are the negative

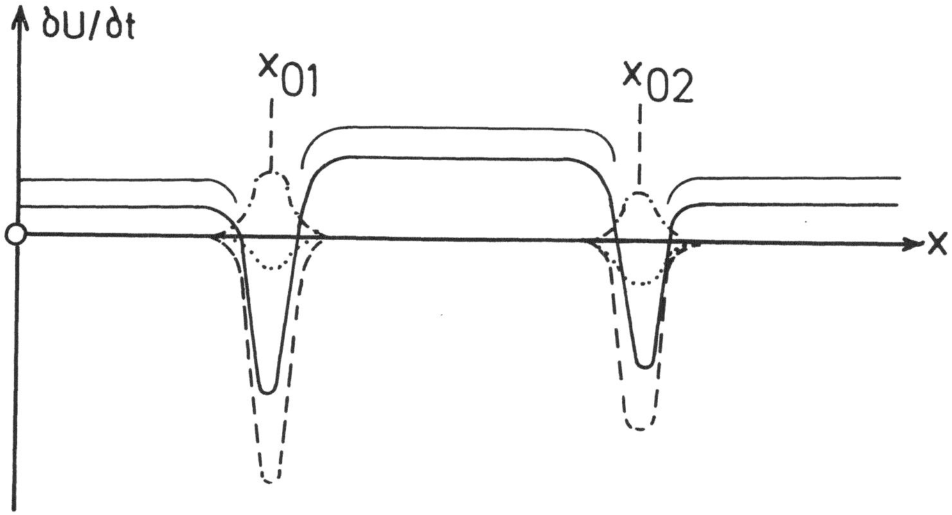

Fig. 22 Rate of change of the internal energy in the presence of an electric current as a function of the cell coordinate x. For further details see text.

terms containing the sum over the $\partial U/\partial m_i$'s (dashed curves), a term proportional to $\rho\,(\partial U/\partial T)_{p,..}$ which we also have chosen to be negative (dotted curves), and the term $j_2\; f(\rho_1, j_2)\cdot \partial\rho_1/\partial x$ (dot-dashed curves) which in this example is assumed to be positive. The term containing the pressure effect may be neglected. Then the total rate of change of internal energy which is the sum of the five contributions is depicted in Fig.22 as the fully drawn curve. If j_2 is increased, the total curve rises and the $f(j_2, \rho_1)$ contribution fills out the two minima to a greater extent.

Next we integrate the left-hand side of eq.(214a) with respect to the cell coordinate x. Let the "end point" of the cell be at $x = \ell$. For simplicity we assume that A, the cross-section of the cell system, is constant. Also, the gradient of the pressure p we may safely set equal to zero. Then we have

$$A\int_0^l \rho \frac{\partial u}{\partial t} dx + p\frac{dV}{dt} + A\int_0^l \operatorname{div} j_Q dx = A\int_0^l \rho \frac{\partial u}{\partial t} dx + p\frac{dV}{dt} + \int_f j_\xi df$$
$$= A\int_0^l \rho \frac{\partial u}{\partial t} dx + p\frac{dV}{dt} - \frac{dQ}{dt}$$

where $\int df$ is the total surface of the system and dQ/dt is the total amount of heat per unit time delivered to the system. When we now integrate the right-hand side of eq.(214a) we obtain

$$A\int_0^l \rho \frac{\partial u}{\partial t} dx + p\frac{dV}{dt} - \frac{dQ}{dt}$$
$$= A E_i j_\xi + A\int_0^l r j_\xi^2 dx + A\int_0^l f(\rho_1, j_\xi)\frac{\partial \rho_1}{\partial x} dx\, j_\xi - A\int_0^l \rho\left(\frac{\partial u}{\partial T}\right)_{p,.} \frac{d\tau_T}{dx} dx\, j_\xi$$
$$-A\int_0^l \rho\left(\frac{\partial u}{\partial p}\right)_{T,.} \frac{d\tau_p}{dx} dx\, j_\xi - A\int_0^l \sum\left(\frac{\partial U}{\partial m_i}\right)_{T,p..} \frac{d\tau_{\rho i}}{dx} dx \cdot j_\xi \qquad (215)$$

where we have introduced the constant of integration E_i in order to be in agreement with eq.(210). However, we may also write this equation, using the left-hand side of eq.(215)

$$A\int_0^l \rho \frac{\partial u}{\partial t} dx + p\frac{dV}{dt} - \frac{dQ}{dt}$$
$$= -(\tau_\varepsilon - \tau_\varepsilon^0) j_\xi A = E j_\xi A = E J_\xi \qquad (216)$$

We say, E is the electrical work delivered to the system per unit time. Comparing eqs.(215) and (216) we have

$$E = \int_0^l r j_\xi dx + \int_0^l f(j_\xi, \rho_1)\frac{\partial \rho_1}{\partial x} dx$$
$$-\int_0^l \rho\left(\frac{\partial u}{\partial T}\right)_{p,..} \frac{d\tau_x}{dx} dx - \int_0^l \rho\left(\frac{\partial u}{\partial p}\right)_{T,.} \frac{d\tau_p}{dx} dx - \int_0^l \sum\left(\frac{\partial U}{\partial m_i}\right) \frac{d\tau_{\rho i}}{dx} dx \qquad (217)$$

where E is the electrical potential of the system. Since E_i is an arbitrary constant, in order to simplify writing, we shall set $E_i = 0$ from now on.

6.4 "Complete" and "truncated" systems.

We have said that E is the electrical potential of the system and it is important now to specify the concept of the "system". So far the "system" has only been formally given by the range of integration $0 \leq x \leq \ell$ that is, each material part of the assembly being located at an x falling in the range $0 \leq x \leq \ell$ belongs to the system, all other parts do not belong to the system. Also for $0 \leq x \leq \ell$ there is a surroundings of the system into which heat flows from the system or from which heat flows into the system. However, this specification still is insufficient. In fact we have to state that in order to make eqs.(215)-(217) valid relations, it is required that the ρ_i are variable parameters of the system and that in the presence of the electric current the law of mass conservation connected with the variation of the ρ_i can be fulfilled within the system. The system has to be considered as a container of extension $0 \leq x \leq \ell$ in which a chemical reaction proceeds and all mass fluxes have to be inside the system. In this sense E, the electrical potential of the system, in the form

$$E = \int_0^\ell r j_Q \, dx + \int_0^\ell f(\rho_1, j_\varrho) \frac{\partial \rho_1}{\partial x} dx$$

$$- \int_0^\ell \rho \left(\frac{\partial u}{\partial T}\right)_{p,..} \frac{d\tau_T}{dx} dx - \int_0^\ell \rho \left(\frac{\partial u}{\partial p}\right)_{T..} \frac{d\tau_p}{dx} dx - \int_0^\ell \sum \left(\frac{\partial U}{\partial m_i}\right)_{T,p..} \frac{d\tau_{\rho_i}}{dx} dx \tag{217a}$$

is a property of the system for a given set of variables T, p, ρ_i, $\dot{j}_\varrho$ (and dQ/dt, see below). From the treatment given in chapter 3 it will be seen that a system in this sense has to have at least two boundaries around which the composition changes. If in such a two boundary system a local change of composition $\partial \rho_i / \partial t$ is not connected with the electric current, - for instance because we have only two different metals in contact - then the last term on the right of eq.(217a) drops out; we are only concerned then with the Peltier effect. Details cannot be given in the present treatment. Also we may assume that the system is constructed such that all electrokinetic effects disappear, then, the term containing $\partial u / \partial p$ can also be omitted.

Now let us imagine that in Fig.22 the system ends between the two boundaries x_{01} and x_{02}, that is, that the system is cut off at x_ℓ with

$x_\ell = \frac{1}{2}(x_{01} + x_{02})$, say. For the reasons explained above, the three last terms of eq.(217a) disappear; the higher power term is connected with the local mass production for the system of interest in the field of electrochemistry and is thus without significance for the "truncated" system. Consequently for the truncated system we are left with the electrical potential

$$E = j_2 \int_0^{x_\ell} r\,dx \qquad (210a)$$

We return to the "complete" system for which eq.(217a) is valid. Then E may have positive or negative values. For example, if we consider Fig.22, and if care is taken that the current density is small enough, then we have $E<0$ and consequently $E J_2\,dt = dW_{el} < 0$. A typical representation of such a situation is a set up in which in the range $0 \leq x < x_{01}$ we have Na(metal), then in the range $x_{01} \leq x \leq x_{02}$ an aqueous NaCl solution is placed and finally, for $x_{02} < x \leq \ell$ Cl_2 is adsorbed on Pt. If the electric current is reversed, then the two minima at x_{01} and x_{02} become maxima and consequently $dW_{el} > 0$. Finally the situation depicted in Fig. 5 may be considered. Now we have a minimum at $x = x_{01}$ and a maximum of equal size at $x = x_{02}$, so the total effect is to produce $E J_2 > 0$, and this is true for both directions of J_2 .

Next we have to deal with the following question. We had to choose a "complete system" in which the law of mass conservation is fulfilled, in order to make eq.(217) a valid relation. We ask then in what way has the system to be constructed in order to be a complete system with respect to energy conservation as well? In other words, what have we to do in order to assure the validity of eq.(216) either with E given by eq.(217a), or again by the reduced form of eq.(210a), depending on the circumstances, with the nontrivial requirement $j_2 \neq 0$. One possible answer is that the system has to be a closed loop, that is $x = \ell$ has to be at the same position as x = 0. Of course, the system may also be extended to include, in addition to the three boundaries of Fig.22, one or several other boundaries around which the composition changes. In these conditions the system is also a "complete" system with respect to energy conservation. Further we see that the "truncated system", when arranged to form a loop is automatically a "complete system". Now eq.(214), written in a slightly simplified form

$$\rho \frac{\partial u}{\partial t}\frac{\partial T}{\partial t} + \rho \frac{\partial u}{\partial p}\frac{\partial p}{\partial t} + \sum \frac{\partial U}{\partial m_i}\frac{\partial \rho_i}{\partial t} + p\frac{dv}{dt}$$

$$= - \mathrm{div}\, \lambda \frac{\partial T}{\partial x} + r j_\xi^2 + f(\rho_1, j_\xi)\frac{\partial \rho_1}{\partial x} j_\xi - \rho \frac{\partial u}{\partial T}\frac{\partial \tau_T}{\partial \rho_1}\frac{\partial \rho_1}{\partial x} j_\xi$$

$$- \rho \frac{\partial u}{\partial p}\frac{\partial \tau_p}{\partial \rho_1}\frac{\partial \rho_1}{\partial x} j_\xi - \sum \frac{\partial U}{\partial m_i}\frac{\partial \tau_{\rho_i}}{\partial \rho_1}\frac{\partial \rho_1}{\partial x} j_\xi$$

is one of the equations which together with the generalized diffusion equation and the equation of motion describes the local change of state of the system.

The integrated form is, according to eq.(216) and (217)

$$A\oint \rho \frac{\partial u}{\partial t} dx = A\oint \left(r j_\xi^2 + f(\rho_1, j_\xi)\frac{\partial \rho_1}{\partial x} j_\xi\right) dx$$

$$- A\oint \left\{\rho \left(\frac{\partial u}{\partial T}\right)\frac{d\tau_T}{dx} - \rho\left(\frac{\partial u}{\partial p}\right)\frac{d\tau_p}{dx} - \sum \left(\frac{\partial U}{\partial m_i}\right)\frac{d\tau_{\rho_i}}{dx}\right\} dx \cdot j_\xi$$

$$= E j_\xi A = 0 \tag{218}$$

where, as a first example we have written the formula for the adiabatic loop with constant volume, i.e. the isolated "complete system" in vacuo. However, the loop could also be placed in an adiabatic calorimeter. Thus we see that eq.(218) when integrated over time is one verification of eq.(197b). The total ring-system has only to be properly divided in two parts, one spanning the range $0 \leq x \leq \ell$, the other one extending from x = ℓ to x = 0. The work is electrical work; it is not yet connected with any non-equivalence of ordinary space. Next we may choose an isothermal loop. Now the complete system has to have a surroundings which is no longer isolated. In this case eq.(216) is valid together with eq.(217a) with $d\tau_T/dx = 0$. The energy production parameters change with time only because the local composition changes; the temperature variation vanishes. In the special case of stationarity, which however cannot be sustained without a movement of parts of the loop relative to another system in which an electric current exists - a magnetic field - we have $dU/dt = dV/dt = 0$, thus

$$-\frac{dQ}{dt} = E j_\xi A$$

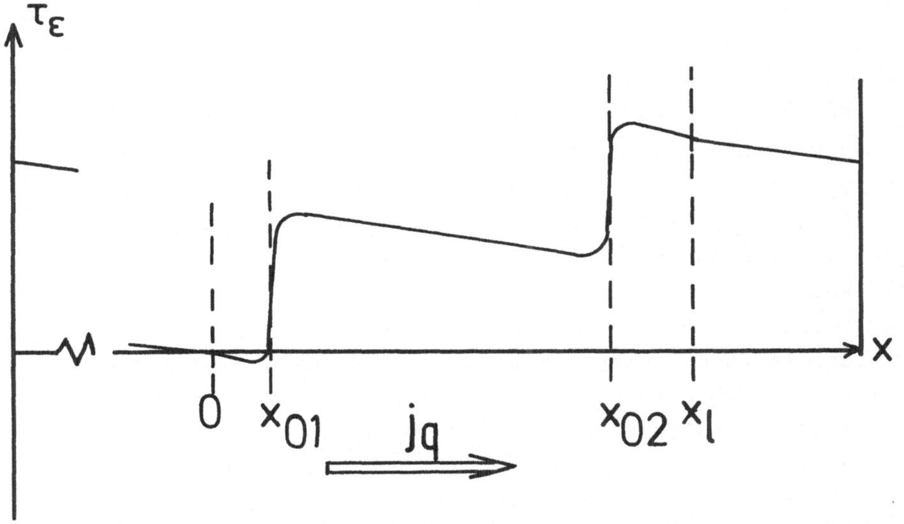

Fig. 23 Schematic representation of the excess energy transport factor as a function of the cell coordinate x. Note that we have $E = -\tau_\varepsilon$.

It is instructive to look at the behaviour of the excess energy transport factor, τ_ε which we have defined in eq.(205), see also eq.(206). In Fig. 23 τ_ε is depicted in a schematic way for the same system as shown in Fig.22. We have set $E_i = 0$. One sees that between x = 0 and $x = x_{01}$ τ_ε decreases with a constant slope, the (negative) divergence of τ_ε is a constant quantity and along the range $0 \leq x < x_{01}$ there is a uniform production of internal energy and heat thus the excess energy flux becomes smaller. However, then, around $x = x_{01}$ there is a large local rate of decrease of internal energy and the consequence is a drastic increase of the excess energy flux. In our Fig.22 we have assumed that the term containing $f(\rho_1, \dot{j}_2)$ is positive. In this case there is an effect acting against the increase of τ_ε around x_{01}. For low current density this term may also add to the increase of τ_ε (see below). Between x_{01} and x_{02} internal energy has to be produced and heat delivered to the surroundings, consequently $\dot{j}_\varepsilon'$ decreases. Around x_{02} there is another increase of $\dot{j}_\varepsilon'$ and finally all energy taken up in the two regions around x_{01} and x_{02} is converted to real, i.e. observable internal energy (via the parameter T) and heat so that at x = 0 the excess energy flux has attained its original value $\dot{j}_\varepsilon' = 0$. This is the feature of a chemical reaction with delocalized mass and energy conservation. It is brought into being by the manner of construction of the material system. The system so constructed could be used to define the electric current $J_2 = A \cdot \dot{j}_2$ in the following way. Assume we have observed the rate of change of the internal energy field, $\partial U/\partial t(x)$, and that we have knowledge of the rate of heat delivery to the surroundings, the instantaneous composition distribution, and the substance parameters entering in the excess energy transport factor τ_ε. Then $\dot{j}_2$ is the physical parameter which ensures that the left-hand and right-hand sides of an equation of type (218) (which is written for the adiabatic case) are equal. It is clear that if there is no chemical reaction proceeding in the closed loop, then the electric current cannot be defined in this way. Now, for definition one has to have recourse to ordinary space; there has to be a movement of the loop relative to another one (or relative to a permanent magnet, which is equivalent). It follows that the general definition of an electric current involves a "movement" in abstract substance space and in ordinary space. There is thus the possibility of a reversal of the unidirectional mapping operations described under eqs.(197) and (197a).

6.5 The reversible electric work and the electromotive force.

As will be shown later (Fig.24), by a proper construction of the system involving more than two boundaries at which the composition changes, we can attain the limiting situation $\dot{j}_2 = 0$ in the loop. This can also be achieved by suitable movement of a part of the loop in a magnetic field.

Then we have in the general case with $|dQ/dt| \geqslant 0$

$$\int_0^\ell \left(\rho \frac{\partial U}{\partial t} + p\frac{dv}{dt} + \mathrm{div}\, j_Q\right) dx = \int_0^\ell \Big(a_0 \frac{d\rho_1}{dx}$$

$$- \rho\left(\frac{\partial U}{\partial T}\right)_{p,..} \frac{d\tau_T}{dx} - \rho\left(\frac{\partial U}{\partial p}\right)_{T,..} \frac{d\tau_p}{dx} + \sum \left(\frac{\partial U}{\partial m_i}\right)_{T,p..} \frac{d\tau_{\rho_i}}{dx}\Big)\, dx \cdot \dot{j}_2$$

$$= E\, \dot{j}_2$$

if $\dot{j}_2$ becomes sufficiently small. "Sufficiently small" means that in eq.(214a) the ratio of the contributions of the quadratic and higher order terms in $\dot{j}_2$ to the one linear in $\dot{j}_2$ can be made as small as we wish.

Next we write:

$$\int_0^{\ell}\Big\{\rho\frac{\partial u}{\partial t}\Big(\frac{\partial T}{\partial t}+\frac{\partial T'}{\partial t}\Big)+\rho\frac{\partial u}{\partial p}\Big(\frac{\partial p}{\partial t}+\frac{\partial p'}{\partial t}\Big)$$

$$+\sum\frac{\partial U}{\partial m_i}\Big(\frac{\partial \rho_i}{\partial t}+\frac{\partial \rho_i'}{\partial t}\Big)+p\frac{dv}{dt}$$

$$+\operatorname{div} j_{Q_{irrev}}+\operatorname{div} j_{Q_{rev}}\Big\}dx = E j_2 \qquad (219)$$

where the unprimed rates of change and $j_{Q_{irrev}}$ correspond to the irreversible processes occurring in the system in the presence and in the absence of an electric current. Since then for the other terms on the left-hand side of eq. (219), we have reversibility, i.e.

$$\Big(\rho\frac{\partial u}{\partial T}\frac{\partial T'}{\partial t}+\rho\frac{\partial u}{\partial p}\frac{\partial p'}{\partial t}+\sum\frac{\partial U}{\partial m_i}\frac{\partial \rho_i}{\partial t}+\operatorname{div} j_{Q_{rev}}\Big)_{-j_2}$$

$$=-\Big(\rho\frac{\partial u}{\partial T}\frac{\partial T'}{\partial t}+\rho\frac{\partial u}{\partial p}\frac{\partial p'}{\partial t}+\sum\frac{\partial U}{\partial m_i}\frac{\partial \rho_i}{\partial t}+\operatorname{div} j_{Q_{rev}}\Big)_{j_2}$$

by definition the negative divergence of the reversible part of the heat flux is connected with S, the entropy per unit mass:

$$-\operatorname{div} j_{Q_{rev}}=\rho T\frac{dS}{dt}$$

Then we obtain

$$\begin{aligned}\frac{\partial T'}{\partial t}&=-\frac{d\tau_T}{dx}j_2\\ \frac{\partial p'}{\partial t}&=-\frac{d\tau_p}{dx}j_2\\ \frac{\partial \rho_i'}{\partial t}&=\mp\frac{d\tau_{\rho_i}}{dx}j_2\\ \rho T\frac{\partial S}{\partial t}&=-a_0\frac{\partial \rho_1}{\partial x}j_2\end{aligned} \qquad (220)$$

and if we neglect the irreversible part of eq.(219) and integrate over the active part of the system we find

$$\int_0^{\ell}\Big\{\rho\frac{\partial u}{\partial t}\frac{d\tau_T}{dx}+\rho\frac{\partial u}{\partial p}\frac{d\tau_p}{dx}+\sum\frac{\partial U}{\partial m_i}\frac{d\tau_{\rho_i}}{\partial x}\Big\}dx\, j_2-\int_0^{\ell}\rho T\frac{\partial S}{\partial t}dx$$

$$+\int_0^{\ell} p \frac{dv}{dt} dx = E \cdot j_\xi \tag{221}$$

with

$$\int_0^{\ell} \rho T \frac{\partial S}{\partial t} dx = \frac{dQ}{dt}$$

The active part is, what is usually considered to be the galvanic cell. Eq.(221) is the electrical analog to eq.(200). Eq.(200) gives the mechanical reversible work (the volume work) which is connected with a change of composition. Correspondingly, eq.(221) gives the reversible electrical work which is needed to change the internal composition variables. Note that we have

$$\frac{d\tau_{\rho_i}}{dx} = \mp \frac{1}{j_\xi} \frac{\partial \rho_i}{\partial t}$$

and that eqs.(200) and (221) refer to unit time.

The entropy S is a function of the state of the system, $S = S(T, p, m_1, m_2 \ldots)$, thus we have for the local rate of change of entropy:

$$\rho \frac{\partial S}{\partial t} = \rho \frac{\partial S}{\partial T} \frac{\partial T'}{\partial t} + \rho \frac{\partial S}{\partial p} \frac{\partial p'}{\partial t} + \sum \frac{\partial S}{\partial m_i} \frac{\partial \rho_i}{\partial t} \tag{222}$$

In eq.(221) dQ/dt may have any value, for instance, the loop may be thermally isolated, then we have dQ/dt = 0. However, another special case is of greater general interest; this is the case of the isothermal loop, or the isothermal galvanic cell under constant pressure. In this situation eq.(221) takes the form

$$\int_0^{\ell} \sum \left(\frac{\partial U}{\partial m_i} - T \frac{\partial S}{\partial m_i} \right) \frac{\partial \rho_i'}{\partial t} dx + \int_0^{\ell} p \frac{dv}{dt} dx = j_\xi E$$

$$= \int_0^{\ell} \sum \frac{\partial U}{\partial m_i} \frac{\partial \rho_i'}{\partial t} dx - \frac{dQ}{dt} + \int_0^{\ell} p \frac{dv}{dt} \qquad j_\xi \to 0$$

where eq.(222) has been used. Since the system is isothermal and isobaric at each internal position, we may also write

$$\int_0^{\ell} \sum_i \mu_i^* \frac{\partial \rho_i'}{\partial t} dx = \frac{1}{A} \frac{dG}{dt} = j_\xi E \tag{223}$$

where μ_i^* is the specific chemical potential of component i and G is the Gibbs free energy. Using eqs.(220), eq.(223) may also be rewritten to yield

$$-\int_0^{\ell} \sum_i \mu_i^* \frac{d\tau_{\varrho_i}}{dx}\, dx \cdot j_z = -\int_0^{\ell} \sum_i \mu_i^* \frac{\partial \tau_{\varrho_i}}{\partial \varrho_i} \frac{\partial \varrho_i}{\partial x}\, dx \cdot j_z = E\, j_z \qquad (224)$$

thus

$$(E)_{j_z \to 0} = -\int_0^{\ell} \sum_i \mu_i^* \frac{\partial \tau_{\varrho_i}}{\partial \varrho_i} \frac{\partial \varrho_i}{\partial x}\, dx \qquad (225)$$

The electric potential of the galvanic cell as given in eq.(225) is also called the electromotive force of the cell. In eq.(225) in a simplifying manner we have assumed that the τ_{ϱ_i} , only depend on the respective ϱ_i . For practical application to be given below, in eqs. (223)-(225) the generalized transport numbers will be replaced by the conventional ones and they will be transformed to the constituent coordinate system. In Fig.24 we visualize the behaviour of the excess energy transport factor in the situation where we have reversibility, i.e. $j_z \to 0$. Between x = ℓ and x = 0 the loop contains a section with a very high resistance. Of course this section may also be a second galvanic cell which is placed against the first one.

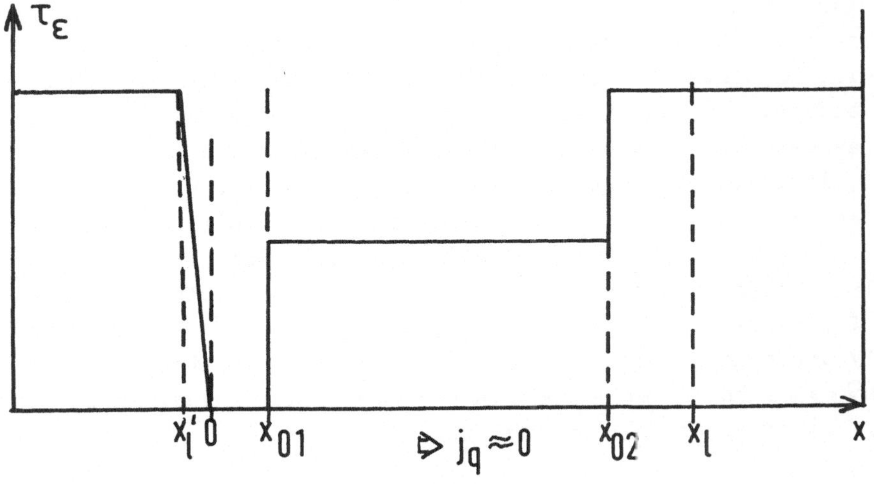

Fig. 24 Excess energy transport factor as a function of the cell coordinate x in the limiting situation $j_z \to 0$.

We turn now to a system of the type shown in Fig.22 which is not a closed loop. Thus we have an isolated open system. After constructing the system, j_z immediately drops to zero, approximately following the law of decay $\exp(t/\mathcal{R}C)$, where the geometrical factor C is the capacitance of the system. Then in eq.(216) $\dot{j}_2$ is exactly zero. The left-hand side of eq.(216) is also exactly zero; this may not be true if irreversible processes occur and the system is surrounded by a heat bath, then $dQ/dt \neq 0$. In any case E is not a well-defined quantity, E is a meaningless parameter and eq.(217) (or (217a)) is not a valid equation. In other words our open system as described so far is not a complete system in the sense of being able to undergo an internal change of state satisfying the condition of mass and energy conservation.

Now the motion of any macroscopic body in the surroundings of the system makes it a complete system with $\dot{j}_2 \neq 0$ and E being a well-defined quantity and also eq.(217a) becomes a valid relation. This macroscopic body may be a metallic system, or it may be a dielectric. In all theses cases the body loses or gains kinetic energy when approaching the cell, and this is always connected with $\dot{j}_2 \neq 0$. Then, since the left-hand side of eq.(216) is measurable, the application of a magnetic needle to measure $\dot{j}_2$ leads to a well-defined result for the electric potential E. We see again that the electric current is defined by a combined movement in abstract substance space and ordinary space. In a particular situation the moving body may be another galvanic cell which, of course has also an internal degree of freedom. Also, the movement of two bodies in contact which do not consist of the same substance causes an electric current in the galvanic cell. We recognize this possibility as being the reversal of the effects described on pages 111-112, in fact, it is the electrostatic charging of the two bodies and their subsequent motion. In the latter situations also eq.(210a) together with eq.(216) becomes a valid equation. In principle it would also be possible to make the capacitance of the system variable, then the shape of the system changes and finally it performes oscillations if properly constructed. In all these cases we have related dU/dt and dQ/dt to the work done in ordinary space involving the non-equivalence of space points, i.e. the forces.

Next assume that we bring a "charged" body from infinity to a given position $\mathbf{r}$ outside the system. We measure the work performed, i.e. measure the potential E. Then we repeat the same process for a smaller body and a smaller charge, we find another E value. Extrapolation to zero charge - i.e. zero electrical work E - and performing the corresponding procedure for all space points outside the system yields the electrostatic potential as a function of the position $\mathbf{r}$.

Reference

1) M. Born, Physik.Zeitschr. 22, 218(1921)

Chapter 7

A comparison of our electrolyte diffusion treatment with the conventional one

7.1 A brief summary: Electric current density and mass flux in our theory.

We have now to compare our treatment of the mass flux, the electric current, and the excess mass and energy fluxes in inhomogeneous systems with the theory generally accepted and found in the literature[1,2,3]. We wish to give as simple an analysis as possible and therefore we consider a binary electrolyte-water system, for instance NaCl + H_2O. Of course the NaCl concentration is not uniform throughout the system. In the conventional treatments the fluxes are the centre of interest, and thus we shall mainly discuss the interconnection between electric current density and mass flux on the one hand and the field properties determining the system on the other hand. Usually the mass fluxes are referred to the solvent fixed coordinate system and instead of the gradient of the solute particle density the gradient (or gradients) of the chemical potential are written in the equations. We assume that the electrodes in the conventional sense - if at all present - are remote and processes at the electrodes do not enter explicitly into the treatment. Comparison of the circumstances at the boundary metal/electrolyte will be postponed to the next chapter.

Then in our approach we have the total solute mass flux in the solvent fixed coordinate system

$$\bar{j}_s = -\mathcal{D}_s' \frac{\partial \rho_s}{\partial x} = -\Omega_{ss}' \frac{\partial \mu^*_{NaCl}}{\partial x} \tag{226}$$

where the diffusion coefficients $\mathcal{D}_s'$ and Ω_{ss}' are related to our previously used diffusion coefficients which are referred to the cell-fixed coordinate system by the following relations:

$$\mathcal{D}_s' = \left(1 + \frac{\rho_s}{\rho_w}\right)\mathcal{D}_s$$

$$\Omega_{ss}' = \left(1 + \frac{\rho_s}{\rho_w}\right)\Omega_{ss}$$

Next we turn to the consideration of the electric current density.

Let us begin with a truncated system (with respect to the delocalized mass conservation). In this situation the excess energy transport factor is

$$\tau_\varepsilon = -\int_0^x r\,dx \cdot j_2 \equiv -E_\Omega$$

(see eq.(210) where we have set $E_i = 0$). This gives

$$\frac{d\tau_\varepsilon}{dx} = -r j_2 = -\frac{dE_\Omega}{dx}$$

or, rewritten in the conventional form of Ohm's law

$$j_2 = \varkappa \frac{dE_\Omega}{dx} \tag{227a}$$

$$= \varkappa \frac{dE}{dx} = -\varkappa \frac{d\boldsymbol{E}}{dx} \tag{227b}$$

where $\varkappa$ is the specific conductance and $\boldsymbol{E} = -E$. In the sense of our treatment E_Ω is the resistive part of the electric potential and the physical meaning of eq.(227a) is the following. If it is known that there is a certain decrease of the excess energy flux through the system per unit length, $-dE_\Omega/dx$, which is observable as an increase of the internal energy per unit time, then there must exist in the system an electric current density j_2 as given by eq.(227a). The electrical current in turn is a consequence of the fact that somewhere outside the truncated system, spatially separated - however coherently proceeding - chemical reactions occur, or that "charged" macroscopic bodies are moving in the surroundings or that a conducting loop not belonging to the system moves in a magnetic field. It should be emphazised that the system is truncated with respect to delocalized mass conservation, however, together with its surroundings it forms a complete system with respect to delocalized energy conservation.

Next we turn to eq.(227b). Here the quantity $\boldsymbol{E} = -E$ is interpreted to be the electrostatic potential as a function of the cell coordinate x. Then $d\boldsymbol{E}/dx$ is the electrical field strength inside the system. We see that, whereas in eq.(227a) the constant quantity j_2 is "the generating function" for the function dE_Ω/dx, $\varkappa$ being a function of x, in eq.(227b) the function $d\boldsymbol{E}/dx$ generates or causes the function j_2. In this view $d\boldsymbol{E}/dx$ is an intrinsic property of the space inside the system and has its continuation outside the system, where the definition of $\boldsymbol{E}$ is given by the procedure described at the end of the preceding chapter. The latter interpretation of E and dE/dt are the conventional ones. We note that in the case of stationarity inside the system, i.e. $dj_{\mathfrak{e}}/dt = 0$, $d(dE/dx)/dt = 0$, the field strength outside the system may be varying with time, depending on the particular conditions chosen.

The mass flux as given by eq.(226) and the electric current density $j_{\mathfrak{e}}$ are independent quantities as long as we are dealing with a truncated system. $\partial\rho_s/dx$ and dE_{Ω}/dx may both have any arbitrary value within the range of validity of the linear relations.

In the next step we consider the complete system in which the delocalized energy and mass conservation laws can be satisfied. Now eq. (227a) is still valid, but eq.(227b) is no longer valid if E is considered to represent any kind of electrical potential. According to our definition of the excess energy transport factor, we have, if we apply eq.(211) and confine ourselves to the isothermal and isobaric case:

$$\begin{aligned} j_{\mathfrak{e}} &= \varkappa \sum \frac{\partial U}{\partial m_i} \frac{d\tau_{\rho_i}}{dx} + \varkappa \frac{d\tau_{\varepsilon,hp}}{dx} - \varkappa \frac{d\tau_{\varepsilon}}{dx} \\ &= \varkappa \sum \frac{\partial U}{\partial m_i} \frac{d\tau_{\rho_i}}{dx} + \varkappa \frac{d\tau_{\varepsilon,hp}}{dx} + \varkappa \frac{dE}{dx} \end{aligned} \tag{228}$$

which, when properly rewritten contains the specific chemical potentials and the gradients of the transport numbers. Furthermore, in general eq. (228) contains terms of higher order in $j_{\mathfrak{e}}$. As we have described at some length, in the complete system with regard to mass and energy conservation, a local real mass production in the inhomogeneous regions is connected with $j_{\mathfrak{e}}$ which modifies the gradients of ρ_i and consequently the mass fluxes j_s. Finally we recall that there are the two excess constituent mass fluxes which are strictly coupled to $j_{\mathfrak{e}}$, which however are not uniquely defined unless a specification of the boundary conditions is given.

7.2 Treatment of electric current density in the conventional theory.

Now we turn to the conventional treatment given in the literature [1,2,3]. One fundamental feature of this treatment, which makes it different from our method, is the absence of the distinction between a truncated and a complete system. This is due to the fact that the conventional theory only contains local quantities. Processes at a given position are never considered to be determined by or occurring coherently with other processes in remote parts of the total system. In the conventional treatment ionic fluxes j_{Na^+} and j_{Cl^-} are introduced, and $M_{Na}j_{Na^+}$, $M_{Cl}j_{Cl^-}$ are the corresponding ionic mass fluxes. We note that these quantities are given by an intrinsic definition, because the divergence of these fluxes does not represent an observable quantity. Let us write these quantities in the usual way:

$$-j_{Na^+} = l_{11}\frac{\partial \tilde{\mu}_{Na^+}}{\partial x} + l_{12}\frac{\partial \tilde{\mu}_{Cl^-}}{\partial x} \tag{229}$$

$$-j_{Cl^-} = l_{21}\frac{\partial \tilde{\mu}_{Na^+}}{\partial x} + l_{22}\frac{\partial \tilde{\mu}_{Cl^-}}{\partial x} \tag{230}$$

They refer to the solvent fixed coordinate system, i.e. $j_w = 0$. The l_{ij} are the constant coefficients of the linear equations (229) and (230); "constant" means that they do not depend on x explicitly. The above equations contain the gradients of the electrochemical potentials

$$\tilde{\mu}_{Na^+} = \mu_{Na^+} + \mathcal{F}\boldsymbol{E} \tag{231}$$

$$\tilde{\mu}_{Cl^-} = \mu_{Cl^-} - \mathcal{F}\boldsymbol{E} \tag{232}$$

μ_{Na^+}, μ_{Cl^-} are the single ion chemical potentials; they have, as is well known, no macroscopic operational meaning. However they are considered to be calculable - at least in principle - within the framework of statistical mechanics. The electrostatic potential $\boldsymbol{E}$ as appearing in eqs.(231) and (232) again is the continuation into the interior of the system of the physical property briefly characterized at the end of the preceding chapter. This justifies the same notation as introduced in eq.(227b). Thus to each x, $0 \leq x \leq l$ inside the system a quantity $\boldsymbol{E} = \boldsymbol{E}(x)$ is assigned representing the electrostatic potential at this position. Combination of eqs.(229) - (232) yields

$$-j_{Na^+} = l_{11}\frac{\partial \mu_{Na^+}}{\partial x} + l_{12}\frac{\partial \mu_{Cl^-}}{\partial x} + (l_{11} - l_{12})\mathcal{F}\frac{d\boldsymbol{E}}{dx} \tag{233}$$

$$-j_{Cl^-} = l_{21}\frac{\partial \mu_{Na^+}}{\partial x} + l_{22}\frac{\partial \mu_{Cl^-}}{\partial x} - (l_{21} - l_{22})\mathcal{F}\frac{d\boldsymbol{E}}{dx} \tag{234}$$

Again the local quantity $dE(x)/dx$ is the electric field strength at a given position x. In general dE/dx defines a vector field which is given by the nature of the system including the boundary conditions. dE/dx is one of the driving "forces" for the fluxes written in eqs. (233) and (234). The other "driving forces" occurring in these equations, $\partial\mu_{Na^+}/\partial x$ and $\partial\mu_{Cl^-}/\partial x$, are considered to be determined by the "chemical nature" of the ionic species. It is worth mentioning that there is a coupling between the effects of the two types of forces. The coefficients of dE/dx are linear functions of the coefficients for the chemical driving forces.

As they do not have a direct operational meaning, the fluxes $\dot{j}_{Na^+}$ and $\dot{j}_{Cl^-}$ have to be considered as auxiliary or model quantities. It has been the aim in the literature to build up physical, i.e. measurable quantities from the $\dot{j}_{Na^+}$ and $\dot{j}_{Cl^-}$ such that the four unknown coefficients l_{ij} are uniquely determined by a suitable set of experimental quantities. We shall compare this building up procedure with the results of our treatment and in particular we ask whether the mass flux and the electric current density (eqs.(226) and (227)) as we have related them to the analytical field can also be built up from the two auxiliary fluxes $\dot{j}_{Na^+}$ and $\dot{j}_{Cl^-}$. If the answer turns out to be positive, then the conventional treatment and our treatment are equivalent.

The first observable property we wish to consider is the electric current density $\dot{j}_Q$. Comparison of eq.(227b) with eqs.(233) and (234) leads us to the proposal that $\dot{j}_Q$ is related to the two quantities $\dot{j}_{Na^+}$ and $\dot{j}_{Cl^-}$ in a linear way:

$$\dot{j}_Q = \alpha_1 \dot{j}_{Na^+} + \alpha_2 \dot{j}_{Cl^-} \tag{234a}$$

with the necessary requirement

$$\alpha_1\left(l_{11}\frac{\partial\mu_{Na^+}}{\partial x} + l_{12}\frac{\partial\mu_{Cl^-}}{\partial x}\right) + \alpha_2\left(l_{21}\frac{\partial\mu_{Na^+}}{\partial x} + l_{22}\frac{\partial\mu_{Cl^-}}{\partial x}\right) = 0 \tag{235}$$

Thus we have

$$\dot{j}_Q = -\left[\alpha_1(l_{11} - l_{12}) + \alpha_2(l_{21} - l_{22})\right]\mathcal{F}\frac{dE}{dx} \tag{236}$$

and considering eq.(227)

$$\mathcal{H} = \left[\alpha_1(l_{11} - l_{12}) + \alpha_2(l_{21} - l_{22})\right]\mathcal{F} \tag{237}$$

Our last equation is the one usually reported if we set $\alpha_1 = +\mathcal{F}$ and $\alpha_2 = -\mathcal{F}$. Apart from this there is one fundamental difference because eq.(235) does not occur in the conventional treatments. This is so because usually the measurement of $\varkappa$ is confined to the situation $\partial\mu_{Na^+}/\partial x = \partial\mu_{Cl^-}/\partial x = 0$. It acts as a kind of boundary condition. Only in this case j_2 can be considered to be connected to dE/dx via $\varkappa$. If $\partial\mu_{Na^+}/\partial x \neq 0$, $\partial\mu_{Cl^-}/\partial x \neq 0$, then in the usual treatment one sets in fact:

$$j_2 = \mathcal{F}\left(j_{Na^+} + (-1)\, j_{Cl^-}\right) \tag{238}$$

with j_{Na^+} and j_{Cl^-} as given by eqs.(233) and (234). As already mentioned the $\partial\mu_{Na^+}/\partial x$ and $\partial\mu_{Cl^-}/\partial x$, in addition to dE/dx are considered to be the driving forces of the real electric current. The contribution of dE/dx has to be known from some suitable source.

From the point of view of our theory, eq.(227b) is always valid for a truncated system. Thus $\varkappa$ has a physical meaning and can be measured for a system in which gradients of the composition do not vanish. In contrast to this, when we are dealing with a complete system with respect to mass and energy conservation, in our approach eq.(238) is clearly not valid. This is because we have found that eq.(228) holds. As we shall see later, according to eq.(238) j_2 is determined by the transport numbers and the gradients of the chemical potentials whereas our eq.(228) implies the chemical potentials themselves and the gradients of the transport numbers. Thus, for a physical arrangement which is to be considered a complete system with respect to mass and energy conservation, the two approaches, the conventional one and the one presented in this book, are not compatible.

There remains the examination of the compatibility for a truncated system. To do this, for $d\mu_{Na^+}/dx \neq 0$, $d\mu_{Cl^-}/dx \neq 0$ we solve eq.(235) which yields:

$$\ell_{21} = -\frac{\alpha_1}{\alpha_2}\ell_{11}$$
$$\ell_{12} = -\frac{\alpha_2}{\alpha_1}\ell_{22}$$

and inserting these relations in eq.(237) gives

$$\varkappa = 0$$

Thus we see that by the requirement that the gradients $\partial\mu_{Na^+}/\partial x$, $\partial\mu_{Cl^-}/\partial x$ do not vanish and still do not contribute to the electric current, the efficiency of dE/dx is also reduced to zero. This is due to the fact that the coefficients of $\partial\mu_{Na^+}/\partial x$, $\partial\mu_{Cl^-}/\partial x$, and dE/dx are coupled by a linear equation.

We conclude that even in the case where the system is a truncated one, the conventional approach and our treatment are not compatible. Given the statement that the current density j_2 is the same physical quantity in both cases, then the quantity dE/dx must have a different physical significance in the two theories. We may also say that in the conventional approach the function j_2 generates another function dE/dx than it does in our treatment. However div j_2 does not strictly vanish in the conventional treatment, thus, even the electric current density is not exactly the same quantity.

7.3 Transport numbers in the conventional treatment

Next we turn to the discussion of transport numbers which in the conventional theory are defined as:

$$j_{Na^+}\mathcal{F} = t_{Na^+} j_2 \tag{239}$$

$$-j_{Cl^-}\mathcal{F} = t_{Cl^-} j_2 \tag{240}$$

Comparison with eq.(238) yields:

$$t_{Na^+} + t_{Cl^-} = 1$$

These relations suggest that our excess constituent mass fluxes as defined in eqs.(110) can be represented by the two ionic fluxes given in eqs.(233) and (234) (after having applied a suitable correction which takes into account solvent motion caused by electrodes fixed in space), i.e.

$$j_{Na^+} = \frac{1}{M_{Na}} j_{Na}^{\,\prime}$$

$$j_{Cl^-} = \frac{1}{M_{Cl}} j_{Cl}^{\,\prime}$$

We see immediately that, as a consequence of eqs.(227), (239), and (240) in a truncated system we must have:

$$j_{Na^+} \sim \frac{dE}{dx} \quad ; \quad -j_{Cl^-} \sim \frac{dE}{dx}$$

which, as we know already, is not guaranteed by eqs.(233) and (234) as they stand. Thus we have to require more than demanded by eq.(235); the constraint is stronger

$$l_{11}\frac{\partial\mu_{Na^+}}{\partial x} + l_{12}\frac{\partial\mu_{Cl^-}}{\partial x} = 0 \qquad (241)$$

$$l_{21}\frac{\partial\mu_{Na^+}}{\partial x} + l_{22}\frac{\partial\mu_{Cl^-}}{\partial x} = 0 \qquad (242)$$

must hold. Then combining eqs.(233), (234) and (234a) with eqs.(239) and (240) and setting $\alpha_1 = \mathcal{F}$, $\alpha_2 = -\mathcal{F}$ gives

$$t_{Na^+} = \frac{l_{11} - l_{12}}{l_{11} - (l_{12} + l_{21}) + l_{22}} \qquad (243)$$

$$t_{Cl^-} = \frac{l_{22} - l_{21}}{l_{11} - (l_{12} + l_{21}) + l_{22}} \qquad (244)$$

These are again the relations usually reported in the literature[1,2], however, the fulfilment of eqs.(241) and (242) is not required.

If we set $\partial\mu_{Na^+}/\partial x \neq 0$ and $\partial\mu_{Cl^-}/\partial x \neq 0$ in eqs.(241) and (242) we obtain $l_{ik} = 0$ (i = 1,2; k = 1, 2) and the quantities t_{Na^+} and t_{Cl^-} in eqs.(243) and (244) remain undetermined. In fact, in our treatment transport numbers are a typical property of a complete system with respect to delocalized mass conservation. Their definition and determination necessarily involves electrodes or similar boundaries with well-defined mass fluxes. In a truncated system transport numbers are not defined and not measurable quantities. In contrast to this, in the conventional treatment the transport numbers are well-defined quantities under all circumstances because the constraints given by eqs.(241) and (342) are released. The question remains, whether the measurement of transport numbers applying either the Hittorf or moving boundary methods strictly occurs within the framework of the conventional theory. Finally, even transport number measurements involving EMF measurements possibly transgress the domain of the conventional theory.

7.4 The mass fluxes in the conventional treatment.

Next we wish to construct the mass flux $\dot{j}_s$ given in eq.(226), from the ionic fluxes $\dot{j}_{Na^+}$ and $\dot{j}_{Cl^-}$ as formulated in eqs.(233) and (234).

The simplest suggestion is to set

$$\dot{j}_s = \beta_1 M_{Na}\dot{j}_{Na^+} + \beta_2 M_{Cl}\dot{j}_{Cl^-} \qquad (245)$$

where β_i, i = 1, 2 are tentative building up parameters. Following our programme to exclude metal/electrolyte and similar boundaries acting as electrodes from the present discussion we state that the system is closed by chemically inactive walls. This means that all mass fluxes vanish at these boundaries. Having established this, the first question concerning the realizibility of the proposal given in eq.(245) has to deal with the electrical potential gradient $d\mathbf{E}/dx$. Comparison of eqs. (226), (233), (234), and (245) tells us that, if we would hope to be successful, then we must have

$$\begin{aligned} \ell_{11} &= \ell_{12} \\ \ell_{22} &= \ell_{21} \end{aligned} \tag{246}$$

or

$$\frac{d\mathbf{E}}{dx} = 0$$

Now it is quite clear that according to our standpoint $d\mathbf{E}/dx = 0$ can never be a condition for a diffusive mass flux to be a well-defined quantity. This holds true for a truncated system as well as for a complete system, when $d\mathbf{E}/dx = -dE_\Omega/dx$. Thus we must accept the validity of eq.(246). Then combination of eqs.(233),(234),(245), and (246) yields:

$$\begin{aligned} -j^s &= \beta_1 \ell_{11}\left(\frac{\partial\mu_{Na^+}}{\partial x} + \frac{\partial\mu_{Cl^-}}{\partial x}\right)M_{Na} \\ &\quad + \beta_2 \ell_{22}\left(\frac{\partial\mu_{Na^+}}{\partial x} + \frac{\partial\mu_{Cl^-}}{\partial x}\right)M_{Cl} \end{aligned}$$

Another requirement is given by the stoichiometry of the dissolved salt:

$$j_{Na^+} = j_{Cl^-}$$

which gives

$$\beta_1 = \beta_2 = 1$$

and

$$\ell_{11} = \ell_{22}$$

Of course we have

$$\frac{\partial\mu_{Na^+}}{\partial x} + \frac{\partial\mu_{Cl^-}}{\partial x} = \frac{\partial\mu_{NaCl}}{\partial x}$$

and consequently the final result is:

$$-j^s = \Omega'_{ss}\frac{\partial\mu_{NaCl}}{\partial x}$$

with

$$\Omega'_{ss} = M_{NaCl}\,\ell_{11} = (M_{Na} + M_{Cl})\,\ell_{11}$$

If we insert eqs.(246) in eq.(237) then again we obtain the result $\varkappa = 0$. Thus, if - as is done in our treatment - the mass flux is considered to be independent of the electric field strength but of course connected with the gradient of the chemical potential via a set of linear coefficients, then an electric current density $j_q \neq 0$ cannot be determined by the same coefficients.

In the conventional theory several authors[1,2] have followed a particular path to represent the diffusion flux of the electrolyte (as a whole) in terms of the set of coefficients l_{ik} occurring in eqs. (229) and (230). The requirement $j_q = 0$ is taken as a criterion for the limiting situation to study and to measure the pure diffusion flux which then yields the mutual diffusion coefficient of the electrolyte. Relation (238) with $j_q = 0$ is used to eliminate dE/dx from one of the equations (233) or (234). This elimination of dE/dx at the same time leads also to the appearance of $\partial\mu_{Na^+}/\partial x$ and $\partial\mu_{Cl^-}/\partial x$ in the form $\partial\mu_{Na^+}/\partial x + \partial\mu_{Cl^-}/\partial x$ which is very suggestive because now

$$\lim_{j_q \to 0} j_{Na^+}$$

only contains macroscopic quantities which are measurable for electrically neutral systems.

This method fixes the value of dE/dx which turns out to be non-zero. Thus we see again that the quantity dE/dx in the conventional theory must be different from the $-dE/dx$ occurring in our eqs.(227a) and (227b) because the mass flux is independent, or better, it is not connected at all with dE/dx. In the usual theories $dE/dx \neq 0$ is taken to be the same as the liquid junction or diffusion potential. As will be shown below, in the framework of our method the diffusion potential can only occur in a complete system with respect to delocalized mass conservation. One point may already be mentioned here. As indicated above the ordinary mutual diffusion process occurs if j_q is exactly zero. The diffusion potential according to our scheme is a meaningful concept only in the limit $j_q \to 0$, that is, in a situation where in the pertinent expression all terms of higher order in j_q are very small compared with the linear one. Moreover the result $dE/dx \neq 0$ in an ordinary diffusion process with $j_q = 0$ raises the question of the continuation of the electric field strength outside the system.

7.5 Summary of comparison between the two approaches.

In summary, we have arrived at the following result. The two treatments of the diffusion problem in the presence of an electric current, the conventional one and our treatment, are incompatible. The former approach mixes the concepts of mass transport and electric current; the latter one provides a strict separation of both these phenomena.

It is of interest to remark that SCHÖNERT[4] has eliminated dE/dx from eq.(233) in the general case i.e. for $j_Q = 0$ and $j_Q \neq 0$. He obtains an expression

$$j_{Na^+} = \frac{t_{Na^+} j_Q}{\mathcal{F}} - D_{sw} \frac{\partial c_{NaCl}}{\partial x} \tag{247}$$

which then is taken to represent the mass flux in the general situation $j_Q \neq 0, \partial c_{NaCl}/\partial x \neq 0$. Similar expressions have also been given by LONGSWORTH[5] and by NEWMAN[6], however the diffusion coefficient occurring in their equations is not explicitly the mutual diffusion coefficient of the neutral salt against water. The divergence of j_{Na^+} in eq.(247) is then set equal to $-\partial c_{NaCl}/\partial t$ and apparently this is the same expression as our eq.(115) if we apply it to a binary system and divide by $M_{Na} + M_I$. In fact, SCHÖNERT has given a correct treatment of the moving boundary phenomenon in a binary system[4], exactly as it would follow from our treatment. However, since eq.(247) has been derived from eqs.(233), (234), and (238) involving coupling between the electric current and mass flow, the apparent agreement must be accidental. Indeed the deviation from the conventional treatment is easily traced: SCHÖNERT used the relation $\mathrm{div}\, j_Q = 0$ which conflicts with eq.(238) together with eqs.(233) and (234). Certainly we have $\partial^2\mu_k/\partial x^2 \neq 0$ (k = Na^+, Cl^-) somewhere in a non-uniform system; the question whether $\partial^2 E/\partial x^2 \neq 0$ has often been discussed in the literature without yielding a very satisfactory answer[7].

Very recently an interesting formulation of the combined diffusion and electric current problem has been given by EKMAN, LIUKKONEN, and KONTTURI[8]. The work of these authors may be considered to lie halfway between the conventional treatment and ours. They write the ionic fluxes in the form:

$$j_{Na^+} = -\, l_{11}^{*}\left(RT \frac{\partial}{\partial x} \ln c_{Na^+} + \mathcal{F}\frac{dE}{dx}\right)$$

$$j_{Cl^-} = -l_{22}^{*}\left(RT\frac{\partial}{\partial x}\ln c_{Cl^-} - \mathcal{F}\frac{dE}{dx}\right)$$

which essentially are the Nernst-Planck equations[9]. They use these equations together with the requirement that the electric current density has to be treated separately and independent of the diffusion flow of the neutral components. Thus they do not calculate the electric current density by the simple formula eq.(238), rather they obtain the expression for $j_{\mathfrak{q}}$ from the properties of a transformation matrix relating the constituent fluxes to the ionic fluxes, and in this matrix scheme the electric current is also incorporated in a special way. Still, the authors arrive at the result that the driving force for the electric current is

$$X_{(j_{\mathfrak{q}})} = -\mathcal{F}\frac{dE}{dx} - RT\left(t_{Na^+}\frac{\partial}{\partial x}\ln c_{Na^+} - t_{Cl^-}\frac{\partial}{\partial x}\ln c_{Cl^-}\right)$$

and the second term on the right according to their statement may be interpreted to represent the diffusion potential and as such it is both approximate and conventional and has no connection with any directly measurable quantity. On the other side, according to this treatment $X_{(j_{\mathfrak{q}})} = j_{\mathfrak{q}}\mathcal{F}/\varkappa$ is a measurable quantity but the value of dE/dx depends on the same conventions as the diffusion potential; clearly here their view deviates strongly from that given in our treatment.

Literature References

1) D.G. Miller, J.physic.Chem. 1966, 70, 2639
2) R. Haase and J. Richter, Z.Naturforsch., 1967, 22a, 1761
3) A. Katchalsky and P.F. Curran, Nonequilibrium Thermodynamics in Biophysics, Harvard University Press, Cambridge Mass. 1975
4) H. Schönert, Ber.Bunsengesellschaft phys.Chem.,1975, 79, 408
5) L.G. Longsworth, J.Amer.Chem.Soc. 1943, 65, 1755
6) J. Newman, in: C.W. Tobias, Advances in Electrochemistry and Electrochemical Engineering, Vol.5, Intersciences, London 1967
7) D.R. Hafemann, J.phys.Chem., 1965, 69, 4226
8) A. Ekman, S. Liukkonen, and K.Kontturi, Electro-Chim. Acta, 1978, 23, 243
9) see e.g. J. Koryta, J.Dvorak and V. Bohackova, Electrochemistry, Chapman/Methuen, London 1970
German translation: Lehrbuch der Elektrochemie, Springer-Verlag Wien 1975

Chapter 8

The electromotive force of a galvanic cell

8.1 The cell Na/NaI/KCl/Cl_2, some general outlines.

As described previously, the galvanic cell is a system constructed from a set of thermodynamic phases such that an electric current can exist and a chemical reaction can proceed in the system. It seems to be a reasonable starting point to describe a system in which two different electrolyte solutions are placed in contact with another. The remaining two boundaries are formed from solid materials and the loop is closed by some inert metal which does not enter in the reaction scheme. Fig. 25 gives a schematic representation of a special system which involves the constituents Na, K, Cl, and I together with water. In Fig.25 the

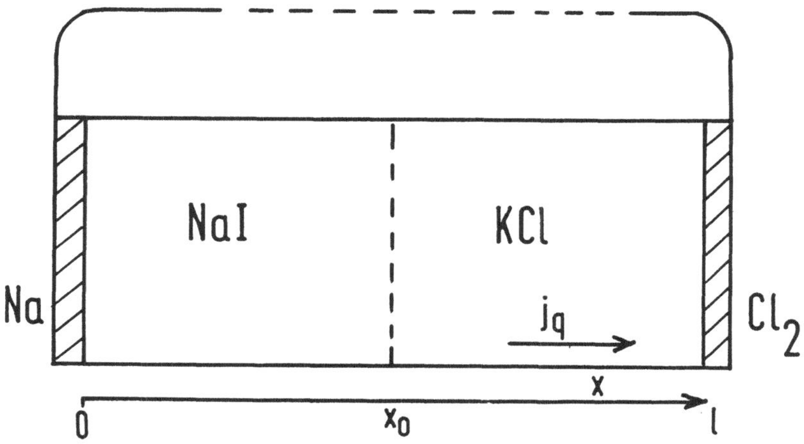

Fig. 25 Schematic representation of the galvanic cell described in this section. Here and in the following figures showing galvanic cells, the set up is really drawn as a closed loop. The portion of the loop drawn as a dashed line represents a device making the electric current sufficiently small.

galvanic cell has indeed been drawn as a closed loop. This is to remind the reader that the galvanic cell is a device shaped in a particular way in which a chemical reaction proceeds. In the following pages we shall repeatedly give this schematic representation of a galvanic cell as a "reaction vessel". This is in contrast to the abbreviated form of representation which is commonly used in conventional treatments and also in the heading of this section.

As has been shown before (chaper 2) in the arrangement shown in Fig.25 the analytical field is given by three freely variable quantities, for instance the partial mass densities

$$\rho_S,\ \rho_{Na},\ \rho_{Cl}$$

Then, the remaining constituent partial densities ρ_K and ρ_I are also uniquely determined. The three independent quantities form the analytical field:

$$\rho_S = \rho_S(x,t) = \left(1+\frac{M_I}{M_{Na}}\right)\rho_{Na}(x,t) + \left(1+\frac{M_I}{M_K}\right)\rho_K(x,t) + \left(1-\frac{M_I}{M_{Cl}}\right)\rho_{Cl}(x,t)$$

$$= (M_{Na}+M_I)\,c_{Na}(x,t) + (M_K+M_I)\,c_K(x,t) + (M_{Cl}-M_I)\,c_{Cl}(x,t)$$

$$\rho_{Na} = \rho_{Na}(x,t) = M_{Na}\,c_{Na}(x,t)$$

$$\rho_{Cl} = \rho_{Cl}(x,t) = M_{Cl}\,c_{Cl}(x,t)$$

For simplicity we can assume that the system has a constant cross-section A. If the cross-section is not constant, then we have to recall that the product $A j_2 = J_2$ is constant throughout the system. As indicated in Fig. 25 we begin with a system for which at t = 0 the liquid electrolyte region has the following partial mass density distribution.

ρ_{Na}	= const.	for	$0 \leq x < x_0$
ρ_{Cl}	= 0	for	$0 \leq x < x_0$
ρ_S	$= \rho_{NaI}$	for	$0 \leq x < x_0$
ρ_{Na}	= 0	for	$x_0 \leq x \leq \ell$
ρ_{Cl}	= const.	for	$x_0 \leq x \leq \ell$
ρ_S	$= \rho_{KCl}$	for	$x_0 \leq x \leq \ell$

All changes $\partial\rho_i/\partial t$ occurring in the following represent reversible changes, for simplicity we omit the prime.

At x = 0 there is a Na electrode, for instance an amalgam electrode. We neglect the fact that $X_{Na} < 1$ in the electrode material (X = mole fraction), the addition of a corresponding term to our final formulas will be straightforward. At $x = \ell$ there is a chlorine electrode. As already mentioned, in this treatment the "electrodes" are the solid materials (in exceptional cases also liquid metals) at the boundaries of the liquid electrolyte, not the "half-cells". At $x = x_0$ there may be a diaphragm, however, this is not essential for our treatment. Now during a time interval dt, dn moles of Na and 1/2 dn moles of Cl_2 are removed from the electrode material and converted to dissolved material. It is not obvious yet what this dissolved material is. This will be shown later. We wish to calculate the electromotive force of the system and according to eq.(223) we need the change of the total Gibbs free energy connected with the process occurring in the system. Through the electrode process the system suffers a decrease of Gibbs free energy which is

$$(dG)_{electrode} \equiv (dG)_{el} = \left(-\mu_{Na'} \frac{j_q}{\mathcal{F}} dt - \frac{1}{2}\mu_{Cl_2} \frac{j_q}{\mathcal{F}} dt\right) A \qquad (248a)$$

where $\mu_{Na'}$ is the molar Gibbs free energy of the metallic sodium and μ_{Cl_2} is the molar Gibbs free energy of the chlorine gas. We can also express these quantities as specific ones:

$$\frac{(dG)_{el}}{A} = -\mu^*_{Na'} M_{Na} \frac{j_q}{\mathcal{F}} dt - \frac{1}{2}\mu^*_{Cl_2} M_{Cl_2} \frac{j_q}{\mathcal{F}} dt \qquad (248)$$

Next we consider the increase of the Gibbs free energy in the solution due to the production of new solute material. Obviously, the local increase of Gibbs-free energy is

$$(d^2G)_{solution} \equiv (d^2G)_s = A\left\{\frac{\partial \rho_s}{\partial t}\mu^*_s + \frac{\partial \rho_{Na}}{\partial t}\mu^*_{Na} + \frac{\partial \rho_{Cl}}{\partial t}\mu^*_{Cl}\right\} dx\, dt \qquad (249)$$

This is so because, as explained previously, the solution is completely defined by the three variables ρ_s, ρ_{Na}, and ρ_{Cl}. We assume that A is constant in the cell. The expression (249) has to be integrated over the entire electrolyte system, thus we have

$$(dG)_s = A\left\{\int_0^{\ell}\frac{\partial \rho_s}{\partial t}\mu^*_s dx + \int_0^{\ell}\frac{\partial \rho_{Na}}{\partial t}\mu^*_{Na} dx + \int_0^{\ell}\frac{\partial \rho_{Cl}}{\partial t}\mu^*_{Cl} dx\right\} dt \qquad (250)$$

Apart from this, there are direct fluxes of solute into the solution from the two electrodes: They cause a change of Gibbs free energy:

$$(dG)_{electrode\ flux} \equiv (dG)_{ef} = A\Big[(\mu^*_s j_s dt + \mu^*_{Na} j_{Na} dt +$$

$$+ \mu^*_{Cl} j_{Cl} dt)_{anode} + (\mu^*_s |j_s| dt + \mu^*_{Na} |j_s| dt + \mu^*_{Cl} |j_{Cl}| dt)_{cathode}\Big] \qquad (251)$$

Now, at the anode, x = 0 by definition, we know for certain that $j_{Cl} = 0$, similarly at the cathode, $x = \ell$, we have $j_{Na} = 0$.

The fluxes at the anode have already been calculated; they are given by eqs.(150) and (151). Now let us consider the fluxes at the cathode. We start from the general equations (112) - (114) and then write the expressions for the gradients of the transport numbers:

$$t_{Cl}(x) = t_{Cl} + \frac{1-t_{Cl}}{\delta x}(x - l + \delta x)$$

$$\frac{dt_{Cl}}{dx} = \frac{1-t_{Cl}}{\delta x}$$

$$t_K(x) = t_K \frac{l-x}{\delta x} \quad ; \quad \frac{dt_K}{dx} = -\frac{t_K}{\delta x}$$

$$t_{Na}(x) = \frac{dt_{Na}}{dx} = 0$$

Thus we obtain from eqs.(112)-(114)

$$\frac{\partial \rho_s}{\partial t} = -\operatorname{div} j_s - \left[-(M_K + M_I)\frac{t_K}{\delta x} - (M_{Cl} - M_I)\frac{1-t_{Cl}}{\delta x}\right]\frac{j_q}{\mathcal{F}}$$

$$\frac{\partial \rho_{Cl}}{\partial t} = -\operatorname{div} j_{Cl} + \frac{M_{Cl}}{\delta x \mathcal{F}} j_q (1-t_{Cl})$$

The integration yields:

$$\int_{l-\delta x}^{l} \frac{\partial \rho_s}{\partial t} dx = -\int_{l-\delta x}^{l} \operatorname{div} j_s dx - \left[-(M_K + M_I)t_K - (M_{Cl} - M_I)(1-t_{Cl})\right]\frac{j_q}{\mathcal{F}}$$

$$\int_{l-\delta x}^{l} \frac{\partial \rho_{Cl}}{\partial t} dx = -\int_{l-\delta x}^{l} \operatorname{div} j_{Cl} dx + \frac{M_{Cl}}{\mathcal{F}} j_q (1-t_{Cl})$$

Now the fluxes j_s and j_{Cl} vanish at the boundary $x = l$, their functional form may be represented by

$$j_i(x) = \frac{j_i(l-x)}{\delta x}$$

$$l - \delta x \leq x \leq l \quad ; \quad i = s, Cl$$

thus we have

$$\operatorname{div} j_i = \frac{\partial j_i}{\partial x} = -\frac{j_i}{\delta x}$$

and consequently

$$\int_{l-\delta x}^{l} \operatorname{div} j_i dx = -j_i \quad , \quad i = s, Cl$$

Finally we observe that $\delta x \to 0$ and we get

$$\int_{\ell-\delta x}^{\ell} \frac{\partial \rho_i}{\partial t}\, dx \longrightarrow 0$$

and the fluxes at the surface of the cathode are

$$(j_s)_{x=\ell} = -\left[(M_K + M_I)t_K + (M_{Cl} - M_I)(1 - t_{Cl})\right]\frac{j_q}{\mathcal{F}}$$

$$(j_{Cl})_{x=\ell} = -\frac{M_{Cl}}{\mathcal{F}} j_q (1 - t_{Cl})$$

In fact, in our special situation we have

$$t_K = 1 - t_{Cl}$$

because $t_{Na} = 0$.

Thus we find

$$(j_s)_{x=\ell} = -(M_K + M_{Cl})(1 - t_{Cl})\frac{j_q}{\mathcal{F}} \tag{252}$$

$$(j_{Cl})_{x=\ell} = -\frac{M_{Cl}}{\mathcal{F}} j_q (1 - t_{Cl}) \tag{253}$$

Now we introduce these fluxes in eq.(251), use also the fluxes given by eqs.(150) and (151) and obtain:

$$\frac{(dG)_{ef}}{A} = \Big[\left(\mu_s^{*(a)} M_{NaI} + \mu_{Na}^{*(a)} M_{Na}\right)(1 - t_{Na}) + \left(\mu_s^{*(c)} M_{KCl} + \mu_{Cl}^{*(c)} M_{Cl}\right)(1 - t_{Cl})\Big]\frac{j_q}{\mathcal{F}}\, dt \tag{254}$$

The superscripts (a) and (c) refer to the anode and cathode, respectively. It is instructive to insert expressions for μ_s^*, μ_{Na}^*, and μ_{Cl}^*, as given in connection with eq.(41) in chapter 2, in this equation. The result is:

$$\frac{(dG)_{ef}}{A} = \frac{1}{\mathcal{F}}\left[M_{NaI}\,\mu_{NaI}^{*(a)}(1 - t_{Na}) + M_{KCl}\,\mu_{KCl}^{*(c)}(1 - t_{Cl})\right] j_q\, dt \tag{254a}$$

Now we turn to the second contribution of the total increase of the Gibbs free energy. We return to eq.(250). Formally we may define a total solute transport number by the relation:

$$j_s' = (M_{K\text{Cl}} + M_{Na\text{Cl}} + M_{NaI})\, t_s \frac{j_2}{\mathcal{F}}$$

Then we have

$$\frac{\partial \rho_s'}{\partial t} = - \operatorname{div} j_s' = -(M_{K\text{Cl}} + M_{Na\text{Cl}} + M_{NaI}) \frac{dt_s}{dx} \frac{j_2}{\mathcal{F}}$$

This may be compared with:

$$\frac{\partial \rho_s'}{\partial t} = - \left\{ (M_{Na} + M_I) \frac{dt_{Na}}{dx} + (M_K + M_I) \frac{dt_K}{dx} - (M_{Cl} - M_I) \frac{dt_{Cl}}{dx} \right\} \frac{j_2}{\mathcal{F}}$$

which gives the result:

$$\frac{dt_s}{dx} = \frac{M_{NaI}\, dt_{Na}/dx + M_{K\text{Cl}}\, dt_K/dx - (M_{Cl} - M_I)\, dt_{Cl}/dx}{M_{NaI} + M_{K\text{Cl}} + M_{Na\text{Cl}}}$$

or

$$t_s = \frac{M_{NaI}\, t_{Na} + M_{K\text{Cl}}\, t_K + (M_I - M_{Cl})\, t_{Cl}}{M_{NaI} + M_{K\text{Cl}} + M_{Na\text{Cl}}} \tag{255}$$

This is the total solute transport number. With this result we can write for $(dG)_s/A$

$$\frac{(dG)_s}{A} = - \left\{ \int_0^l \mu_s \frac{dt_s}{dx} dx + \int_0^l \mu_{Na} \frac{dt_{Na}}{dx} dx - \int_0^l \mu_{Cl} \frac{dt_{Cl}}{dx} dx \right\} \frac{j_2}{\mathcal{F}} dt \tag{256}$$

with

$$\mu_s = \mu_s^* (M_{K\text{Cl}} + M_{NaI} + M_{Na\text{Cl}})$$

$$\mu_{Na} = \mu_{Na}^* M_{Na} \; ; \quad \mu_{Cl} = \mu_{Cl}^* M_{Cl}$$

The sum of eqs. (248), (254) and (256) when divided by dt represent the left-hand side of eq.(223):

$$\frac{1}{A} \left\{ (dG)_{el} + (dG)_s + (dG)_{ef} \right\} = j_2 E\, dt \tag{223a}$$

Thus we have achieved the formal solution of our problem by computing the electromotive force of a fairly general type of galvanic cell. It is expressed in the "asymmetric" total solute-constituents language. If the μ_i^*'s and the transport numbers $t_i = t_i(\rho_s, \rho_{Na}, \rho_{Cl})$ were given, then they could be introduced, integrated and the final result would be available. Details will be worked out below. However, it is instructive and more convenient for a comparison with conventional results to introduce in these equations the chemical potentials expressed in the construction coordinate system and the transport numbers in the full consituents language.

8.2 The cell $Na/NaI/KCl/Cl_2$, explicit formulas for electromotive force

Let us start with the computation of the free energy change in the electrolyte solution. Then we have (using the first part of eq.(35)) when we consider the free energy change per unit volume

$$\frac{(dG)_s}{V\,dt} = \mu^*_{KCl} M_{KCl} \frac{\partial \rho_K}{\partial t}\frac{1}{M_K}$$
$$+ \mu^*_{NaCl} M_{NaCl}\left(\frac{\partial \rho_{Cl}}{\partial t}\frac{1}{M_{Cl}} - \frac{\partial \rho_K}{\partial t}\frac{1}{M_K}\right)$$
$$+ \mu^*_{NaI} M_{NaI}\left(\frac{\partial \rho_{Na}}{\partial t}\frac{1}{M_{Na}} + \frac{\partial \rho_K}{\partial t}\frac{1}{M_K} - \frac{\partial \rho_{Cl}}{\partial t}\frac{1}{M_{Cl}}\right)$$

Now we replace the $\partial \rho_i/\partial t$ by the corresponding gradients of the transport numbers (fundamental law of electrochemistry, see page 58)

$$\frac{(dG)_s}{dV} = \left\{ -\mu^*_{KCl} M_{KCl}\frac{dt_K}{dx} + \mu^*_{NaCl} M_{NaCl}\left(\frac{dt_{Cl}}{dx} + \frac{dt_K}{dx}\right)\right.$$
$$\left. - \mu^*_{NaI} M_{NaI}\left(\frac{dt_K}{dx} + \frac{dt_{Na}}{dx} + \frac{dt_{Cl}}{dx}\right)\right\}\frac{j_e}{\mathcal{F}}\,dt$$

This gives then for the contribution of the electrolyte to the change in free energy:

$$\frac{(dG)_s}{A} = -\left\{\int_0^l \mu^*_{NaI} M_{NaI}\left(\frac{dt_{Na}}{dx} + \frac{dt_K}{dx} + \frac{dt_{Cl}}{dx}\right)dx\right.$$
$$- \int_0^l \mu^*_{NaCl} M_{NaCl}\left(\frac{dt_{Cl}}{dx} + \frac{dt_K}{dx}\right)dx$$
$$\left. + \int_0^l \mu^*_{KCl} M_{KCl}\frac{dt_K}{dx}\,dx\right\}\frac{j_e}{\mathcal{F}}\,dt \qquad (257)$$

The boundary between the two solutions is at x_o, and has a width of δx. Then we write for the transport numbers:

$$t_{Na}(x) = \frac{t_{Na}}{\delta x}(x_o + \delta x - x) \; ; \qquad \frac{dt_{Na}}{dx} = -\frac{t_{Na}}{\delta x}$$

$$t_K(x) = \frac{t_K}{\delta x}(x - x_o) \; ; \qquad \frac{dt_K}{dx} = \frac{t_K}{\delta x}$$

$$t_{Cl}(x) = \frac{t_{Cl}}{\delta x}(x - x_o) \; ; \qquad \frac{dt_{Cl}}{dx} = \frac{t_{Cl}}{\delta x}$$

This gives in eq.(257)

$$\frac{(dG)_s}{A} = \left\{ \int_{x_o}^{x_o+\delta x} \mu^*_{NaI} M_{NaI} \frac{1}{\delta x}[t_{Na} - t_{Cl} - t_K]\, dx + \frac{1}{\delta x}\int_{x_o}^{x_o+\delta x} \mu^*_{NaCl} M_{NaCl}(t_{Cl} + t_K)\, dx - \frac{1}{\delta x}\int_{x_o}^{x_o+\delta x} \mu^*_{KCl} M_{KCl}\, t_K\, dx \right\} \frac{j_q}{F} dt$$

$$= \left[\frac{t_{Na} - t_K - t_{Cl}}{\delta x} \int_{x_o}^{x_o+\delta x} \mu_{NaI}\, dx + \frac{t_{Cl} + t_K}{\delta x}\int_{x_o}^{x_o+\delta x} \mu_{NaCl}\, dx - \frac{t_K}{\delta x}\int_{x_o}^{x_o+\delta x} \mu_{KCl}\, dx \right] \frac{j_q}{F} dt$$

$$= \left[\frac{t_{Na} - t_K - t_{Cl}}{\delta x} \left\{ \int_{x_o}^{x_o+\delta x} \mu^o_{NaI}\, dx + RT \int_{x_o}^{x+\delta x} \ln a_{NaI}\, dx \right\} + \frac{t_{Cl} + t_K}{\delta x} \left\{ \int_{x_o}^{x_o+\delta x} \mu^o_{NaCl}\, dx + RT \int_{x_o}^{x_o+\delta x} \ln a_{NaCl}\, dx \right\} - \frac{t_K}{\delta x} \left\{ \int_{x_o}^{x_o+\delta x} \mu^o_{KCl}\, dx + RT \int_{x_o}^{x_o+\delta x} \ln a_{KCl}\, dx \right\} \right] \frac{j_q}{F} dt$$

$$= \left[(t_{Na} - t_K - t_{Cl})\mu^o_{NaI} + \frac{t_{Na} - t_K - t_{Cl}}{\delta x} RT \int_{x_o}^{x_o+\delta x} \ln a_{NaI}\, dx + (t_K + t_{Cl})\mu^o_{NaCl} + \frac{t_K + t_{Cl}}{\delta x} RT \int_{x_o}^{x_o+\delta x} \ln a_{NaCl}\, dx - t_K \mu^o_{KCl} - \frac{t_K}{\delta x} RT \int_{x_o}^{x_o+\delta x} \ln a_{KCl}\, dx \right] \frac{j_q}{F} dt$$

Next we use linear x dependences of the activities (which are given in the language of construction space):

$$\left.\begin{aligned} a_{NaI}(x) &= \frac{a_{NaI}}{\delta x}(x_o + \delta x - x) \\ a_{KCl}(x) &= \frac{a_{KCl}}{\delta x}(x - x_o) \end{aligned}\right\} \quad \text{for } x_o \leq x \leq x_o + \delta x$$

Of course, this is an approximation.

In the above equations also, a_{NaCl} the activity of NaCl occurs. As has been outlined in chapter 2, when the diffusion process between KCl and NaI is considered in the construction coordinate system, then NaCl appears during the process of intermixing. Also, the chemical reaction, coherent with j_2 produces NaCl. The same is true for KI, however, this component does not appear explicitly in our equations. Of course, the production of NaCl only occurs around the boundary x_o. Thus for NaCl we choose the following approximation:

$$a_{NaCl}(x) = \frac{a_{NaCl}}{\delta x/2}(x - x_o) \qquad \text{for} \quad x_o \leq x \leq x_o + \delta x/2$$

$$a_{NaCl}(x) = \frac{a_{NaCl}}{\delta x/2}(x_o + \delta x - x) \qquad \text{for} \quad x_o + \delta x/2 \leq x \leq x_o + \delta x$$

Then the integrals are

$$\frac{1}{\delta x}\int_{x_o}^{x_o+\delta x} \ln a_{NaI}(x)\,dx = \ln a_{NaI} - 1$$

$$\frac{1}{\delta x}\int_{x_o}^{x_o+\delta x} \ln a_{KCl}(x)\,dx = \ln a_{KCl} - 1$$

$$\frac{1}{\delta x}\int_{x_o}^{x_o+\delta x} \ln a_{NaCl}(x)\,dx = \ln a_{NaCl} - 1$$

With these approximations one arrives at the equations:

$$\frac{(dG)_s}{A} = \Big\{ t_{Na}\mu^o_{NaI} + (t_K + t_{Cl})(\mu^o_{NaCl} - \mu^o_{NaI}) - t_K\mu^o_{KCl}$$
$$+ (t_{Na} - t_K - t_{Cl})RT(\ln a_{NaI} - 1)$$

$$+ (t_K + t_{Cl}) RT(\ln a_{NaCl} - 1) - t_K RT(\ln a_{KCl} - 1)\Big\} \frac{j_q}{\mathcal{F}} dt \tag{258}$$

Now we add this contribution to that from the electrode fluxes. The result is :

$$\frac{(dG)_{s+ef}}{A} = \Big\{ \mu^o_{NaI}(1-t_{Na}) + RT \ln a_{NaI}(1-t_{Na}) + \mu^o_{KCl}(1-t_{Cl})$$
$$+ RT \ln a_{KCl}(1-t_{Cl})$$
$$+ t_{Na}\mu^o_{NaI} + (t_K + t_{Cl})(\mu^o_{NaCl} - \mu^o_{NaI}) - t_K \mu^o_{KCl}$$
$$+ t_{Na} RT(\ln a_{NaI} - 1) - (t_K + t_{Cl}) RT(\ln a_{NaI} - 1)$$
$$+ (t_K + t_{Cl}) RT(\ln a_{NaCl} - 1) - t_K RT(\ln a_{KCl} - 1)\Big\} \frac{j_q}{\mathcal{F}} dt$$

and after some rearrangements one finds :

$$\frac{(dG)_{s+ef}}{A} = \Big\{ \mu^o_{NaI} + \mu^o_{KCl} + RT \ln a_{NaI} + RT \ln a_{KCl}$$
$$- RT(t_{Na} - t_K) + (t_K + t_{Cl})\big[\mu^o_{NaCl} - \mu^o_{NaI} - \mu^o_{KCl}$$
$$+ RT(\ln a_{NaCl} - \ln a_{NaI} - \ln a_{KCl})\big]\Big\} \frac{j_q}{\mathcal{F}} dt \tag{259}$$

Now the addition of eqs.(259) and (248) gives the total change of Gibbs free energy in the system

$$\frac{dG}{A} = \Big\{ \mu^o_{NaI} + \mu^o_{KCl} + RT \ln a_{NaI} + RT \ln a_{KCl} - \mu_{Na'}$$
$$- \frac{1}{2}\mu_{Cl_2} - RT(t_{Na} - t_K)$$
$$+ (t_K + t_{Cl})\big[\mu^o_{NaCl} - \mu^o_{NaI} - \mu^o_{KCl} + RT(\ln a_{NaCl}$$
$$- \ln a_{NaI} - \ln a_{KCl})\big]\Big\} \frac{j_q}{\mathcal{F}} dt$$

Now we introduce the definition

$$\mathcal{F} E_{Nernst} = \Big(E_o + \frac{RT}{\mathcal{F}}(\ln a_{NaI} + \ln a_{KCl})\Big)\mathcal{F}$$

with

$$\mathcal{F}E_o = \mu^o_{NaI} + \mu^o_{KCl} - \mu_{Na'} - \frac{1}{2}\mu_{Cl_2}$$

Then, using eq.(223), our final result is:

$$E = E_{Nernst} - \frac{RT}{\mathcal{F}}\Big[(t_{Na} - t_K)$$

$$- (t_K + t_{Cl})\Big(\frac{1}{RT}(\mu^o_{NaCl} - \mu^o_{NaI} - \mu^o_{KCl})$$

$$+ \ln a_{NaCl} - \ln a_{NaI} - \ln a_{KCl}\Big)\Big] \qquad (260a)$$

$$= E'_{Nernst} - \frac{RT}{\mathcal{F}}(t_{Na} - t_K) \qquad (260b)$$

with

$$E'_{Nernst} = E'_o + \frac{RT}{\mathcal{F}}\ln a_{NaCl} \qquad (261a)$$

$$E'_o = \mu^o_{NaCl} - \mu_{Na'} - \frac{1}{2}\mu_{Cl_2} \qquad (261b)$$

since in the present arrangement we have $t_K + t_{Cl} = 1$.

One recognizes from the nature of eq.(261a) and (261b) the chemical reaction in the galvanic cell which is:

$$Na + 1/2\ Cl_2 = NaCl\ .$$

The product of the reaction appears only around the boundary x_o, not at the metal and gas electrode. Note that we have indeed a delocalized mass conservation. The activity of NaCl is not a known quantity because it is not defined directly by the construction operation; in the first moment ideally we should have $a_{NaCl} = 0$, however immediately after placing the NaI and KCl solution in contact, diffusion begins to produce NaCl (and KI), see chapter 2. It may be seen that our result does not depend on the width δx of the boundary around x_o. Thus the regions where all four constituents are present may be as large as we wish, provided that at the two electrodes we have "pure" NaI and pure "KCl" and the linear approximations for the t_i 's and a_i 's are valid. Both these restrictions will be removed in the following sections.

It is important now to devote some words to the liquid junction or diffusion potential. Obviously, the second term on the right-hand side of eq.(260a) is the liquid junction potential. From the derivation of this formula it may be seen that it belongs wholly to the reversible part of the electric potential of the cell. We can even rewrite the final result - as we did in eq.(260b)- so that, apart from the small concentration dependence of $t_{Na} - t_K$, the entire composition dependent part of the electromotive force is due to the diffusion potential. Diffusion, i.e. the irreversibility only enters in so far as in our result a_{NaCl} is partly given as the solution of the diffusion equations, in fact, as the solution of the diffusion equations in the construction coordinate system as shown by the formulas (55)-(57). So, the "reversible" EMF slowly changes with time. For reasons of completeness we should mention that there is also a rate of change dG/dt due to the irreversible diffusion process. This contribution corresponds to the term we neglected when proceeding from eq.(219) to eq.(221) in chapter 6. However, from our point of view these irreversible local changes of the μ_i^* 's with time are only "transformations" of the local changes of the partial mass densities and have no further physical significance.

8.3 Consideration of a more genral case: The galvanic cell $Na/NaCl^{(a)}, NaI^{(a)}, KCl^{(a)}/NaCl^{(c)}, NaI^{(c)}, KCl^{(c)}/Cl_2$

We are now going to discuss the extension of our method to a system in which all four constituent concentrations are $\neq 0$ everywhere. The schematic representation is given in Fig.26. First we consider the change with time of the Gibbs-free energy in the electrolyte solution. We have to start from eq.(257).

Fig. 27 describes the situation at the liquid junction. It will be seen that linear approximations with respect to the cell coordinate x, are used. It follows that in the region $x_0 \leq x \leq x_0 + \delta x$ we have the slopes

$$\frac{dt_{Na}}{dx} = \frac{t_{Na}^{(c)} - t_{Na}^{(a)}}{\delta x} \quad ; \quad \frac{dt_K}{dx} = \frac{t_K^{(c)} - t_K^{(a)}}{\delta x}$$

$$\frac{dt_{Cl}}{dx} = \frac{t_{Cl}^{(c)} - t_{Cl}^{(a)}}{\delta x}$$

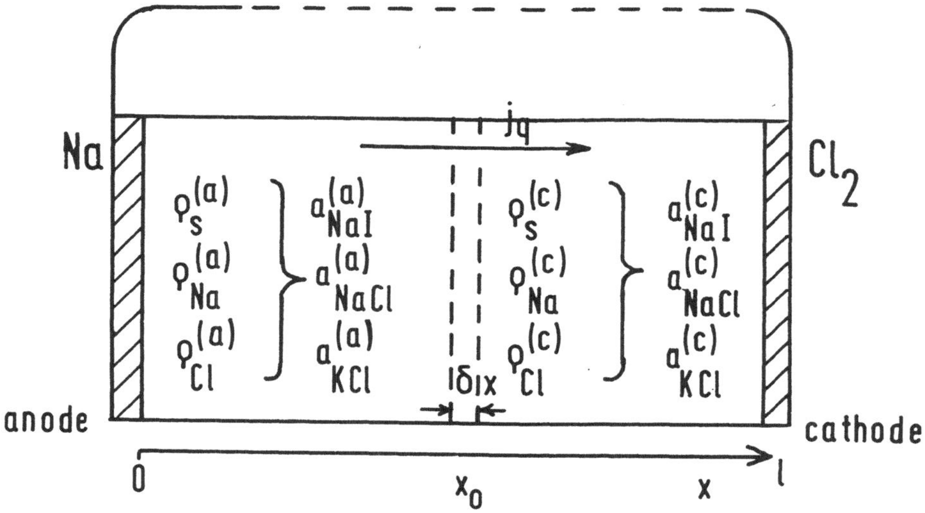

Fig. 26 Galvanic cell in which both electrode compartments contain the four constituents Na, K, Cl and I

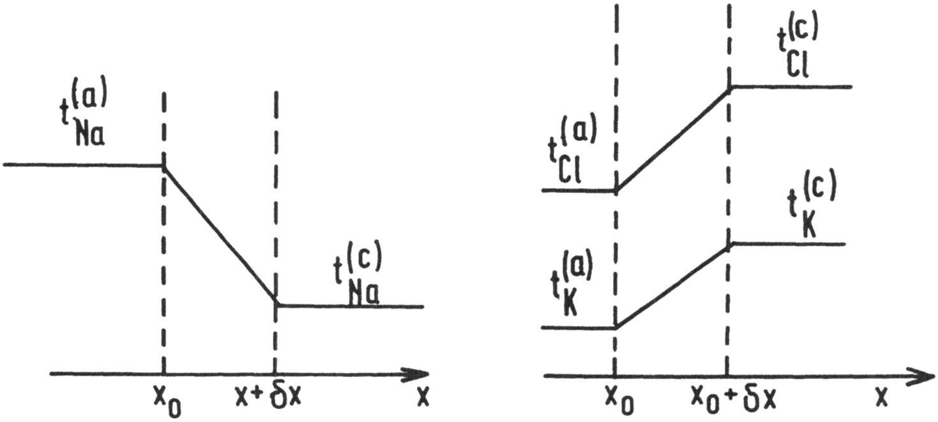

Fig. 27 Schematic representaiton of the x-dependences of various transport numbers at the liquid junction assuming a linear approximation.

These relations we introduce in eqs.(257) to give

$$\frac{(dG)_s}{A} = \Big\{ -\int_{x_0}^{x_0+\delta x} \mu_{NaI} \frac{1}{\delta x} (\Delta t_{Na} + \Delta t_K + \Delta t_{Cl}) dx + \frac{1}{\delta x} \int_{x_0}^{x_0+\delta x} \mu_{NaCl} (\Delta t_{Cl} + \Delta t_K) dx - \frac{1}{\delta x} \int_{x_0}^{x_0+\delta x} \mu_{KCl} \Delta t_K dx \Big\} \frac{j_q}{F} dt$$

with

$$\Delta t_{Na} = t_{Na}^{(c)} - t_{Na}^{(a)} \; ; \quad \Delta t_{Cl} = t_{Cl}^{(c)} - t_{Cl}^{(a)} \; ; \quad \Delta t_K = t_K^{(c)} - t_K^{(a)}$$

Further

$$\frac{(dG)_s}{A} = \Big[-(\Delta t_{Na} + \Delta t_{Cl} + \Delta t_K)\mu_{NaI}^{o} - \frac{\Delta t_{Na} + \Delta t_{Cl} + \Delta t_K}{\delta x} RT \int_{x_0}^{x_0+\delta x} \ln a_{NaI}\, dx$$

$$+ (\Delta t_K + \Delta t_{Cl})\mu_{NaCl}^{o} + \frac{\Delta t_{Cl} + \Delta t_K}{\delta x} RT \int_{x_0}^{x_0+\delta x} \ln a_{NaCl}\, dx$$

$$- \Delta t_K \mu_{KCl}^{o} - \frac{\Delta t_K}{\delta x} RT \int_{x_0}^{x_0+\delta x} \ln a_{KCl}\, dx \Big] \frac{j_2}{\mathcal{F}} dt \qquad (262)$$

We introduce linear approximations for the activities, e.g.

$$a_{NaI}(x) = a_{NaI}^{(a)} + \frac{a_{NaI}^{(c)} - a_{NaI}^{(a)}}{\delta x}(x - x_0)$$

and corresponding expressions for a_{NaCl} and a_{KCl} .
These expressions are inserted in eq.(262) and the integration over x gives

$$\frac{(dG)_s}{A} = \Big\{ -(\Delta t_{Na} + \Delta t_{Cl} + \Delta t_K)\mu_{NaI}^{o} + (\Delta t_K + \Delta t_{Cl})\mu_{NaCl}^{o}$$

$$-(\Delta t_{Na} + \Delta t_K + \Delta t_{Cl})RT \left[\frac{a_{NaI}^{(c)} \ln a_{NaI}^{(c)} - a_{NaI}^{(a)} \ln a_{NaI}^{(a)}}{\Delta a_{NaI}} - 1 \right]$$

$$+ (\Delta t_K + \Delta t_{Cl})RT \left[\frac{a_{NaCl}^{(c)} \ln a_{NaCl}^{(c)} - a_{NaCl}^{(a)} \ln a_{NaCl}^{(a)}}{\Delta a_{NaCl}} - 1 \right]$$

$$- \Delta t_K \mu_{KCl}^{o} - \Delta t_K RT \left[\frac{a_{KCl}^{(c)} \ln a_{KCl}^{(c)} - a_{KCl}^{(a)} \ln a_{KCl}^{(a)}}{\Delta a_{KCl}} - 1 \right] \Big\} \frac{j_2}{\mathcal{F}} dt \qquad (263)$$

Now we turn to the contribution to dG stemming from the electrode fluxes. We have to use eq.(251). The fluxes at the anode are given by the expressions eq.(150), (151), and (152). Corresponding expressions for the cathode are not yet available to us. However eqs.(252) and (253) give us a partial answer and we have to complete these relations:

$$(j_s)_{x=\ell} = -\left[(M_{Na}+M_I)t_{Na}^{(c)} + (M_K + M_I)t_K^{(c)} + (M_{Cl}-M_I)(1-t_{Cl}^{(c)})\right]\frac{j_2}{\mathcal{F}} \tag{264}$$

$$(j_{Na})_{x=\ell} = -\frac{M_{Na}}{\mathcal{F}} j_2 t_{Na}^{(c)} \tag{265}$$

$$(j_{Cl})_{x=\ell} = -\frac{M_{Cl}}{\mathcal{F}} j_2 (1-t_{Cl}^{(c)}) \tag{266}$$

These equations together with eq.(251) and (150)-(152) yield the desired result.

$$\begin{aligned}\frac{(dG)_{ef}}{A} = \Big[& \mu_s^{*(a)} M_{NaI}(1-t_{Na}^{(a)}) - \mu_s^{*(a)} M_{KI} t_K^{(a)} \\ & + \mu_s^{*(a)}(M_{Cl}-M_I)t_{Cl}^{(a)} + \mu_{Na}^{*(a)} M_{Na}(1-t_{Na}^{(a)}) \\ & + \mu_{Cl}^{*(a)} M_{Cl} t_{Cl}^{(a)} + \mu_{Na}^{*(c)} M_{Na} t_{Na}^{(c)} + \mu_{Cl}^{*(c)}(1-t_{Cl}^{(c)})M_{Cl} \\ & + \mu_s^{*(c)} M_{NaI} t_{Na}^{(c)} + \mu_s^{*(c)} M_{KI} t_K^{(c)} + \mu_s^{*(c)}(M_{Cl}-M_I)(1-t_{Cl}^{(c)})\Big]\frac{j_2}{\mathcal{F}}\,dt\end{aligned} \tag{267}$$

Now for the μ_s^*, μ_{Na}^*, μ_{Cl}^* we use the expressions given as eqs.(41) - (41c). We find:

$$\begin{aligned}\frac{(dG)_{ef}^{(a)}}{A} = \Big\{ & \mu_{NaI}^* M_{NaI}\left[(1-t_{Na}^{(a)}) - t_K^{(a)} - t_{Cl}^{(a)}\right] - \mu_{KCl}^* M_{KCl} t_K^{(a)} \\ & + \mu_{NaCl}^* M_{NaCl}\left[t_{Cl}^{(a)} + t_K^{(a)}\right]\Big\}\frac{j_2}{\mathcal{F}}\,dt\end{aligned}$$

This formula we identify with the first part of eq.(35), and we see that the changes of the constituent masses, m_i, i = Na, K, Cl, are just those given by the transport numbers, with the signs suitably reversed. Therefore we can also write the corresponding expression for the cathode and we obtain:

$$\frac{(dG)_{ef}^{(a)}}{A} = \left\{ \mu_{NaI}^{(a)} \left[(1 - t_{Na}^{(a)}) - t_K^{(a)} - t_{Cl}^{(a)} \right] - \mu_{KCl}^{(a)} t_K^{(a)} + \mu_{NaCl}^{(a)} (t_K^{(a)} + t_{Cl}^{(a)}) \right\} \frac{j_2}{F} dt \qquad (268)$$

and

$$\frac{(dG)_{ef}^{(c)}}{A} = \left\{ \mu_{NaI}^{(c)} \left[t_{Na}^{(c)} + t_K^{(c)} - (1 - t_{Cl}^{(c)}) \right] + \mu_{KCl}^{(c)} t_K^{(c)} + \mu_{NaCl}^{(c)} \left[(1 - t_{Cl}^{(c)}) - t_K^{(c)} \right] \right\} \frac{j_2}{F} dt \qquad (269)$$

Then, returning to the toal electrode flux contribution

$$\frac{(dG)_{ef}}{A} = \frac{(dG)_{ef}^{(a)}}{A} + \frac{(dG)_{ef}^{(c)}}{A} \qquad (270)$$

and splitting the chemical potentials into a standard term and a composition dependent term, one derives :

$$\begin{aligned} \frac{(dG)_{ef}}{A} = \Big\{ & \mu_{NaI}^{(a)} + \mu_{NaCl}^{(c)} - \mu_{NaI}^{(c)} - \mu_{NaI}^{o} (t_{Na}^{(a)} + t_K^{(a)} + t_{Cl}^{(a)}) \\ & - RT (t_{Na}^{(a)} + t_K^{(a)} + t_{Cl}^{(a)}) \ln a_{NaI}^{(a)} - \mu_{KCl}^{o} t_K^{(a)} - t_K^{(a)} RT \ln a_{KCl}^{(a)} \\ & + \mu_{NaCl}^{o} (t_K^{(a)} + t_{Cl}^{(a)}) + RT (t_K^{(a)} + t_{Cl}^{(a)}) \ln a_{NaCl}^{(a)} \\ & + \mu_{NaI}^{o} (t_{Na}^{(c)} + t_K^{(c)} + t_{Cl}^{(c)}) + RT (t_{Na}^{(c)} + t_K^{(c)} + t_{Cl}^{(c)}) \ln a_{NaI}^{(c)} \\ & + \mu_{KCl}^{o} t_K^{(c)} + RT t_K^{(c)} \ln a_{KCl}^{(c)} \\ & - \mu_{NaCl}^{o} (t_K^{(c)} + t_{Cl}^{(c)}) - RT (t_K^{(c)} + t_{Cl}^{(c)}) \ln a_{NaCl}^{(c)} \Big\} \frac{j_2}{F} dt \end{aligned} \qquad (271)$$

In order to get the total change of the Gibbs-free energy due to the production of electrolyte, we have to add eqs.(263) and (271). After some fairly lengthy algebraic operations one finds the result:

$$\frac{(dG)_{ef+s}}{A} = \Big\{ \mu_{NaI}^{(a)} - \mu_{NaI}^{(c)} + \mu_{NaCl}^{(c)}$$

$$+RT\left[(t_{Na}^{(c)} + t_{Cl}^{(c)} + t_{K}^{(c)}) \left(\frac{a_{NaI}^{(a)} \ln a_{NaI}^{(a)}/a_{NaI}^{(c)}}{a_{NaI}^{(c)} - a_{NaI}^{(a)}} + 1 \right) \right.$$

$$\left. - (t_{Na}^{(a)} + t_{Cl}^{(a)} + t_{K}^{(a)}) \left(\frac{a_{NaI}^{(c)} \ln a_{NaI}^{(a)}/a_{NaI}^{(c)}}{a_{NaI}^{(c)} - a_{NaI}^{(a)}} + 1 \right) \right]$$

$$-RT\left[(t_{Cl}^{(c)} + t_{K}^{(c)}) \left(\frac{a_{NaCl}^{(a)} \ln a_{NaCl}^{(a)}/a_{NaCl}^{(c)}}{a_{NaCl}^{(c)} - a_{NaCl}^{(a)}} + 1 \right) \right.$$

$$\left. -(t_{Cl}^{(a)} + t_{K}^{(a)}) \left(\frac{a_{NaCl}^{(c)} \ln a_{NaCl}^{(a)}/a_{NaCl}^{(c)}}{a_{NaCl}^{(c)} - a_{NaCl}^{(a)}} + 1 \right) \right] + RT\left[t_{K}^{(c)} \cdot \right.$$

$$\left. \left(\frac{a_{KCl}^{(a)} \ln a_{KCl}^{(a)}/a_{KCl}^{(c)}}{a_{KCl}^{(c)} - a_{KCl}^{(a)}} + 1 \right) - t_{K}^{(a)} \left(\frac{a_{KCl}^{(c)} \ln a_{KCl}^{(a)}/a_{KCl}^{(c)}}{a_{KCl}^{(c)} - a_{KCl}^{(a)}} + 1 \right) \right] \Big\} \frac{I_2}{F}\, dt \tag{272}$$

Now it only remains to determine the Nernst potential. We develop the first terms of eq.(272):

$$\mu_{NaI}^{(a)} - \mu_{NaI}^{(c)} + \mu_{NaCl}^{(c)}$$

$$= \mu_{NaI}^{0} + RT \ln a_{NaI}^{(a)} - \mu_{NaI}^{0} - RT \ln a_{NaI}^{(c)}$$

$$+ \mu_{NaCl}^{0} + RT \ln a_{NaCl}^{(c)}$$

$$= \mu_{NaCl}^{0} + RT \ln \frac{a_{NaI}^{(a)} a_{NaCl}^{(a)}}{a_{NaI}^{(c)}} \tag{273}$$

Then the Nernst potential is

$$E_{Nernst} = \frac{1}{\mathcal{F}}\left[\mu^{o}_{NaCl} - \mu_{Na'} - \frac{1}{2}\mu_{Cl_2} + RT\ln\frac{a^{(a)}_{NaI}\,a^{(c)}_{NaCl}}{a^{(c)}_{NaI}}\right]$$

or with

$$E_o = \frac{1}{\mathcal{F}}\left(\mu^{o}_{NaCl} - \mu_{Na'} - \frac{1}{2}\mu_{Cl_2}\right)$$

we have:

$$E_{Nernst} = E_o + \frac{RT}{\mathcal{F}}\ln\frac{a^{(a)}_{NaI}\,a^{(c)}_{NaCl}}{a^{(c)}_{NaI}} \qquad (274)$$

Using eq.(272) together with eqs.(273),(274) and (223) we arrive at the final result:

$$E = E_o + \frac{RT}{\mathcal{F}}\ln a^{(c)}_{NaCl}\frac{a^{(a)}_{NaI}}{a^{(c)}_{NaI}} + E_{diff} \qquad (275)$$

with

$$\begin{aligned} E_{diff} = \frac{RT}{\mathcal{F}}\Bigg\{ & \left(t^{(c)}_{Na} + t^{(c)}_{K} + t^{(c)}_{Cl}\right)\left(\frac{a^{(a)}_{NaI}\ln a^{(a)}_{NaI}/a^{(c)}_{NaI}}{a^{(c)}_{NaI} - a^{(a)}_{NaI}} + 1\right) \\ & - \left(t^{(a)}_{Na} + t^{(a)}_{K} + t^{(a)}_{Cl}\right)\left(\frac{a^{(c)}_{NaI}\ln a^{(a)}_{NaI}/a^{(c)}_{NaI}}{a^{(c)}_{NaI} - a^{(a)}_{NaI}} + 1\right) - \left(t^{(c)}_{K} + t^{(c)}_{Cl}\right)\left(\frac{a^{(a)}_{NaCl}\ln a^{(a)}_{NaCl}/a^{(c)}_{NaCl}}{a^{(c)}_{NaCl} - a^{(a)}_{NaCl}} + 1\right) \\ & + \left(t^{(a)}_{K} + t^{(a)}_{Cl}\right)\left(\frac{a^{(c)}_{NaCl}\ln a^{(a)}_{NaCl}/a^{(c)}_{NaCl}}{a^{(c)}_{NaCl} - a^{(a)}_{NaCl}} + 1\right) + t^{(c)}_{K}\left(\frac{a^{(a)}_{KCl}\ln a^{(a)}_{KCl}/a^{(c)}_{KCl}}{a^{(c)}_{KCl} - a^{(a)}_{KCl}} + 1\right) \\ & - t^{(a)}_{K}\left(\frac{a^{(c)}_{KCl}\ln a^{(a)}_{KCl}/a^{(c)}_{KCl}}{a^{(c)}_{KCl} - a^{(a)}_{KCl}} + 1\right)\Bigg\} \end{aligned} \qquad (275a)$$

As an application of this formula let us consider one special case. The anode compartment only contains NaI, the cathode compartment contains mainly KCl, but with a very small amount of NaCl, so that $c^{(c)}_{NaCl} \approx 0$, and also $a_{NaCl}^{(c)} \approx 0$. The anode compartment may contain as well a small amount of NaCl, i.e. $a_{NaCl}^{(a)} \approx 0$, but this is of minor importance. Thus we have

$$t_{Na}^{(c)} = 0 \;, \quad t_{K}^{(a)} = 0, \quad t_{Cl}^{(a)} = 0$$

where, in fact we should have written $t_{Na}^{(c)} \approx 0$. Then since the cathode compartment does not contain iodine, we have also

$$t_{K}^{(c)} + t_{Cl}^{(c)} \approx 1$$

We introduce all these equations in eq.(275) and obtain:

$$E = E_0 + \frac{RT}{\mathcal{F}} \ln a_{NaCl}^{(c)} + \frac{RT}{\mathcal{F}} \ln a_{NaI}^{(a)} - \frac{RT}{\mathcal{F}} \ln a_{NaI}^{(c)}$$

$$+ \frac{RT}{\mathcal{F}} \Big\{ 1 \cdot \Big(\frac{a_{NaI}^{(a)} \ln a_{NaI}^{(a)}/a_{NaI}^{(c)}}{a_{NaI}^{(c)} - a_{NaI}^{(a)}} + 1 \Big) - t_{Na}^{(a)} (0 + 1)$$

$$- 1(0 + 1) + 0 + t_{K}^{(c)} (0 + 1) - 0 \Big\}$$

which gives:

$$E = E_0 + \frac{RT}{\mathcal{F}} \ln a_{NaCl}^{(c)} - \frac{RT}{\mathcal{F}} (t_{Na}^{(a)} - t_{K}^{(c)}) \qquad (276)$$

We compare this result with eq.(260b). We see that both equations are very similar. The only difference is that now we have $\ln a_{NaCl}^{(c)}$ instead of $\ln a_{NaCl}$. This is due to the other concentration distribution of NaCl which in the former case was a triangular function (see Fig.28). This was a consequence of the free diffusion process which did not permit us direct knowledge of $a_{NaCl} = a_{NaCl}(x)$ whereas now we have essentially a step function which is determined by the construction operation with a fixed value of $a_{NaCl}^{(c)}$.

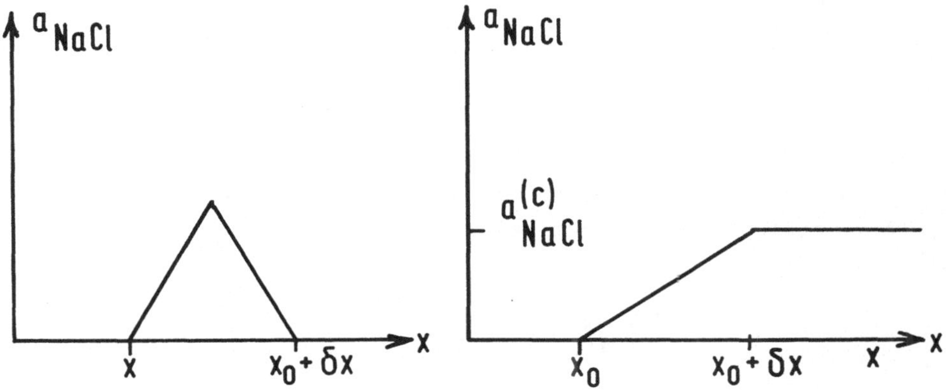

Fig. 28 a) A "triangle function" as an approximation to the NaCl component distribution around the liquid junction between NaI and KCl solution.
b) "Step function" when the system was constructed with small amount of NaCl in the cathode compartment.

8.4 Coordinate system transformations.

We return now to eqs.(254) and (256) where the total change of the Gibbs free energy of the solution was given in the coordinate system; total solute - two constituents. We wish now to extend this formulation in order to represent dG in the three coordinate system representations we have introduced and used in the chapters 1,2 and 3. Then,including our previous approach we have:

(1) Use of the construction coordinate system or "construction space"
(2) Use of the constituents coordinate system or "constituents space".
(3) Use of the total solute mass - two constitutents coordinate system or space.

It should be emphasized that the final results for the electromotive force of a galvanic cell so far derived (eqs.(260),(275), and (276)) were expressed in a mixed language, because the transport numbers were given in the constituents coordinate system whereas the chemical potentials were in the construction coordinate system. We shall now express both the transport numbers and the chemical potentials in the same property space.

a) Construction space.
According to eqs.(16)-(18) we have:

$$\frac{\partial c^*_{KCl}}{\partial t} = \frac{\partial c_K}{\partial t} \quad , \quad \frac{\partial c^*_{NaCl}}{\partial t} = \frac{\partial c_{Cl}}{\partial t} - \frac{\partial c_K}{\partial t}$$

$$\frac{\partial c^*_{NaI}}{\partial t} = \frac{\partial c_{Na}}{\partial t} + \frac{\partial c_K}{\partial t} - \frac{\partial c_{Cl}}{\partial t}$$

With our fundamental electrochemical equations this gives (see eqs. (112)-(114) neglecting the term referring to the irreversible mixing process)

$$\frac{\partial c^*_{KCl}}{\partial t} = -\frac{dt_K}{dx}\frac{j_q}{\mathcal{F}} \quad , \quad \frac{\partial c^*_{NaCl}}{\partial t} = \frac{j_q}{\mathcal{F}}\left(\frac{dt_{Cl}}{dx} + \frac{dt_K}{dx}\right)$$

$$\frac{\partial c^*_{NaI}}{\partial t} = -\frac{j_q}{\mathcal{F}}\left(\frac{dt_{Na}}{dx} + \frac{dt_K}{dx} + \frac{dt_{Cl}}{dx}\right)$$

Now we define the three component transport numbers:

$$t_{KCl} = t_K \quad , \quad t_{NaCl} = t_{Cl} + t_K \tag{277}$$

$$t_{NaI} = t_{Na} + t_K + t_{Cl}$$

Then we have

$$\frac{\partial c^*_{KCl}}{\partial t} = -\frac{j_q}{\mathcal{F}}\frac{dt_{KCl}}{dx} \quad , \quad \frac{\partial c^*_{NaCl}}{\partial t} = \frac{j_q}{\mathcal{F}}\frac{dt_{NaCl}}{dx}$$

$$\frac{\partial c^*_{NaI}}{\partial t} = -\frac{j_q}{\mathcal{F}}\frac{dt_{NaI}}{dx} \tag{278}$$

Now we consider the equations for the change of G due to the electrode fluxes. We obtain with eqs.(278) and (268)-(270)

$$\frac{(dG)_{ef}}{A} = \Big\{\mu^{(a)}_{NaI}(1 - t^{(a)}_{NaI}) - \mu^{(a)}_{KCl}\, t^{(a)}_{KCl} + \mu^{(a)}_{NaCl}\, t^{(a)}_{NaCl}$$

$$+ \mu^{(c)}_{NaI}(t^{(c)}_{NaI} - 1) + \mu^{(c)}_{KCl}\, t^{(c)}_{KCl} + \mu^{(c)}_{NaCl}(1 - t^{(c)}_{NaCl})\Big\}\frac{j_q}{\mathcal{F}}\,dt$$

likewise for the solution we have according to eq.(257)

$$\frac{(dG)_s}{A} = -\left\{ \int_0^l \mu_{NaI} \frac{dt_{NaI}}{dx} dx - \int_0^l \mu_{NaCl} \frac{dt_{NaCl}}{dx} dx + \int_0^l \mu_{KCl} \frac{dt_{KCl}}{dx} dx \right\} \frac{j_q}{\mathcal{F}} dt \tag{279}$$

Again we apply the approximation of a linear x dependence of the t_i and a_i, i = NaI, NaCl, KCl. This after incorporation of the Na(metal) and Cl_2 contribution leads to the total change of Gibbs free energy:

$$\frac{dG}{A} = j_q \, dt \left\{ E_0 + \frac{RT}{\mathcal{F}} \ln a_{NaCl}^{(c)} \cdot \frac{a_{NaI}^{(a)}}{a_{NaI}^{(c)}} + E_{diff} \right\}$$

which gives

$$E = E_0 + \frac{RT}{\mathcal{F}} \ln a_{NaCl}^{(c)} \frac{a_{NaI}^{(a)}}{a_{NaI}^{(c)}} + E_{diff} \tag{280}$$

with

$$\begin{aligned} E_{diff} = \frac{RT}{\mathcal{F}} \Bigg\{ & t_{NaI}^{(c)} \left(\frac{a_{NaI}^{(a)} \ln a_{NaI}^{(a)}/a_{NaI}^{(c)}}{a_{NaI}^{(c)} - a_{NaI}^{(a)}} + 1 \right) \\ & - t_{NaI}^{(a)} \left(\frac{a_{NaI}^{(c)} \ln a_{NaI}^{(a)}/a_{NaI}^{(c)}}{a_{NaI}^{(c)} - a_{NaI}^{(a)}} + 1 \right) - t_{NaCl}^{(c)} \left(\frac{a_{NaCl}^{(a)} \ln a_{NaCl}^{(a)}/a_{NaCl}^{(c)}}{a_{NaCl}^{(c)} - a_{NaF}^{(a)}} + 1 \right) \\ & + t_{NaCl}^{(a)} \left(\frac{a_{NaCl}^{(c)} \ln a_{NaCl}^{(a)}/a_{NaCl}^{(c)}}{a_{NaCl}^{(c)} - a_{NaCl}^{(a)}} + 1 \right) + t_{KCl}^{(c)} \left(\frac{a_{KCl}^{(a)} \ln a_{KCl}^{(a)}/a_{KCl}^{(c)}}{a_{KCl}^{(c)} - a_{KCl}^{(a)}} + 1 \right) \\ & - t_{KCl}^{(a)} \left(\frac{a_{KCl}^{(c)} \ln a_{KCl}^{(a)}/a_{KCl}^{(a)}}{a_{KCl}^{(c)} - a_{KCl}^{(a)}} + 1 \right) \Bigg\} \end{aligned} \tag{280a}$$

The correspondence between eqs.(275a) and (280a) can be directly seen. In the special case $a_{NaCl}^{(c)} \approx 0$, $a_{NaI}^{(a)} = a_{KCl}^{(a)} = 0$ one obtains

$$E = E_0 + \frac{RT}{F} \ln a^{(c)}_{NaCl} - \frac{RT}{F}(t^{(a)}_{NaI} - t^{(c)}_{KCl})$$

which is the same as eq.(276) (see eqs(277) and note that $t^{(a)}_{KCl} = t^{(a)}_{NaCl} = 0$).

b) Full constituents space.

By inverting eqs.(37)-(39) and taking into account eq.(40), we obtain the transformation relations to transform the chemical potentials from the constituent to the construction space

$$\begin{aligned} \mu_{NaI} &= \mu_{Na} \\ \mu_{NaCl} &= \mu_{Na} + \mu_{Cl} \\ \mu_{KCl} &= \mu_K + \mu_{Cl} \end{aligned} \tag{281}$$

It is once again emphasized that the μ_{Na}, μ_K , and μ_{Cl} do not represent "single ion" chemical potentials. Considering the transformation relations eq.(37)-(39) we can also write

$$\mu_i = \mu_i^{\circ} + RT \ln a_i \qquad i = Na, K, Cl \tag{282}$$

where the constituent activity a_i is a function of the constituent concentrations c_{Na}, c_K, c_{Cl}.

With eqs.(281) and (282) and eqs.(286) - (270) we obtain for the contribution of the electrode fluxes to dG

$$\frac{(dG)_{ef}}{A} = \Big[\mu^{(a)}_{Na} + \mu^{(c)}_{Cl}$$

$$+ \mu^{\circ}_{Na} \Delta t_{Na} - RT t^{(a)}_{Na} \ln a^{(a)}_{Na} + RT t^{(c)}_{Na} \ln a^{(c)}_{Na} - \mu^{\circ}_{Cl} \Delta t_{Cl}$$

$$- RT t^{(c)}_{Cl} \ln a^{(c)}_{Cl} + RT t^{(a)}_{Cl} \ln a^{(a)}_{Cl} + \mu^{\circ}_K \Delta t_K$$

$$- RT t^{(a)}_K \ln a^{(a)}_K + RT t^{(c)}_K \ln a^{(c)}_K \Big] \frac{j}{F} dt \tag{283}$$

Next we consider the change of free energy in the electrolyte solution. Again the starting point is eq.(257). One finds:

$$\frac{(dG)_s}{A} = -\left\{ \int_0^l \mu_{Na} \frac{dt_{Na}}{dx} dx - \int_0^l \mu_{Cl} \frac{dt_{Cl}}{dx} dx + \int_0^l \mu_K \frac{dt_K}{dx} dx \right\} \frac{j_2}{\mathcal{F}} dt$$

$$= \left\{ -\Delta t_{Na} \mu^0_{Na} + \Delta t_{Cl} \mu^0_{Cl} - \Delta t_K \mu^0_K \right.$$

$$-\Delta t_{Na} RT \left[\frac{a^{(c)}_{Na} \ln a^{(c)}_{Na} - a^{(a)}_{Na} \ln a^{(a)}_{Na}}{\Delta a_{Na}} - 1 \right]$$

$$+\Delta t_{Cl} RT \left[\frac{a^{(c)}_{Cl} \ln a^{(c)}_{Cl} - a^{(a)}_{Cl} \ln a^{(a)}_{Cl}}{\Delta a_{Cl}} - 1 \right]$$

$$\left. -\Delta t_K RT \left[\frac{a^{(c)}_K \ln a^{(c)}_K - a^{(a)}_K \ln a^{(a)}_K}{\Delta a_K} - 1 \right] \right\} \frac{j_2}{\mathcal{F}} dt \qquad (284)$$

with

$$\Delta t = t^{(c)}_i - t^{(a)}_i$$

and the approximation of linear x dependence of the t_i and the a_i, i = Na, Cl, K.

Now we add eqs.(283) and (284) to obtain after the usual reformulations

$$E = E_0^{const} + \frac{RT}{\mathcal{F}} \ln a^{(a)}_{Na} a^{(c)}_{Cl} + E_{diff} \qquad (285)$$

with

$$E_0^{const} = \mu^0_{Na} + \mu^0_{Cl} - \mu_{Na'} - \frac{1}{2}\mu_{Cl_2}$$

and

$$E_{diff} = \frac{RT}{\mathcal{F}} \left[t^{(c)}_{Na} \left(\frac{a^{(a)}_{Na} \ln a^{(a)}_{Na}/a^{(c)}_{Na}}{a^{(c)}_{Na} - a^{(a)}_{Na}} + 1 \right) - t^{(a)}_{Na} \left(\frac{a^{(c)}_{Na} \ln a^{(a)}_{Na}/a^{(c)}_{Na}}{a^{(c)}_{Na} - a^{(a)}_{Na}} + 1 \right) \right.$$

$$- t^{(c)}_{Cl} \left(\frac{a^{(a)}_{Cl} \ln a^{(a)}_{Cl}/a^{(c)}_{Cl}}{a^{(c)}_{Cl} - a^{(a)}_{Cl}} + 1 \right) + t^{(a)}_{Cl} \left(\frac{a^{(c)}_{Cl} \ln a^{(a)}_{Cl}/a^{(c)}_{Cl}}{a^{(c)}_{Cl} - a^{(a)}_{Cl}} + 1 \right)$$

$$\left. + t^{(c)}_K \left(\frac{a^{(a)}_K \ln a^{(a)}_K/a^{(c)}_K}{a^{(c)}_K - a^{(a)}_K} + 1 \right) - t^{(a)}_K \left(\frac{a^{(c)}_K \ln a^{(a)}_K/a^{(c)}_K}{a^{(c)}_K - a^{(a)}_K} + 1 \right) \right] \qquad (285a)$$

In this formula - as already mentioned - the constituent chemical potentials have the form (as may be derived from eqs.(35)-(39))

$$\mu_{Na} = \mu_{NaI}$$
$$\mu_{Cl} = \mu_{NaCl} - \mu_{NaI}$$
$$\mu_K = \mu_{KCl} + \mu_{NaI} - \mu_{NaCl}$$

For the special situation $a_{Na}^{(c)} \approx 0$, $a_K^{(a)} = a_{Cl}^{(a)} = 0$ we derive

$$E = E_0^{const} + \frac{RT}{F} \ln a_{Na}^{(a)} a_{Cl}^{(c)} - \frac{RT}{F}(t_{Na}^{(a)} + t_{Cl}^{(c)} - t_K^{(c)}) \tag{286}$$

which is different from (276) because now the a_i, i = Na, Cl, K, are assumed to be linear functions of x whereas previously the a_{NaI}, a_{KCl}, and a_{NaCl} were taken as linear functions of x.

c) The total solute-two constituents property space.

According to eq.(255) the total solute transport number is defined as

$$t_s = \frac{M_{NaI} t_{Na} + M_{KCl} t_K - (M_{Cl} - M_I) t_{Cl}}{M_{NaI} + M_{NaCl} + M_{KCl}} \tag{255}$$

Then with eq.(255) and eqs.(150)-(152) the anode flux contribution to the change of Gibbs free energy is:

$$\frac{(dG)_{ef}^{(a)}}{A} = \left\{ \mu_s^{*(a)}(M_{NaI} - t_s^{(a)} \sum M) + \mu_{Na}^{(a)}(1 - t_{Na}^{(a)}) + \mu_{Cl}^{(a)} t_{Cl}^{(a)} \right\} \frac{j_q dt}{F}$$

where we have set:

$$M_{NaI} + M_{KCl} + M_{NaCl} = \sum M.$$

Likewise, at the cathode the flux contribution is

$$\frac{(dG)_{ef}^{(c)}}{A} = \left\{ \mu_s^{*(c)}((M_{Cl} - M_I) + t_s^{(c)} \sum M) + \mu_{Na}^{(c)} t_{Na}^{(c)} + \mu_{Cl}^{(c)}(1 - t_{Cl}^{(c)}) \right\} \frac{j_q dt}{F}$$

This gives for the total electrode flux effect

$$\begin{aligned} &\frac{1}{A}\left[(dG)_{ef}^{(a)} + (dG)_{ef}^{(c)}\right] \\ &= \Big\{ \mu_s^{*(a)} M_{NaI} + \mu_{Na}^{(a)} + \mu_s^{*(c)}(M_{Cl} - M_I) + \mu_{Cl}^{(c)} \\ &\quad + \Delta t_s \mu_s^{\circ} - t_s^{(a)}(\textstyle\sum M)/M_{KI} \cdot RT \ln a_s^{(a)} + t_s^{(c)}(\sum M)/M_{KI}\, RT \ln a_s^{(c)} \\ &\quad - \Delta t_{Cl} \mu_{Cl}^{\circ} - t_{Cl}^{(c)} RT \ln a_{Cl}^{(c)} + t_{Cl}^{(a)} RT \ln a_{Cl}^{(a)} \\ &\quad + \Delta t_{Na} \mu_{Na}^{\circ} - t_{Na}^{(a)} RT \ln a_{Na}^{(a)} + t_{Na}^{(c)} RT \ln a_{Na}^{(c)} \Big\} \frac{j_q dt}{F} \end{aligned} \tag{287}$$

with

$$\mu_s^{\circ} = (M_{KCl} + M_{NaCl} + M_{NaI})\mu_s^{*\circ} = \sum M \cdot \mu_s^{*\circ}$$

For the solution we have according to eq.(256)

$$\frac{(dG)_s}{A} = -\left\{ \int_0^l \mu_s \frac{dt_s}{dx} dx - \int_0^l \mu_{Cl} \frac{dt_{Cl}}{dx} dx + \int_0^l \mu_K \frac{dt_K}{dx} dx \right\} \frac{j_2}{\mathcal{F}} dt$$

$$= \Big\{ -\Delta t_s \mu_s^{\circ} + \Delta t_{Cl} \mu_{Cl}^{\circ} - \Delta t_{Na} \mu_{Na}^{\circ}$$

$$- \frac{\sum M}{M_{KI}} \Delta t_s RT \left[\frac{a_s^{(c)} \ln a_s^{(c)} - a_s^{(a)} \ln a_s^{(a)}}{\Delta a_s} - 1 \right] + \Delta t_{Cl} RT$$

$$\cdot \left[\frac{a_{Cl}^{(c)} \ln a_{Cl}^{(c)} - a_{Cl}^{(a)} \ln a_{Cl}^{(a)}}{\Delta a_{Cl}} - 1 \right] - \Delta t_{Na} RT \left[\frac{a_{Na}^{(c)} \ln a_{Na}^{(c)} - a_{Na}^{(a)}}{a_{Na}^{(c)} -} \right.$$

$$\left. \frac{\ln a_{Na}^{(a)}}{- a_{Na}^{(a)}} - 1 \right] \Big\} \frac{j_2}{\mathcal{F}} dt \qquad (288)$$

Finally we have to add the two contributions, one from the electrode fluxes, and one from the electrolyte solution, and we substract the electrode material parts. The result is:

$$E = E_0^{s,Na,Cl} + \frac{RT}{\mathcal{F}} \left(\frac{M_{NaI}}{M_{KI}} \ln a_s^{(a)} + \ln a_{Na}^{(a)} \right.$$

$$\left. + \frac{(M_{Cl} - M_I)}{M_{KI}} \ln a_s^{(c)} + \ln a_{Cl}^{(c)} \right) + E_{diff} \qquad (289)$$

with

$$E_0^{s,Na,Cl} = M_{NaI} \mu_s^{*\circ} + \mu_{Na}^{\circ} + (M_{Cl} - M_I)\mu_s^{*\circ} + \mu_{Cl}^{\circ}$$

$$- \mu_{Na'} - \frac{1}{2}\mu_{Cl_2}$$

$$= M_{NaCl} \mu_s^{*\circ} + \mu_{Na}^{\circ} + \mu_{Cl}^{\circ} - \mu_{Na'} - \frac{1}{2}\mu_{Cl_2}$$

$$E_{diff} = \frac{RT}{\mathcal{F}}\left[\frac{\Sigma M}{M_{KI}} t_s^{(c)}\left(\frac{a_s^{(a)} \ln a_s^{(a)}/a_s^{(c)}}{a_s^{(c)} - a_s^{(a)}} + 1\right)\right.$$

$$- \frac{\Sigma M}{M_{KI}} t_s^{(a)}\left(\frac{a_s^{(c)} \ln a_s^{(a)}/a_s^{(c)}}{a_s^{(c)} - a_s^{(a)}} + 1\right) - t_{Cl}^{(c)}\left(\frac{a_{Cl}^{(a)} \ln a_{Cl}^{(a)}/a_{Cl}^{(c)}}{a_{Cl}^{(c)} - a_{Cl}^{(a)}} + 1\right)$$

$$+ t_{Cl}^{(a)}\left(\frac{a_{Cl}^{(c)} \ln a_{Cl}^{(a)}/a_{Cl}^{(c)}}{a_{Cl}^{(c)} - a_{Cl}^{(a)}} + 1\right) + t_{Na}^{(c)}\left(\frac{a_{Na}^{(a)} \ln a_{Na}^{(a)}/a_{Na}^{(c)}}{a_{Na}^{(c)} - a_{Na}^{(a)}} + 1\right)$$

$$\left. - t_{Na}^{(a)}\left(\frac{a_{Na}^{(a)} \ln a_{Na}^{(a)}/a_{Na}^{(c)}}{a_{Na}^{(c)} - a_{Na}^{(a)}} + 1\right)\right] \qquad (289a)$$

with the following denotations:

$$\mu_s^* = \mu_s^{*o} + \frac{RT}{M_{KI}} \ln a_s^*$$

$$\mu_s = (\Sigma M)\mu_s^* = \mu_s^o + \frac{RT\Sigma M}{M_{KI}} \ln a_s$$

$$= (\Sigma M_s)\mu_s^{*o} + \frac{RT\Sigma M}{M_{KI}} \ln a_s$$

8.5. Simplifications by introduction of suitable approximations.

It will be seen that the results, eqs.(280), (280a), (285), (285a), and (289), (289a) all have the same formal structure. So the simplifications and approximations to be introduced now are in principle valid for each case; however, it may occur that the conditions to be fulfilled in a particular property space are a better representation of reality than in the other two.

Let us first consider the case that all three transport numbers are essentially constant along the diffusion path x. This will be approximately true in the system which is built up from solutions having the same relative composition at each position x, however the total solute concentration varies. Taking as an example the construction components representation, then we deduce from eqs.(280) and (280a):

$$E = E_0 + \frac{RT}{F} \ln a^{(c)}_{NaCl} \frac{a^{(a)}_{NaI}}{a^{(c)}_{NaI}} + E_{diff} \qquad (290)$$

with E_o as given in eq.(274) and

$$E_{diff} = \frac{RT}{F}\left[t_{NaI} \ln \frac{a^{(c)}_{NaI}}{a^{(a)}_{NaI}} - t_{NaCl} \ln \frac{a^{(c)}_{NaCl}}{a^{(a)}_{NaCl}} + t_{KCl} \ln \frac{a^{(c)}_{KCl}}{a^{(a)}_{KCl}} \right] \qquad (290a)$$

t_{NaI} = const., t_{NaCl} =const., t_{KCl} = const.

As will be seen below, the diffusion potential in this situation is simply a superposition of the electromotive forces of three concentration cells with transference, equipped with a suitable sign.

Furthermore, if the ratios of the activities in the cathode compartment to those in the anode compartment are equal for all three construction components, i.e.

$$\frac{a^{(c)}_{NaI}}{a^{(a)}_{NaI}} = \frac{a^{(c)}_{NaCl}}{a^{(a)}_{NaCl}} = \frac{a^{(c)}_{KCl}}{a^{(a)}_{KCl}}$$

then it follows:

$$E_{diff} = \frac{RT}{F}\left\{ (t_{Na} + t_K) \ln \frac{a^{(c)}_{NaI}}{a^{(a)}_{NaI}} \right\}$$

and finally, if $a^{(c)}_{NaI} = a^{(a)}_{NaI}$

$$E_{diff} = 0$$

then we have a galvanic cell without transference.

The other type of simplification becomes valid if the cell is constructed in such a manner that the total constituents concentration remains essentially constant along the diffusion path, i.e.

$$c_{Na} + c_K = c_I + c_{Cl} = c_{const}$$

with

$$\frac{dc_{const}}{dx} \approx 0$$

Then we may introduce a proportionality approximation with respect to the constituent concentration. We assume the validity of the relations:

$$t_{Na} \approx t_{Na}^{(a)} \frac{c_{Na}}{c_{Na}^{(a)}} \quad ; \quad t_{Cl} \approx t_{Cl}^{(c)} \frac{c_{Cl}}{c_{Cl}^{(c)}} \tag{291}$$

$$t_K \approx t_K^{(c)} \frac{c_K}{c_K^{(c)}}$$

Of course these equations imply certain features of the mode of preparation of the cell, for example we must have $c_{Na}^{(a)}, c_{Cl}^{(c)}, c_K^{(c)} \neq 0$.

Next we introduce proportionality approximations for the activities, as well:

$$a_{Na} \approx a_{Na}^{(a)} \frac{c_{Na}}{c_{Na}^{(a)}} \; ; \; a_{Cl} \approx a_{Cl}^{(c)} \frac{c_{Cl}}{c_{Cl}^{(c)}} \tag{292}$$

$$a_K \approx a_K^{(c)} \frac{c_K}{c_K^{(c)}}$$

Next we consider the integrals occurring in eq.(279). The parts of these integrals which contain the concentration dependence of the chemical potentials have the form

$$RT \int_0^l \frac{dt_i}{dx} \ln a_i(x)\, dx \qquad i = Na, Cl, K$$

which may be integrated by parts to give:

$$\int_0^l \frac{dt_i}{dx} \ln a_i(x)\, dx = t_i \ln a_i \Big|_0^l - \int_0^l t_i \frac{da_i/dx}{a_i(x)} dx \tag{293}$$

Now we introduce eqs. (291) and (292) in the integral on the right-hand side of eq.(293). The result is

$$\int_0^l t_i \frac{da_i/dx}{a_i(x)} dx = \frac{t_{Na}^{(a)}}{c_{Na}^{(a)}} \int_0^l \frac{dc_{Na}}{dx} dx = \frac{t_{Na}^{(a)}}{c_{Na}^{(a)}} \left(c_{Na}^{(c)} - c_{Na}^{(a)} \right)$$

$$= \frac{t_i^{(c)}}{c_i^{(c)}} \int_0^l \frac{dc_i}{dx} dx = \frac{t_i^{(c)}}{c_i^{(c)}} \left(c_i^{(c)} - c_i^{(a)} \right)$$

$$i = Cl, K. \tag{294}$$

Then with eqs.(293) and (294) we have

$$\frac{(dG)_s}{A} = -\left\{ \int_0^l \mu_{Na} \frac{dt_{Na}}{dx} dx - \int_0^l \mu_{Cl} \frac{dt_{Cl}}{dx} dx + \int_0^l \mu_K \frac{dt_K}{dx} dx \right\} \frac{j_e\, dt}{\mathcal{F}} \tag{295}$$

$$= \Big\{ -\Delta t_{Na} \mu^o_{Na} + \Delta t_{Cl} \mu^o_{Cl} - \Delta t_K \mu^o_K$$

$$- RT t^{(c)}_{Na} \ln a^{(c)}_{Na} + RT t^{(a)}_{Na} \ln a^{(a)}_{Na} + \frac{t^{(a)}_{Na}}{c^{(a)}_{Na}} (c^{(c)}_{Na} - c^{(a)}_{Na}) RT$$

$$+ RT t^{(c)}_{Cl} \ln a^{(c)}_{Cl} - RT t^{(a)}_{Cl} \ln a^{(a)}_{Cl} - \frac{t^{(c)}_{Cl}}{c^{(c)}_{Cl}} (c^{(c)}_{Cl} - c^{(a)}_{Cl}) RT$$

$$- RT t^{(c)}_K \ln a^{(c)}_K - RT t^{(a)}_K \ln a^{(a)}_K + \frac{t^{(c)}_K}{c^{(c)}_K} (c^{(c)}_K - c^{(a)}_K) RT \Big\} \frac{j_e}{\mathcal{F}} dt \tag{295a}$$

This can be combined with eq.(283) and the result is:

$$E = E_o^{const} + \frac{RT}{\mathcal{F}} \ln a^{(a)}_{Na} a^{(c)}_{Cl} + E_{diff} \tag{296}$$

with

$$E_{diff} = \frac{RT}{\mathcal{F}} \left[t^{(a)}_{Na} \left(\frac{c^{(c)}_{Na}}{c^{(a)}_{Na}} - 1 \right) - t^{(c)}_{Cl} \left(1 - \frac{c^{(a)}_{Cl}}{c^{(c)}_{Cl}} \right) + t^{(c)}_K \left(1 - \frac{c^{(a)}_K}{c^{(c)}_K} \right) \right] \tag{296a}$$

We note that this formula holds for any arbitrary concentration profile as long as the proportionality relations, eqs.(291) and (292), are sufficiently well fulfilled. Also, given the transport numbers and the chemical potentials in one of the other two coordinate systems, then these quantities may be transformed to the three constituents coordinate system, if eqs.(291), (292), and (294) are good approximations. Then eqs.(296) and (296a) give a simple expression for the EMF.

Finally for the special case which we have considered repeatedly, namely $a^{(c)}_{NaCl} \approx 0$, $a^{(a)}_K = a^{(a)}_{Cl} = 0$ we derive

$$E_{diff} = -\frac{RT}{\mathcal{F}}\left(t_{Na}^{(a)} + t_{Cl}^{(c)} - t_{K}^{(c)}\right)$$

which is the same as eq.(286).

8.6 The galvanic cell containing two metal electrodes, for example Na(metal and K(metal).

Now the chemical process is:

Na + electrolyte solution ⟶

K + electrolyte solution

At the cathode we have the deposition of K(metal), but no longer the consumption of Cl_2(gas). This means that at the surface layer of the cathode ($x = \ell$) our transport numbers have the properties:

$$t_{Na}(x) = \frac{t_{Na}(\ell - x)}{\delta x} \quad , \quad \frac{dt_{Na}}{dx} = -\frac{t_{Na}}{\delta x}$$

$$t_{Cl}(x) = \frac{t_{Cl}(\ell - x)}{\delta x} \quad , \quad \frac{dt_{Cl}}{dx} = -\frac{t_{Cl}}{\delta x}$$

$$t_K(x) = t_K + \frac{1 - t_K}{\delta x}(x - \ell + \delta x)$$

$$\frac{dt_K}{dx} = \frac{1 - t_K}{\delta x}$$

This gives for the integral of $\partial \rho_s/\partial t$ ......
in the layer $\ell - \delta x < x < \ell$

$$\int_{\ell-\delta x}^{\ell} \frac{\partial \rho_s}{\partial t} dx = -\int_{\ell-\delta x}^{\ell} \operatorname{div} j_s \, dx - \Big[-(M_{Na} + M_I)t_{Na} + (1 - t_K)$$
$$\cdot (M_K + M_I) + (M_{Cl} - M_I)t_{Cl}\Big]\frac{j_e}{\mathcal{F}}$$

As previously (see page 142) we consider the limit $\delta x \to 0$. Then the left-hand side of the above equation vanishes and it follows for the total solute mass flux at the cathode ($x = \ell$):

$$(j_s)_{x=l} = -\left[(M_{Na}+M_I)t_{Na} - (1-t_K)(M_K+M_I) - (M_{Cl}-M_I)t_{Cl}\right]\frac{j_2}{\mathcal{F}} \quad (297)$$

The first special case to be considered is that corresponding to the simple cell described on page 144 where the anode and cathode compartments contain "pure" NaI and KCl solutions, respectively. Then we have $t_K = 0$ at the anode and $t_{Na} = 0$ at the cathode. Thus

$$(j_s)_{x=l} = \left\{(1-t_K)(M_K+M_I) + (M_{Cl}-M_I)t_{Cl}\right\}\frac{j_2}{\mathcal{F}}$$

and since $$1 = t_K + t_{Cl}$$

$$(j_s)_{x=l} = M_{KCl}(1-t_K)\frac{j_2}{\mathcal{F}}$$

It should be noted that this is a mass flux having a direction out of the electrolyte solution. Thereby we obtain for the total contribution to the change of Gibbs free energy from the two electrodes

$$\frac{(dG)_{ef}}{A} + \frac{(dG)_{el}}{A}$$

$$= \left[\mu^*_{NaI}M_{NaI}(1-t_{Na}) - \mu^*_{KCl}M_{KCl}(1-t_K)\right]\frac{j_2}{\mathcal{F}}dt$$

$$+ \frac{\mu_{K'} - \mu_{Na'}}{\mathcal{F}} j_2\, dt$$

$$(298)$$

$$= \left[\mu_{NaI}(1-t_{Na}) - \mu_{KCl}(1-t_K)\right]\frac{j_2}{\mathcal{F}}dt + (\mu_{K'} - \mu_{Na'})\frac{j_2}{\mathcal{F}}dt$$

Of course the contribution from the electrolyte solution is the same as given previously because this part of the galvanic cell has not been changed. Thus we have to add eqs.(258) and (298). After some rearrangements one finds a total change of Gibbs free energy in the cell:

$$\frac{dG}{A} = j_2\, dt\left\{E_{Nernst} - \frac{RT}{\mathcal{F}}\left[(t_{Na}-t_K) - (t_K+t_{Cl})\right.\right.$$

$$\left.\cdot\left(\frac{\mu^0_{NaCl}}{RT} - \frac{\mu^0_{NaI}}{RT} + \ln a_{NaCl} - \ln a_{NaI}\right)\right]$$

with

$$\mathcal{F} E_{Nernst} = \mathcal{F}(E_0 + \frac{RT}{\mathcal{F}} \ln \frac{a_{NaI}}{a_{KCl}})$$

and

$$\mathcal{F} E_0 = \mu^0_{NaI} - \mu^0_{KCl} + \mu_{K'} - \mu_{Na'}$$

which can be rewritten to give ($t_K + t_{Cl} = 1$)

$$E = E_0' + \frac{RT}{\mathcal{F}} \ln \frac{a_{NaCl}}{a_{KCl}} - \frac{RT}{\mathcal{F}}(t_{Na} - t_K) \tag{299}$$

with

$$E_0' = \mu^0_{NaCl} - \mu^0_{KCl} + \mu_{K'} - \mu_{Na'}$$

This is the same result as eqs.(260b)-(261b), the only difference being that $1/\mathcal{F}(\mu_{KCl} - \frac{1}{2}\mu_{Cl_2} - \mu_{K'})$ is subtracted. It may be seen that, in agreement with the statement given in the beginning of this section, the chemical reaction which proceeds in the system is

$$Na + KCl(aq.) \longrightarrow NaCl(aq.) + K$$

and it should be noticed that the NaCl activity occurring in eq.(299) is determined by the NaCl production connected with the diffusion process or with the current which has already passed through the system. This is the same situation as has been described in connection with eq.(260).

We turn now to the general configuration of the cell:

Na(metal) / non-uniform mixed electrolyte / K(metal)

We consider the system shown in Fig.26, and only replace the Cl_2 cathode by a K(metal) electrode. dG at the anode is the same as given in eq. (268) because here nothing has changed. However at the cathode some modifications occur. The total solute mass flux is given by eq.(297) and the constituent mass fluxes are:

$$(j_K)_{x=l} = (1 - t_K) M_K \frac{j_q}{\mathcal{F}}$$

$$(j_{Na})_{x=l} = - t_{Na} M_{Na} \frac{j_q}{\mathcal{F}}$$

$$(j_{Cl})_{x=l} = t_{Cl} M_{Cl} \frac{j_q}{\mathcal{F}}$$

These fluxes have to be multiplied by the respective specific chemical potentials. Translation of the μ_i^*'s from the ρ_s, ρ_{Na}, ρ_{Cl} - language into the construction language and change to molar chemical potentials yields the flux contribution to dG from the cathode:

$$\frac{(dG)_{ef}^{(c)}}{A} = \Big\{\mu_{NaI}^{(c)}\left[t_{Na}^{(c)} + t_{K}^{(c)} + t_{Cl}^{(c)} - 1\right] - \mu_{KCl}^{(c)}\left(1 - t_{K}^{(c)}\right) + \mu_{NaCl}^{(c)}\left(1 - t_{Cl}^{(c)} - t_{K}^{(c)}\right)\Big\}\frac{jq}{\mathcal{F}}\,dt$$

It is the same expression as eq.(269), apart from an additional term $-\mu_{KCl}^{(c)}$ which mow appears. This is the only modification of the entire treatment of the galvanic cell. As a consequence,we can immediately write the final result for the electromotive force:

$$E = E_o' + \frac{RT}{\mathcal{F}} \ln \frac{a_{NaCl}^{(c)}}{a_{KCl}^{(c)}} \cdot \frac{a_{NaI}^{(a)}}{a_{NaI}^{(c)}} + E_{diff} \tag{300}$$

with

$$E_o' = \frac{1}{\mathcal{F}}\left(\mu_{NaCl}^{o} - \mu_{KCl}^{o} - \mu_{Na'} + \mu_{K'}\right)$$

and E_{diff} exactly the same as given in eq.(275a).

When we consider the special case: $t_{Na}^{(c)} \approx 0$, $t_{Cl}^{(a)} = t_{K}^{(a)} = 0$ then the formula eq.(276) remains valid, we only have to subtract $\mu_{KCl}^{(c)}$. Thus the result is

$$E = E_o' + \frac{RT}{\mathcal{F}} \ln \frac{a_{NaCl}^{(c)}}{a_{KCl}^{(c)}} - \frac{RT}{\mathcal{F}}\left(t_{Na}^{(a)} - t_{K}^{(c)}\right)$$

with E_o' as given in eq.(300).
Of course, this is also the same formal result as eq.(299), only the analytical nature of a_{NaCl} is different as has been explained above.

In the construction representation NaI, NaCl, KCl, we have at the cathode the electrode flux contribution

$$\frac{(dG)_{ef}^{(c)}}{A} = \Big\{\mu_{NaI}^{(c)}\left(-\left(1 - t_{K}^{(c)}\right) + t_{Na}^{(c)} + t_{Cl}^{(c)}\right) - \mu_{KCl}^{(c)}\left(1 - t_{K}^{(c)}\right) + \mu_{NaCl}^{(c)}\left(-t_{Cl}^{(c)} + \left(1 - t_{K}^{(c)}\right)\right)\Big\}\frac{jq}{\mathcal{F}}\,dt$$

$$= \left\{ \mu_{NaI}^{(c)} (t_{NaI}^{(c)} - 1) + \mu_{KCl}^{(c)} t_{KCl}^{(c)} - \mu_{KCl}^{(c)} + \mu_{NaCl}^{(c)} (t_{NaCl}^{(c)} - 1) \right\} \frac{j_q}{\mathcal{F}} dt$$

(see definition on page 158).

Thus we obtain the same expression as before (eq.(280)) only μ_{KCl} is to be subtracted. This is the formula

$$E = E_0' + \frac{RT}{\mathcal{F}} \ln \frac{a_{NaCl}^{(c)}}{a_{KCl}^{(c)}} \cdot \frac{a_{NaI}^{(a)}}{a_{NaI}^{(c)}} + E_{diff}$$

with E_{diff} as given in eq.(280a) and

$$E_0' = \mu_{NaCl}^{o} - \mu_{KCl}^{o} - \mu_{Na'} + \mu_{K'}$$

The Na, K, Cl-three constituents representation leads to the cathode contribution

$$\frac{(dG)_{ef}^{(c)}}{A} = \left\{ \mu_{Na}^{(c)} t_{Na}^{(c)} - (1 - t_K^{(c)}) \mu_K^{(c)} - t_{Cl}^{(c)} \mu_{Cl}^{(c)} \right\} \frac{j_q}{\mathcal{F}} dt$$

$$= \left\{ \mu_{Na}^{(c)} t_{Na}^{(c)} + t_K^{(c)} \mu_K^{(c)} - t_{Cl}^{(c)} \mu_{Cl}^{(c)} - \mu_K^{(c)} \right\} \frac{j_q}{\mathcal{F}} dt$$

This is the same as before, however $\mu_K + \mu_{Cl}$ are to be subtracted. So we get instead of eq.(285)

$$E = E_0^{const'} + \frac{RT}{\mathcal{F}} \ln \frac{a_{Na}^{(a)}}{a_K^{(c)}} + E_{diff}$$

with

$$E_0^{const'} = \mu_{Na}^{o} - \mu_K^{o} + \mu_{K'} - \mu_{Na'}$$

and E_{diff} unchanged as compared with eq.(285a). Also the simpler expressions (290a) and (296a) remain valid if the necessary conditions justifying the respective approximations are fulfilled. Finally, in the same way, in the $\rho_s, \rho_{Na}, \rho_{Cl}$ representation the Nernst potential is identical also, only the contribution due to KCl has to be subtracted. The diffusion potential is the same as before (eq.(289a)).

8.6 Treatment of the galvanic cell in which the electrolyte has arbitrary composition, but the cathode is an I_2- electrode, $Na/NaCl^{(a)}, NaI^{(a)}, KCl^{(a)}/NaCl^{(c)}, NaI^{(c)}, KCl^{(c)}/I_2$.

The last example of this group of systems is the galvanic cell as depicted in Fig.28. It is the same system as we have treated on page 157, with merely the Cl_2 electrode replaced by an iodine electrode or some electrode of a second kind which is equivalent to an iodine electrode.

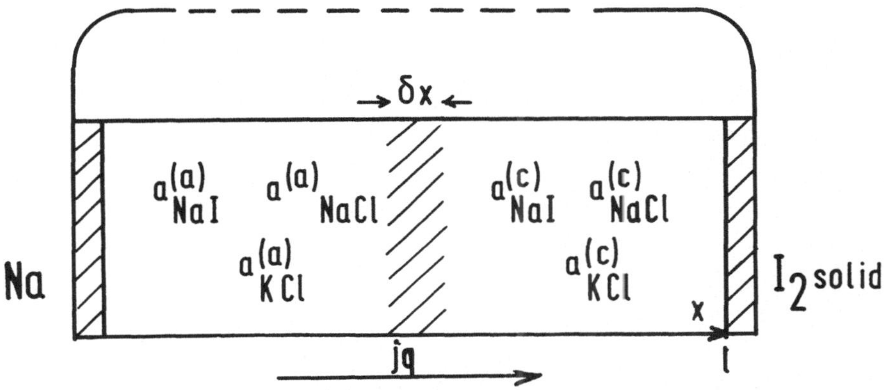

Fig. 29 Schematic representation of galvanic cell in which a non-uniform electrolyte solution containing the independent constituents Na, K, and Cl is placed between a Na and an I_2 electrode.

We have to consider the mass fluxes at the cathode. In order to demonstrate clearly the general nature of the procedure, we rewrite the total solute mass density such that it contains the iodine constituent contribution explicitly: We have

$$\rho_s = M_{Na} c_{Na} + M_K c_K + M_{Cl} c_{Cl} + M_I c_I$$

$$c_K = c_{Cl} + c_I - c_{Na}$$

thus

$$\rho_s = M_{Na} c_{Na} + M_K (c_{Cl} + c_I - c_{Na}) + M_{Cl} c_{Cl} + M_I c_I$$

$$= (M_{Na} - M_K) c_{Na} + (M_K + M_{Cl}) c_{Cl} + (M_K + M_I) c_I$$

Now we obtain for the rate of change of the total solute mass density

$$\frac{\partial \rho_s}{\partial t} = - \operatorname{div} j_s - \left[(M_{Na} - M_K) \frac{dt_{Na}}{dx} - M_{KCl} \frac{dt_{Cl}}{dx} - M_{KI} \frac{dt_I}{dx} \right] \frac{j_q}{\mathcal{F}}$$

which, when integrated at the iodine-cathode gives

$$\int_{\ell-\delta x}^{\ell} \frac{\partial g_s}{\partial t} dx = j^s - \left[-(M_{Na} - M_K) t_{Na}^{(c)} + M_{KCl} t_{Cl}^{(c)} - (1 - t_I^{(c)}) M_{KI} \right] \frac{j_q}{\mathcal{F}}$$

Thus we get

$$(j^s)_{x=\ell} = -\left[(M_{Na} - M_K) t_{Na}^{(c)} - M_{KCl} t_{Cl}^{(c)} + (1 - t_I^{(c)}) M_{KI} \right] \frac{j_q}{\mathcal{F}}$$

$$= -\left[(M_{Na} - M_K) t_{Na}^{(c)} - M_{KCl} t_{Cl}^{(c)} + (t_{Na}^{(c)} + t_{Cl}^{(c)} + t_K^{(c)}) M_{KI} \right] \frac{j_q}{\mathcal{F}}$$

$$= -\left[M_{NaI} t_{Na}^{(c)} - (M_{Cl} - M_I) t_{Cl}^{(c)} + M_{KI} t_K^{(c)} \right] \frac{j_q}{\mathcal{F}} \qquad (301)$$

and correspondingly we have for the other two constituent fluxes:

$$(j_{Na})_{x=\ell} = -\frac{M_{Na}}{\mathcal{F}} j_q t_{Na}^{(c)}$$

$$(j_{Cl})_{x=\ell} = \frac{M_{Cl}}{\mathcal{F}} j_q t_{Cl}^{(c)}$$

which are the same expressions as we had before. Eq.(301) is the same as eq.(264) except that $1 - t_{Cl}^{(c)}$ in the latter formula is replaced by $-t_{Cl}^{(c)}$. Thus, for the cathode flux contribution to dG we also get eq.(269), but with $1 - t_{Cl}^{(c)}$ replaced by $-t_{Cl}^{(c)}$ i.e.

$$\frac{(dG)_{ef}^{(c)}}{A} = \left\{ \mu_{NaI}^{(c)} \left[t_{Na}^{(c)} + t_K^{(c)} + t_{Cl}^{(c)} \right] + \mu_{KCl}^{(c)} t_K^{(c)} + \mu_{NaCl}^{(c)} \left[-t_{Cl}^{(c)} - t_K^{(c)} \right] \right\} \frac{j_q}{\mathcal{F}} dt$$

So we find that we have to add $\mu_{NaI}^{(c)}$ and to subtract $\mu_{NaCl}^{(c)}$ from our previous expression which was valid for the system:

Na(metal)/electrolyte, arbitrary composition/Cl_2(gas)

otherwise this expression remains unchanged. For instance, instead of the result eq.(275), we now have:

$$E = E_0 + \frac{RT}{\mathcal{F}} \ln a_{NaI}^{(a)} + E_{diff} = E_{Nernst} + E_{diff}$$

with

$$E_0 = \frac{1}{\mathcal{F}} \left(\mu_{NaI}^{0} - \mu_{Na'} - \frac{1}{2} \mu_{I_2} \right)$$

and E_{diff} exactly as given in eq.(275a).
It will be seen that the expression for the Nernst potential becomes particularly simple when one of the electrodes is formed from that constituent material which does not belong to the set of the three constituents which describes the electrolyte solution. So, the transport number of this fourth constituent does not appear explicitly.

8.7 A comparison: The conventional treatment of the galvanic cell

Again we consider a galvanic cell which in the electrolyte phase contains the four constituents Na, K, Cl, and I. Of course, these constituents are now ionic species and we shall emphasize this by proper denotation. As before, we assume that the anode is a Na electrode, we follow our previous scheme and treat the two possible cases separately. The cathode may be a Cl_2(gas) electrode or it may be a K(metal) electrode. We begin with an electrolyte phase which may be inhomogeneous, e.g. the system may incorporate a liquid junction.

According to the conventional treatment, in an arbitrary non-equilibrium state there are four ionic fluxes in the solution: These are given by the generalization of our eqs.(300) and (301):

$$\begin{aligned} -j_{Na^+} &= l_{11} \nabla\mu_{Na^+} + l_{12} \nabla\mu_{K^+} + l_{13} \nabla\mu_{Cl^-} + l_{14} \nabla\mu_{I^-} \\ &\quad + (l_{11} + l_{12} - l_{13} - l_{14}) \mathcal{F} \nabla E \\ &\cdots\cdots\cdots\cdots \\ -j_{I^-} &= l_{41} \nabla\mu_{Na^+} + l_{42} \nabla\mu_{K^+} + l_{43} \nabla\mu_{Cl^-} + l_{44} \nabla\mu_{I^-} \\ &\quad - (l_{41} + l_{42} - l_{43} - l_{44}) \mathcal{F} \nabla E \end{aligned}$$

$$\nabla Y = \frac{\partial Y}{\partial x}, \quad Y = \mu_{Na^+} \ldots E$$

The $\nabla\mu_i$, $i = Na^+, K^+, Cl^-, I^-$ are the gradients of the single ion chemical potentials and the l_{ik} are the coefficients of the respective linear relations. Then in analogy to eq.(238) the electric current density is:

$$j_Z = \mathcal{F}(j_{Na^+} + j_{K^+} - j_{Cl^-} - j_{I^-})$$

$$= -\mathcal{F}\{ l_{11}\nabla\mu_{Na^+} + l_{12}\nabla\mu_{K^+} + l_{13}\nabla\mu_{Cl^-} + l_{14}\nabla\mu_{I^-}$$

$$+ l_{21}\nabla\mu_{Na^+} + l_{22}\nabla\mu_{K^+} + l_{23}\nabla\mu_{Cl^-} + l_{24}\nabla\mu_{I^-}$$

$$- l_{31}\nabla\mu_{Na^+} - l_{32}\nabla\mu_{K^+} - l_{33}\nabla\mu_{Cl^-} - l_{34}\nabla\mu_{I^-}$$

$$- l_{41}\nabla\mu_{Na^+} - l_{42}\nabla\mu_{K^+} - l_{43}\nabla\mu_{Cl^-} - l_{44}\nabla\mu_{I^-}$$

$$+ \tilde{\alpha}\mathcal{F}\nabla E$$

with

$$\tilde{\alpha} = l_{11} + 2l_{12} - 2l_{13} - 2l_{14} + l_{22} - 2l_{23} - 2l_{24} + l_{33} + 2l_{34} + l_{44}$$

where the relations $l_{ik} = l_{ki}$ have been used. We have also:

$$j_2 = -\mathcal{F}\{ (l_{11} + l_{21} - l_{31} - l_{41})\nabla\mu_{Na^+}$$

$$+ (l_{12} + l_{22} - l_{32} - l_{42})\nabla\mu_{K^+}$$

$$+ (l_{13} + l_{23} - l_{33} - l_{43})\nabla\mu_{Cl^-}$$

$$+ (l_{14} + l_{24} - l_{34} - l_{44})\nabla\mu_{I^-}$$

$$+ \tilde{\alpha}\mathcal{F}\nabla E \qquad (302)$$

Generalization of eqs.(243) and (244) gives the transport numbers:

$$t_{Na^+} = (l_{11} + l_{12} - l_{13} - l_{14})(\tilde{\alpha})^{-1}$$

$$t_{K^+} = (l_{21} + l_{22} - l_{23} - l_{24})(\tilde{\alpha})^{-1}$$

$$t_{Cl^-} = (l_{33} + l_{34} - l_{31} - l_{32})(\tilde{\alpha})^{-1} \qquad (303)$$

$$t_{I^-} = (l_{44} + l_{43} - l_{41} - l_{42})(\tilde{\alpha})^{-1}$$

Of course, between the t_i the relation holds:

$$t_{Na^+} + t_{K^+} + t_{Cl^-} + t_{I^-} = 1$$

Now, in order to calculate the electromotive force of a galvanic cell, in the conventional treatment we have the condition

$$j_2 = 0 \qquad (304)$$

and eq.(302) is applied to this special situation. As has already been briefly mentioned this step definitely differs from our treatment. In the conventional theory $\dot{j}_Q$ is considered to be exactly zero, whereas in our treatment $\dot{j}_Q$ always has a finite value $\neq 0$, and all our expressions for the electrical work involve the formulation

$$dG = A(\ldots\ldots.)\frac{\dot{j}_Q}{\mathcal{F}}dt \qquad (305)$$

For the calculation of the reversible electromotive force $\dot{j}_Q$ was allowed to have any value $\neq 0$, provided all terms of higher powers in $\dot{j}_Q$ are small compared with the linear ones. Thus, eq.(305) becomes physically meaningless if $\dot{j}_Q = 0$, and consequently, strictly speaking, the conventional treatment involving eq.(304) is also physically meaningless in the framework of our theory. We shall briefly return to this point below.

With eqs.(304) and (303), eq.(302) becomes:

$$-\frac{d\mathbf{E}}{dx}\mathcal{F} = t_{Na^+}\frac{\partial \mu_{Na^+}}{\partial x} + t_{K^+}\frac{\partial \mu_{K^+}}{\partial x} - t_{Cl^-}\frac{\partial \mu_{Cl^-}}{\partial x} - t_{I^-}\frac{\partial \mu_{I^-}}{\partial x} \qquad (306)$$

We integrate this relation over the whole extension of the electrolyte phase:

$$-\mathbf{E}\Big|_0^{\ell} = \frac{1}{\mathcal{F}}\Big\{\int_0^{\ell} t_{Na^+}\frac{\partial \mu_{Na^+}}{\partial x}dx + \int_0^{\ell} t_{K^+}\frac{\partial \mu_{K^+}}{\partial x}dx - \int_0^{\ell} t_{Cl^-}\frac{\partial \mu_{Cl^-}}{\partial x}dx - \int_0^{\ell} t_{I^-}\frac{\partial \mu_{I^-}}{\partial x}dx\Big\} \qquad (307)$$

Obviously, according to the conventional treatment this is the electric potential difference between the two boundaries of the electrolyte phase, i.e. $0 \leq x \leq \ell$.

In the next step the thermodynamic equilibrium at the electrodes is considered which involves the equality of the electrochemical potentials. As a consequence, at the anode we have

$$\mu_{Na'} = \mu_{Na^+} + \mathcal{F}\mathbf{E}^{(a)} \qquad (308)$$

and at the cathode we quote the two alternative forms

$$\frac{1}{2}\mu_{Cl_2} = \mu_{Cl^-} - \mathcal{F}E^{(c)} \tag{309}$$

$$\mu_{K'} = \mu_{K^+} + \mathcal{F}E^{(c)} \tag{310}$$

where the chemical potentials of the elements, $\mu_{Na'}$, $\mu_{K'}$, and μ_{Cl_2} have the same physical significance as in our treatment. If we write eqs.(308)-(310) in such a form that they fit into the scheme of eq.(306) i.e. that on the right-hand side the difference

$$(\mu_i)_{x+\delta x} - (\mu_i)_x$$

appears, where $\delta x > 0$, then we get:

$$-E^{(a)} = (\mu_{Na^+} - \mu_{Na'})\frac{1}{\mathcal{F}} \tag{308a}$$

$$-E^{(c)} = (\tfrac{1}{2}\mu_{Cl_2} - \mu_{Cl^-})\frac{1}{\mathcal{F}} \tag{309a}$$

$$E^{(c)} = (\mu_{K'} - \mu_{K^+})\frac{1}{\mathcal{F}} \tag{310a}$$

Next we add the right-hand side of eq.(308a) to the right-hand side of eq.(307) and subtract the right-hand side of eq.(309a) from the right-hand side of eq.(307). We have to subtract the latter term because in eq.(307) the term referring to the Cl^- ion also carries a minus sign. We also introduce the usual formulas for the chemical potentials. The result is:

$$-E = -E^{(a)} - E\Big|_0^l + E^{(c)}$$

$$= \frac{1}{\mathcal{F}}\Big\{\mu_{Na^+} + \mu_{Cl^-} - \mu_{Na'} - \frac{1}{2}\mu_{Cl_2} + RT\int_0^l t_{Na^+}\frac{d\ln a_{Na^+}}{dx}dx + RT\int_0^l t_{K^+}\frac{d\ln a_{K^+}}{dx}dx - RT\int_0^l t_{Cl^-}\frac{d\ln a_{Cl^-}}{dx}dx - RT\int_0^l t_{I^-}\frac{d\ln a_{I^-}}{dx}dx\Big\} \tag{311}$$

E is the (total) electromotive force of the galvanic cell. Attempts to get information about the non-measurable quantities $E^{(a)}$ and $E^{(c)}$ are well-known.

Now eq.(311) has to be compared with the corresponding result which follows from our treatment. It is natural to choose the three constituents representation for such a comparison. The starting point, as always, is to consider the electric work $E j_2 A\,dt$ done on the system. As we have seen before and applied already several times, the electric work should be divided into three contributions:

$$E j_2 A\,dt = (dG)_{ef} + (dG)_s + (dG)_{el} = (dG)_{\text{electrode flux}} + (dG)_{\text{electrolyte solution}} + (dG)_{\text{electrode material}}$$

The first term is given by eq.(283), the second term is given by eq. (295). We apply eq.(293) in the latter expression. The third term is simply $-(\mu_{Na'}+\frac{1}{2}\mu_{Cl_2})\mathcal{F}^{-1}$. For convenience we rewrite all three contributions as given by the respective equations:

$$\begin{aligned}
E j_2 A\,dt = \frac{j_2 A}{\mathcal{F}} dt \Big\{ & \mu_{Na}^{(a)} + \mu_{Cl}^{(c)} + \mu_{Na}^{o}\Delta t_{Na} + RT(t_{Na}^{(c)}\ln a_{Na}^{(c)} - t_{Na}^{(a)}\ln a_{Na}^{(a)}) \\
& - \mu_{Cl}^{o}\Delta t_{Cl} - RT(t_{Cl}^{(c)}\ln a_{Cl}^{(c)} - t_{Cl}^{(a)}\ln a_{Cl}^{(a)}) \\
& + \mu_{K}^{o}\Delta t_{K} + RT(t_{K}^{(c)}\ln a_{K}^{(c)} - t_{K}^{(a)}\ln a_{K}^{(a)}) \Big\}_{ef} \\
+ \frac{j_2 A}{\mathcal{F}} dt \Big\{ & -\mu_{Na}^{o}\Delta t_{Na} + \mu_{Cl}^{o}\Delta t_{Cl} - \mu_{K}^{o}\Delta t_{K}^{o} \\
& - RT(t_{Na}^{(c)}\ln a_{Na}^{(c)} - t_{Na}^{(a)}\ln a_{Na}^{(a)}) \\
& + RT(t_{Cl}^{(c)}\ln a_{Cl}^{(c)} - t_{Cl}^{(a)}\ln a_{Cl}^{(a)}) \\
& - RT(t_{K}^{(c)}\ln a_{K}^{(c)} - t_{K}^{(a)}\ln a_{K}^{(a)}) \\
& + \int_0^l t_{Na}\frac{d\ln a_{Na}}{dx}dx - \int_0^l t_{Cl}\frac{d\ln a_{Cl}}{dx}dx + \int_0^l t_{K}\frac{d\ln a_{K}}{dx}dx \Big\}_s \\
+ \frac{j_2 A}{\mathcal{F}} dt \Big\{ & -\mu_{Na'} - \frac{1}{2}\mu_{Cl_2} \Big\}_{el}
\end{aligned} \tag{312}$$

It will be seen that apart from the two terms $\mu_{Na}^{(a)}$ and $\mu_{Cl}^{(c)}$ <u>all</u> terms of the electrode flux contribution cancel with the corresponding terms in the electrolyte contribution. Thus we are left with the simple formula

$$E = \frac{1}{\mathcal{F}}\Big\{ \mu_{Na}^{(a)} + \mu_{Cl}^{(c)} - \mu_{Na'} - \frac{1}{2}\mu_{Cl_2}$$

$$+RT\int_0^\ell t_{Na}\frac{d\ln a_{Na}}{dx}dx - RT\int_0^\ell t_{Cl}\frac{d\ln a_{Cl}}{dx}dx + RT\int_0^\ell t_K\frac{d\ln a_K}{dx}dx\Big\} \qquad (313)$$

It will be seen that eqs.(311) and (313) have a very similar formal structure. In contrast to this apparent similarity, the physical content they imply is fundamentally different. The first contribution of eq.(312), i.e. $\{\ldots\}_{\text{electrode flux}}$, represents the effect on the system of the various mass fluxes into (or in other cases out of) the electrolyte solution. As an example the mass fluxes at the anode are depicted schematically in Fig.13. They are the consequence of a real or observable mass production (or mass depletion), as has been outlined in chapter 5, and they are connected with large changes of the excess energy flux (see chapter 6). In the conventional treatment this system of mass fluxes does not exist or, more precisely, it does not exist as a constituting factor of the quantitative derivations.It is replaced by the two abrupt changes of the electrocal potential given by the "thermodynamic equilibria" according to eqs.(308) and (309). They are electrical potential jumps which are intrinsic properties of space in the system. The positive electric current, if present when the equilibrium is disturbed, inside the system, has the same direction as that in which the potential increases. In this way the galvanic cell is considered as an "electrically active" capacitor. However, the electric current which in our treatment is tightly coupled with the mass fluxes at the boundaries, in the conventional treatment of the "static" EMF does not play any explicit role. Correspondingly, in our theory a jump of electric potential inside the system does not exist; at no stage of the development of the theory was it necessary to introduce or to define intrinsic electric potential differences which would be preserved when the electric current vanishes. Only one external parameter, the electric potential E, occurs if $\dot{J}_q \neq 0$ and this quantity characterizes the total system (the resistive contribution does not need to be mentioned here). Still,in spite of the fact that all the mass fluxes at the electrodes are taken into account to form the first term of eq.(312), in the final result they appear only in a very rudimentary form. This is because they are partly balanced by terms of opposite sign occurring in the second parentheses $\{\ldots\}$ of eq.(312), which is the part of the free energy change occurring in the bulk of the electrolyte. They are simply the remaining parts of the mass fluxes which, when multiplied by the corresponding chemical potentials, are exactly equal to the jumps of the electrical potentials at the boundaries as postulated by the conventional theory.

Let us now list the features which equations (311) and (313) have in common. The chemical potentials of the electrode materials, $\mu_{Na'}$ and μ_{Cl_2} are the same physical quantities and they are numerically identical. Likewise, the transport numbers are numerically the same quantities

$$t_{Na} = t_{Na^+} \, , \quad t_{Cl} = t_{Cl^-} \, , \quad t_K = t_{K^+}$$

although their physical significance is different. The fourth transport number in eq.(311) is not independent. In contrast to this, the two equations differ in the following respect. They are based on a different sign convention which however is not a very material distinction. In eq.(311) the activities are single ion quantities, which, as is well known, are only auxiliary properties without direct operational meaning. In practical application mean ionic activities have to be introduced. The corresponding a_i 's in eq.(313) are constituent activities which are experimentally well-defined (see chapter 1) although in general numerical values are not available in the literature. In eq.(311) a fourth integral appears which is absent in eq.(313). This fourth integral only contains dependent quantities.

Finally, it should be noted that in the last four terms of eq.(311) all the classical methods to evaluate the liquid junction potential can be applied, the most well-known being the HENDERSON or the PLANCK treatment, for more details see e.g. the textbook of McINNES[1].

For completeness we have to demonstrate correspondence of final results for the galvanic cell having a K(metal) cathode instead of the Cl_2(gas) cathode. Now we have to add the right-hand sides of eqs.(308a) and (310a) to the right-hand side of eq.(307). It is obvious from eqs.(307) and (310a) that no change of sign is required. The result is:

$$-E = -E^{(a)} - E\Big|_0^{\ell} + E^{(c)}$$

$$= \frac{1}{F}\Big\{\mu_{Na^+} - \mu_{K^+} - \mu_{Na'} + \mu_{K'}$$

$$+ RT\Big(\int_0^{\ell} t_{Na^+}\frac{d\ln a_{Na^+}}{dx}dx + \int_0^{\ell} t_{K^+}\frac{d\ln a_{K^+}}{dx}dx - \int_0^{\ell} t_{Cl^-}\frac{d\ln a_{Cl^-}}{dx}dx - \int_0^{\ell} t_{I^-}\frac{d\ln a_I}{dx}dx\Big)\Big\} \qquad (314)$$

When the same procedures as described for the Na/electrolyte/Cl_2 system are followed it is easily shown that for the galvanic cell to be considered here our theory gives the result:

$$E = \frac{1}{\mathcal{F}}\left\{\mu_{Na}^{(a)} - \mu_{K}^{(c)} + \mu_{K'} - \mu_{Na'}\right.$$

$$\left. + RT\int_0^l t_{Na}\frac{d\ln a_{Na}}{dx}dx - RT\int_0^l t_{Cl}\frac{d\ln a_{Cl}}{dx}dx + RT\int_0^l t_K\frac{d\ln a_K}{dx}dx\right\} \tag{315}$$

which again has the same general structure as eq.(314) and the common and different featues, respectively, as has been outlined in detail above.

Reference

1) D.A. McInnes, The Principles of Electrochemistry, Reinhold, New York 1961

Chapter 9

Galvanic cells containing only one type of anion (or correspondingly, one type of cation).

9.1 The cell configuration $Na/\rho^{(a)}_{NaCl}, \rho^{(a)}_{KCl}/\rho^{(c)}_{NaCl}, \rho^{(c)}_{KCl}/Cl_2$

We now turn to the class of simpler galvanic cells which are built up of only three constituents. Thus we may have two different cationic constituents and only one anionic constituent. As a first example, we choose the NaCl + KCl system. The galvanic cell with an electrolyte containing two different anionic species and only one cationic constituent should be treated in an analogous way.

Of course, the description of the three constituents galvanic cell is simpler than that given in the previous chapters. Still, the formulas given for the four constituent systems cannot in all cases be transformed into those which are valid for the three constituents cell by simply dropping the terms corresponding to the fourth constituent. Therefore we shall give a very brief description of the path from the first equations to the final results. The details of the calculations are analogous to those given for the more general four constituent galvanic cells. Figure 30 depicts schematically the galvanic cell which will be treated.

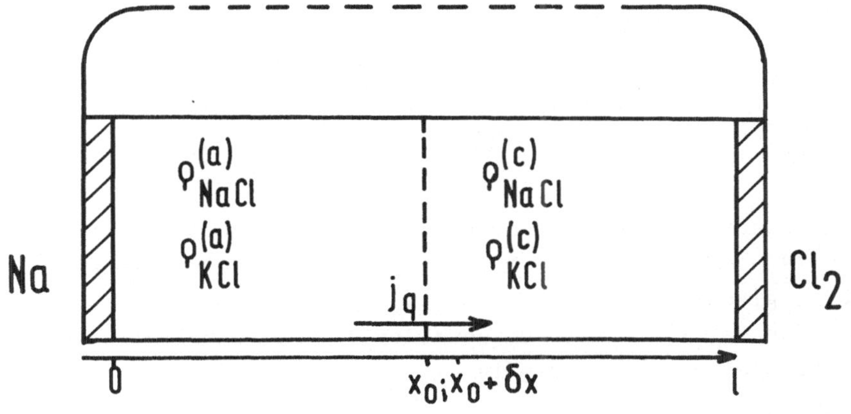

Fig.30 A galvanic cell consisting of a non-uniform NaCl, KCl, H_2O mixture between a Na and a Cl_2 electrode.

Let us begin to describe the system in the total-solute-one constituent property space and let Na be the one independent constituent. The fluxes at the electrodes are

$$(j_s)_{x=0} = M_{NaCl}(1 - t_{Na}^{(a)}) \frac{j_q}{\mathcal{F}} - M_{KCl}\, t_K^{(a)} \frac{j_q}{\mathcal{F}}$$

$$(j_{Na})_{x=0} = (1 - t_{Na}^{(a)}) \frac{M_{Na}}{\mathcal{F}} j_q$$

and correspondingly at the cathode we have :

$$(j_s)_{x=\ell} = -\left[t_K^{(c)} M_{KCl} + M_{NaCl}\, t_{Na}^{(c)} \right] \frac{j_q}{\mathcal{F}}$$

$$(j_{Na})_{x=\ell} = - t_{Na}^{(c)} \frac{M_{Na}}{\mathcal{F}} j_q$$

We convince ourselves that the total solute mass fluxes are written correctly. We have

$$\frac{\partial \rho_s}{\partial t} = M_{Na} \frac{\partial c_{Na}}{\partial t} + M_K \frac{\partial c_K}{\partial t} + M_{Cl} \frac{c_{Cl}}{\partial t}$$

$$c_{Na} + c_K = c_{Cl}$$

$$\frac{\partial c_{Cl}}{\partial t} = \frac{\partial c_{Na}}{\partial t} + \frac{\partial c_K}{\partial t}$$

$$\frac{\partial \rho_s}{\partial t} = (M_{Na} + M_{Cl}) \frac{\partial c_{Na}}{\partial t} + (M_{Cl} + M_K) \frac{\partial c_K}{\partial t}$$

Then the application of the fundamental law of electrochemistry (eqs. (112) and (113), replace M_I by M_{Cl}) yields the total solute mass fluxes j_s as given above. Note that now, apart from j_s, only one constitutent mass flux has to be considered. The contribution to dG from the anode electrode fluxes is:

$$\frac{(dG)_{ef}^{(a)}}{A} = \left\{ -\mu_s^{*(a)} M_{KCl}\, t_K^{(a)} + \mu_s^{*(a)} M_{NaCl} (1 - t_{Na}^{(a)}) + \mu_{Na}^{*(a)} M_{NaCl} (1 - t_{Na}^{(a)}) \right\} \frac{j_q}{\mathcal{F}} dt$$

$$= \left\{ -\mu_{KCl}^{*(a)} M_{KCl}\, t_K^{(a)} + \mu_{NaCl}^{*(a)} M_{NaCl} (1 - t_{Na}^{(a)}) \right\} \frac{j_q}{\mathcal{F}} dt \qquad (316)$$

where eq.(42) has been used to transform the μ_i^* to the construction space. Correspondingly, for the cathode contribution we can immediately write

$$\frac{(dG)_{ef}^{(c)}}{A} = \left\{ \mu_{KCl}^{*(c)} M_{KCl}\, t_K^{(c)} + \mu_{NaCl}^{*(c)} M_{NaCl}\, t_{Na}^{(c)} \right\} \frac{j_2}{F}\, dt \qquad (317)$$

In the bulk of the electrolyte we have the following change of Gibbs free energy :

$$\frac{(dG)_s}{A} = \left\{ - \int_0^{\ell} \mu_{KCl}^{*} M_{KCl} \frac{dt_K}{dx}\, dx - \int_0^{\ell} \mu_{NaCl}^{*} M_{NaCl} \frac{dt_{Na}}{dx}\, dx \right\} \frac{j_2}{F}\, dt$$

We apply the linear approximations of the t_i and a_i with respect to x in the interval $x_0 \leq x \leq x + \delta x$, then with

$$\frac{dt_{Na}}{dx} = \frac{\Delta t_{Na}}{\delta x}, \qquad \frac{dt_K}{dx} = \frac{\Delta t_K}{\delta x}$$

$$\Delta t_{Na} = t_{Na}^{(c)} - t_{Na}^{(a)}, \qquad \Delta t_K = t_K^{(c)} - t_K^{(a)}$$

$$\Delta a_i = a_i^{(c)} - a_i^{(a)}$$

we obtain

$$\frac{(dG)_s}{A} = \frac{j_2}{F} dt \Big\{ - \Delta t_K \mu_{KCl}^{o} - \Delta t_{Na} \mu_{NaCl}^{o}$$

$$- \Delta t_K RT \left[\frac{a_{KCl}^{(c)} \ln a_{KCl}^{(c)} - a_{KCl}^{(a)} \ln a_{KCl}^{(a)}}{\Delta a_{KCl}} - 1 \right]$$

$$- \Delta t_{Na} RT \left[\frac{a_{NaCl}^{(c)} \ln a_{NaCl}^{(c)} - a_{NaCl}^{(a)} \ln a_{NaCl}^{(a)}}{\Delta a_{NaCl}} - 1 \right] \Big\} \qquad (318)$$

As before, in the next step we have to add eqs.(316),(317) and (318), the result is:

$$\frac{(dG)_{ef+s}}{A} = \Big[\mu_{NaCl} +$$

$$RT \left(t_K^{(c)} \left(\frac{a_{KCl}^{(a)} \ln a_{KCl}^{(a)} / a_{KCl}^{(c)}}{a_{KCl}^{(c)} - a_{KCl}^{(a)}} + 1 \right) - t_K \left(\frac{a_{KCl}^{(c)} \ln a_{KCl}^{(a)} / a_{KCl}^{(c)}}{a_{KCl}^{(c)} - a_{KCl}^{(a)}} + 1 \right) \right.$$

$$+ t_{Na}^{(c)}\left(\frac{a_{NaCl}^{(a)} \ln a_{NaCl}^{(a)}/a_{NaCl}^{(c)}}{a_{NaCl}^{(c)} - a_{NaCl}^{(a)}} + 1\right) - t_{Na}^{(a)}\left(\frac{a_{NaCl}^{(c)} \ln a_{NaCl}^{(a)}/a_{NaCl}^{(a)}}{a_{NaCl}^{(c)} - a_{NaCl}^{(c)}} + 1\right)\Bigg]\oint \frac{j_2\,dt}{\mathcal{F}}$$

Thus we arrive at the final formula for the EMF:

$$E = E_0 + \frac{RT}{\mathcal{F}} \ln a_{NaCl}^{(a)} + E_{diff} \tag{319}$$

with

$$E_0 = \frac{1}{\mathcal{F}}\left(\mu_{NaCl}^{o} - \mu_{Na'} - \frac{1}{2}\mu_{Cl_2}\right)$$

and

$$E_{diff} = \frac{RT}{\mathcal{F}}\Bigg\{ t_K^{(c)}\left(\frac{a_{KCl}^{(a)} \ln a_{KCl}^{(a)}/a_{KCl}^{(c)}}{a_{KCl}^{(c)} - a_{KCl}^{(a)}} + 1\right)$$

$$- t_K^{(a)}\left(\frac{a_{KCl}^{(c)} \ln a_{KCl}^{(a)}/a_{KCl}^{(c)}}{a_{KCl}^{(c)} - a_{KCl}^{(a)}} + 1\right) + t_{Na}^{(c)}\left(\frac{a_{NaCl}^{(a)} \ln a_{NaCl}^{(a)}/a_{NaCl}^{(c)}}{a_{NaCl}^{(c)} - a_{NaCl}^{(a)}} + 1\right)$$

$$- t_{Na}^{(a)}\left(\frac{a_{NaCl}^{(c)} \ln a_{NaCl}^{(a)}/a_{NaCl}^{(c)}}{a_{NaCl}^{(c)} - a_{NaCl}^{(a)}} + 1\right)\Bigg\} \tag{319a}$$

Of course, the chemical reaction is:

$$Na + 1/2\ Cl_2 \longrightarrow NaCl\ (aq.)$$

Consideration of some special situations:

a) In the anode compartment there is "pure" NaCl solution, and the cathode compartment contains "pure" KCl solution.

Now we find from eq.(319a)

$$E_{diff} = \frac{RT}{\mathcal{F}}\left\{t_K^{(c)}(0+1) - 0 + 0 - t_{Na}^{(a)}(0+1)\right\}$$

$$= \frac{RT}{\mathcal{F}}\left(t_K^{(c)} - t_{Na}^{(a)}\right)$$

b) Both transport numbers are constant throughout the total system

$$t_K^{(a)} = t_K^{(c)} ; \quad t_{Na}^{(c)} = t_{Na}^{(a)}$$

Then we get from eq.(319a)

$$E_{diff} = \frac{RT}{F}\left(t_K \ln a_{KCl}^{(c)}/a_{KCl}^{(a)} - t_{Na} \ln a_{NaCl}^{(c)}/a_{NaCl}^{(a)}\right)$$

and if the ratios of the activities are also equal we have:

$$E_{diff} = \frac{RT}{F}(t_K - t_{Na}) \ln \frac{a_{KCl}^{(c)}}{a_{KCl}^{(a)}}$$

c) Use of the approximations:

$$t_{Na} = t_{Na}^{(a)} \frac{c_{NaCl}}{c_{NaCl}^{(a)}} , \quad t_K = t_K^{(c)} \frac{c_{KCl}}{c_{KCl}^{(c)}}$$

$$a_{NaCl} = a_{NaCl}^{(a)} \frac{c_{NaCl}}{c_{NaCl}^{(a)}} , \quad a_{KCl} = a_{KCl}^{(c)} \frac{c_{KCl}}{c_{KCl}^{(c)}}$$

With this approximation, as outlined on page 166 one finds:

$$E_{diff} = \frac{RT}{F}\left(t_K^{(c)}\left(1 - \frac{c_{KCl}^{(a)}}{c_{KCl}^{(c)}}\right) - t_{Na}^{(a)}\left(1 - \frac{c_{NaCl}^{(c)}}{c_{NaCl}^{(a)}}\right)\right)$$

9.2 The galvanic cell $Na/\rho_{NaCl}^{(a)}, \rho_{KCl}^{(a)} / \rho_{NaCl}^{(c)}, \rho_{KCl}^{(c)} / K$

The chemical reaction is

Na + KCl(aq.) ⟶ K + NaCl(aq.)

Now we have to replace the chlorine electrode at $x = \ell$ by a K(metal) electrode.

Then instead of eq.(317) we have:

$$\frac{(dG)_{ef}^{(c)}}{A} = \left\{\mu_{NaCl}^{*(c)} M_{NaCl} t_{Na}^{(c)} - (1 - t_K^{(c)}) \mu_{KCl}^{*(c)} M_{KCl}\right\} j_q \frac{dt}{F}$$

$$= \left\{\mu_{NaCl}^{(c)} t_{Na}^{(c)} + \mu_{KCl}^{(c)} t_K^{(c)} - \mu_{KCl}^{(c)}\right\} j_q \frac{dt}{F}$$

All other details remain the same, thus we have only to subtract $\mu_{KCl}^{(c)}$

from our final result (eq.(319):

$$E = E_0 + \frac{RT}{\mathcal{F}} \ln \frac{a^{(a)}_{NaCl}}{a^{(c)}_{KCl}} + E_{diff}$$

with

$$E_0 = \frac{1}{\mathcal{F}}(\mu^0_{NaCl} - \mu^0_{KCl} + \mu_{K'} - \mu_{Na'})$$

and E_{diff} unchanged as before.

Thus all the simplifications considered, retain their particular formulas, only the Nernst potential $RT/\mathcal{F} \ln a^{(a)}_{NaCl}/a^{(c)}_{KCl}$ and a modified standard potential have to be introduced.

9.3 The galvanic cell in which one of the constituents is given by a polyvalent cation.

The transition from the systems so far treated in this chapter, to a galvanic cell containing an electrode which is electrochemically active with respect to a constituent with $z^+ > 1$ (see page 55) is very simple. As an example we choose a system which is equipped with a Ca(metal) electrode instead of the Na(metal) electrode. The electrolyte solution now contains the two components $CaCl_2$ and KCl. Let the partial mass densities in the anode compartment be $\rho^{(a)}_{CaCl_2}$ and $\rho^{(a)}_{KCl}$; in the cathode compartment the partial mass densities are $\rho^{(c)}_{CaCl_2}$ and $\rho^{(c)}_{KCl}$. As before, we consider two different cell configurations. a) The cathode is a chloride(gas)electrode, b) the cathode is a K(metal) electrode.

First some general comments have to be made: The total solute partial mass density is

$$\rho_s = \rho_{Ca} + \rho_K + \rho_{Cl}$$

Application of eq.(2 b) with $c_I = 0$ yields

$$\rho_s = \frac{\rho_{Ca}}{M_{Ca}}(M_{Ca} + 2M_{Cl}) + \frac{\rho_K}{M_K}(M_K + M_{Cl})$$

Combination of eq.(104) and (109) gives

$$\frac{\partial \rho'_{Ca}}{\partial t} = - M_{Ca} \frac{j_q}{2\mathcal{F}} \frac{dt_{Ca}}{dx}$$

Thus, the fundamental equations of electrochemistry are

$$\frac{\partial \rho_s}{\partial t} = - \operatorname{div} j_s - \frac{j_2}{F}\left[\frac{1}{2} M_{CaCl_2} \frac{dt_{Ca}}{dx} + M_{KCl} \frac{dt_K}{dx}\right] \tag{320}$$

$$\frac{\partial \rho_{Ca}}{\partial t} = - \operatorname{div} j_{Ca} - \frac{M_{Ca}}{2F} j_2 \frac{dt_{Ca}}{dx} \tag{321}$$

These are equations (112a) and (113a) if one replaces M_I by M_{Cl}.

Furthermore, we have also:

$$\rho_s = \rho^*_{KCl} + \rho^*_{CaCl}$$

$$\rho_{Ca} = \rho^*_{CaCl_2} \frac{M_{Ca}}{M_{CaCl_2}}$$

and consequently

$$\rho^*_{CaCl_2} = \rho_{Ca} \frac{M_{CaCl_2}}{M_{Ca}}$$

$$\rho^*_{KCl} = \rho_s - \rho_{Ca} \frac{M_{CaCl_2}}{M_{Ca}}$$

These equations are the anologs of eqs. (32) and (33), consequently the analogs of eqs.(42a) and (42b) are also valid

$$\mu^*_s = \mu^*_{KCl} \tag{322}$$

$$\mu^*_{Ca} = \frac{1}{M_{Ca}} M_{CaCl_2}(\mu^*_{CaCl_2} - \mu^*_{KCl}) \tag{323}$$

Now we compare eqs.(320) and (321) with eqs.(112) and (113) where we replace M_I by M_{Cl}. We see that the only modification we have to introduce in the fundamental laws of electrochemistry is to replace M_{NaCl} in the equations for $\partial \rho_s / \partial t$ by $1/2 \cdot M_{CaCl_2}$, and in the equation referring to the constituent mass flux we have to replace M_{Na} by $1/2\ M_{Ca}$. It is clear that in the general case the respective molecular masses M_i have to be replaced by $1/z^+ M_i$. For the chemical potentials we have to use the direct analogs of our previous expressions as shown in eqs.(322) and (323). Following these instructions we shall now write down the most important formulas which must be used to obtain the EMF of the galvanic cells considered below:

a) The cathode is a chlorine(gas) electrode.
The mass fluxes at the electrodes are:

$$(j_s)_{x=0} = \left[\frac{M_{CaCl_2}}{2}(1 - t_{Ca}^{(a)}) - M_{KCl}t_K^{(a)}\right]\frac{j_q}{\mathcal{F}}$$

$$(j_{Ca})_{x=0} = \frac{M_{Ca}}{2}(1 - t_{Ca}^{(a)})\frac{j_q}{\mathcal{F}}$$

$$(j_s)_{x=\ell} = -\left[\frac{M_{CaCl_2}}{2}t_{Ca}^{(c)} + M_{KCl}t_K^{(c)}\right]\frac{j_q}{\mathcal{F}}$$

$$(j_{Ca})_{x=\ell} = -\frac{M_{Ca}}{2}t_{Ca}^{(c)}\frac{j_q}{\mathcal{F}}$$

The contributions to dG from the anode and cathode electrode fluxes are (see eqs.(316) and (317))

$$\frac{(dG)_{ef}^{(a)}}{A} = \left\{-\mu_{KCl}^{*(a)}M_{KCl}t_K^{(a)} + \frac{1}{2}\mu_{CaCl_2}^{*(a)}M_{CaCl_2}(1 - t_{Ca}^{(a)})\right\}\frac{j_q\,dt}{\mathcal{F}}$$

$$\frac{(dG)_{ef}^{(c)}}{A} = \left\{\mu_{KCl}^{*(c)}M_{KCl}t_K^{(c)} + \frac{1}{2}\mu_{CaCl_2}^{*(c)}M_{CaCl_2}t_{Ca}^{(c)}\right\}\frac{j_q\,dt}{\mathcal{F}}$$

In the bulk of the electrolyte solution the change of Gibbs free energy is:

$$\frac{(dG)_s}{A} = \left\{-\int_0^\ell \mu_{KCl}^{*}M_{KCl}\frac{dt_K}{dx}dx - \frac{1}{2}\int_0^\ell \mu_{CaCl_2}^{*}M_{CaCl_2}\frac{dt_{Ca}}{dx}dx\right\}\frac{j_q\,dt}{\mathcal{F}}$$

$$= \frac{j_q}{\mathcal{F}}dt\left\{-\Delta t_K\mu_{KCl}^{o} - \frac{\Delta t_{Ca}}{2}\mu_{CaCl_2}^{o}\right.$$

$$- \Delta t_K RT\left[\frac{a_{KCl}^{(c)}\ln a_{KCl}^{(c)} - a_{KCl}^{(a)}\ln a_{KCl}^{(a)}}{\Delta a_{KCl}} - 1\right]$$

$$\left.- \frac{1}{2}\Delta t_{Ca}RT\left[\frac{a_{CaCl_2}^{(c)}\ln a_{CaCl_2}^{(c)} - a_{CaCl_2}^{(a)}\ln a_{CaCl_2}^{(a)}}{\Delta a_{CaCl_2}} - 1\right]\right\}$$

(see eq.(318)). This gives a total electrolyte contribution

$$\frac{(dG)_{ef+s}}{A} = \Big[\frac{1}{2}\mu_{CaCl_2} + RT\Big(t_K^{(c)}\Big(\frac{a_{KCl}^{(a)} \ln a_{KCl}^{(a)}/a_{KCl}^{(c)}}{a_{KCl}^{(c)} - a_{KCl}^{(a)}} + 1\Big) - t_K^{(a)} \Big(\frac{a_{KCl}^{(c)} \ln a_{KCl}^{(a)}/a_{KCl}^{(c)}}{a_{KCl}^{(c)} - a_{KCl}^{(a)}} + 1\Big) + \frac{t_{Ca}^{(c)}}{2}\Big(\frac{a_{CaCl_2}^{(a)} \ln a_{CaCl_2}^{(a)}/a_{CaCl_2}^{(c)}}{a_{CaCl_2}^{(c)} - a_{CaCl_2}^{(a)}} + 1\Big) - \frac{t_{Ca}^{(a)}}{2}\Big(\frac{a_{CaCl_2}^{(c)} \ln a_{CaCl_2}^{(a)}/a_{CaCl_2}^{(c)}}{a_{CaCl_2}^{(c)} - a_{CaCl_2}^{(a)}} + 1\Big)\Big)\Big] \frac{j_2\, dt}{F}$$

and finally it follows:

$$E = E_0 + \frac{RT}{2F} \ln a_{CaCl_2}^{(a)} + E_{diff} \tag{324}$$

with

$$E_0 = \frac{1}{2F}\left(\mu_{CaCl_2}^{\circ} - \mu_{Ca'} - \mu_{Cl_2}\right)$$

and

$$E_{diff} = \frac{RT}{F}\Big\{t_K^{(c)}\Big(\frac{a_{KCl}^{(a)} \ln a_{KCl}^{(a)}/a_{KCl}^{(c)}}{a_{KCl}^{(c)} - a_{KCl}^{(a)}} + 1\Big) - t_K^{(a)}\Big(\frac{a_{KCl}^{(c)} \ln a_{KCl}^{(a)}/a_{KCl}^{(c)}}{a_{KCl}^{(c)} - a_{KCl}^{(a)}} + 1\Big) - \frac{1}{2}t_{Ca}^{(c)}\Big(\frac{a_{CaCl_2}^{(a)} \ln a_{CaCl_2}^{(a)}/a_{CaCl_2}^{(c)}}{a_{CaCl_2}^{(c)} - a_{CaCl_2}^{(a)}} + 1\Big) - \frac{1}{2}t_{Ca}^{(a)}\Big(\frac{a_{CaCl_2}^{(c)} \ln a_{CaCl_2}^{(a)}/a_{CaCl_2}^{(c)}}{a_{CaCl_2}^{(c)} - a_{CaCl_2}^{(a)}} + 1\Big)\Big\} \tag{324a}$$

(see eq.(319) and (319a).

The chemical reaction is

$$1/2\ Ca + 1/2\ Cl_2 \longrightarrow 1/2\ CaCl_2(aq.)$$

If the anode and cathode compartments contain "pure" $CaCl_2$ and KCl solutions, respectively, we have a diffusion potential

$$E_{diff} = \frac{RT}{F}\left(t_K^{(c)} - \frac{1}{2}t_{Ca}^{(a)}\right)$$

b) The cathode is a K(metal) electrode.
Now the cathode flux contribution to the change of G is:

$$\frac{(dG)_{ef}^{(c)}}{A} = \left\{\frac{1}{2}\mu_{CaCl_2}^{(c)} t_{Ca}^{(c)} - (1 - t_K^{(c)})\mu_{KCl}^{(c)}\right\} \frac{j_2}{F} dt$$

and the anode and the bulk solution contributions remain the same. Thus again, we have only to subtract $\mu_{KCl}^{(c)}$ from our eqs.(324) and (324a). One finds:

$$E = E_o + \frac{RT}{2\mathcal{F}} \ln \frac{a_{CaCl_2}^{(a)}}{(a_{KCl}^{(c)})^2} + E_{diff}$$

with

$$E_o = \frac{1}{\mathcal{F}}\left(\frac{1}{2}\mu^o_{CaCl_2} - \mu^o_{KCl} + \mu_{K'} - \frac{1}{2}\mu_{Ca'}\right)$$

and E_{diff} as given by eq.(324a). The chemical reaction is

$$1/2\ Ca + KCl(aq.) \longrightarrow 1/2\ CaCl_2(aq.) + K$$

9.4 The cell which contains only one kind of binary electrolyte solution, but with varying concentration.

Let the solute in the binary electrolyte solution be NaCl. Now the system is fully described by $\rho_s = \rho_{NaCl}$ and we have

$$\mu_s^* = \mu_{NaCl}^*$$

in analogy to the formula(42a).
Furthermore

$$\frac{\partial c_s}{\partial t} = \frac{\partial c_{Na}}{\partial t} \; ; \quad \frac{\partial c_{Na}}{\partial x} = \frac{M_{Na}}{M_{NaCl}} \frac{\partial c_{Na}}{\partial x}$$

$$\frac{\partial \rho_s}{\partial t} = M_{NaCl} \frac{\partial c_{NaCl}}{\partial t}$$

First we consider the galvanic cell which is shown in Fig.31:

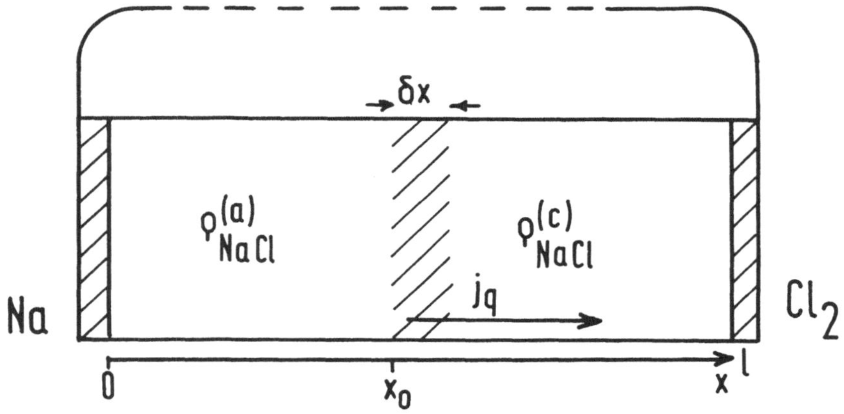

Fig.31 A simple galvanic cell consisting of a non-uniform NaCl solution between a Na(metal) and a Cl_2(gas) electrode.

It is immediately seen that the chemical reaction is:

$$Na + 1/2\ Cl_2 \longrightarrow NaCl(aq.)$$

Following our previous procedure we first consider the case for which $\rho_{NaCl}^{(a)} \neq \rho_{NaCl}^{(c)}$.
Then at the anode the change of Gibbs free energy is:

$$\frac{(dG)_{ef}^{(a)}}{A} = \mu_{NaCl}^{(a)}\left(1 - t_{Na}^{(a)}\right)\frac{j_Q}{F}\,dt \tag{325}$$

and likewise at the cathode we have:

$$\frac{(dG)_{ef}^{(c)}}{A} = \mu_{NaCl}^{(c)}\, t_{Na}^{(c)}\,\frac{j_Q}{F}\,dt \tag{326}$$

For the bulk electrolyte solution we have to set:

$$\frac{(dG)_s}{A} = -\int_0^{\ell} \mu_{NaCl}\,\frac{dt_{Na}}{dx}\,dx\cdot\frac{j_Q}{F}\,dt$$

which with $dt_{Na}/dx = \Delta t_{Na}/\delta x$ in the region $x_0 \leq x \leq x_0 + \delta x$ gives

$$\frac{(dG)_s}{A} = \frac{j_Q}{F}\,dt\left\{-\Delta t_{Na}\,\mu_{NaCl}^{o} - \frac{\Delta t_{Na}}{\delta x}RT\int_{x_0}^{x_0+\delta x}\ln a_{NaCl}\,dx\right\}$$

and using the approximation of linear x dependence of a_{NaCl} together with the abbreviation $\Delta a_{NaCl} = a_{NaCl}^{(c)} - a_{NaCl}^{(a)}$ one finds:

$$\frac{(dG)_s}{A} = \frac{j_Q}{F}\,dt\left\{-\Delta t_{Na}\,\mu_{NaCl}^{o} - RT\Delta t_{Na}\left[\frac{a_{NaCl}^{(c)}\ln a_{NaCl}^{(c)} - a_{NaCl}^{(a)}\ln a_{NaCl}^{(a)}}{\Delta a_{NaCl}} - 1\right]\right\} \tag{327}$$

For the sum of eqs.(325), (326), and (327) we get:

$$\frac{(dG)_{ef+s}}{A} = \frac{j_Q}{F}\,dt\left\{\mu_{NaCl}^{(a)} + RT\left(t_{Na}^{(c)}\left(\frac{a_{NaCl}^{(a)}\ln a_{NaCl}^{(a)}/a_{NaCl}^{(c)}}{a_{NaCl}^{(c)} - a_{NaCl}^{(a)}} + 1\right) - t_{Na}^{(a)}\left(\frac{a_{NaCl}^{(c)}\ln a_{NaCl}^{(a)}/a_{NaCl}^{(c)}}{a_{NaCl}^{(c)} - a_{NaCl}^{(a)}} + 1\right)\right)\right\}$$

The general result follows :

$$E = E_0 + \frac{RT}{\mathcal{F}} \ln a^{(a)}_{NaCl} + E_{diff} \qquad (328)$$

with

$$E_0 = \frac{1}{\mathcal{F}}\left(\mu^0_{NaCl} - \mu_{Na'} - \frac{1}{2}\mu_{Cl_2}\right)$$

and

$$E_{diff} = \frac{RT}{\mathcal{F}}\left(t^{(c)}_{Na}\left[\frac{a^{(a)}_{NaCl} \ln a^{(a)}_{NaCl}/a^{(c)}_{NaCl}}{a^{(c)}_{NaCl} - a^{(a)}_{NaCl}} + 1\right]\right.$$

$$\left. - t^{(a)}_{Na}\left[\frac{a^{(c)}_{NaCl} \ln a^{(a)}_{NaCl}/a^{(c)}_{NaCl}}{a^{(c)}_{NaCl} - a^{(a)}_{NaCl}} + 1\right]\right) \qquad (328a)$$

Special cases are evident. If the difference of the transport numbers may be neglected, we have

$$E_{diff} = \frac{RT}{\mathcal{F}} t_{Na} \ln \frac{a^{(c)}_{NaCl}}{a^{(a)}_{NaCl}}$$

And, if the concentrations are equal in both compartments we have

$$E = E_0 + \frac{RT}{\mathcal{F}} \ln a_{NaCl}$$

i.e. of course, $E_{diff} = 0$, we have a cell without transference.

9.5 The concentration cell with transference.

The concentration cell with transference is obtained if in Fig.31 the chlorine electrode is replaced by a sodium electrode (see Fig.32). The only change is that at the cathode now we must write

$$\frac{(d\Theta)^{(c)}_{ef}}{A} = -\left(1 - t^{(c)}_{Na}\right)\mu^{(c)}_{NaCl} \frac{\dot{q}}{\mathcal{F}} dt$$

$$= \left(-\mu^{(c)}_{NaCl} + t^{(c)}_{Na}\mu^{(c)}_{NaCl}\right)\frac{\dot{q}}{\mathcal{F}} dt$$

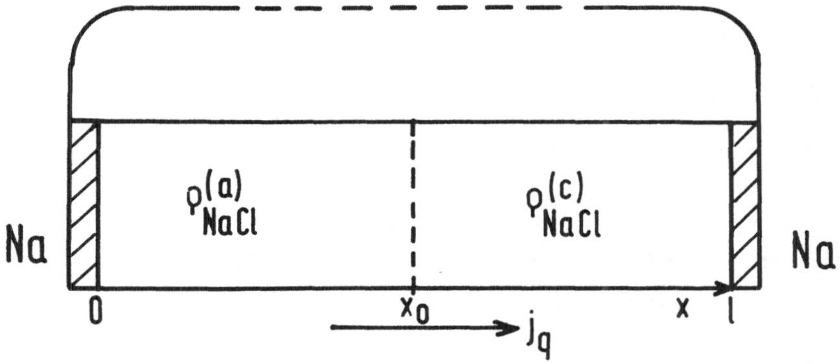

Fig. 32 A concentration cell with transference.

Thus everything remains unchanged as compared with the previous example, only $\mu^{(c)}_{NaCl}$ has to be subtracted: The final result is

$$E = \frac{RT}{\mathcal{F}} \ln \frac{a^{(a)}_{NaCl}}{a^{(c)}_{NaCl}} + E_{diff}$$

with E_{diff} as given by eq.(328a).

If we make the approximation t_{Na} = const. (which a fortiori must be an approximation) we get:

$$E = \frac{RT}{\mathcal{F}}(1 - t_{Na}) \ln \frac{a^{(a)}_{NaCl}}{a^{(c)}_{NaCl}}$$

In fact, this is a well-known formula. a_{NaCl} , the activity of the solute is usually written as the square of the mean ionic activity

$$a_{NaCl} = (a^{\pm}_{NaCl})^2$$

Chapter 10

The galvanic cell with a redox electrode.

10.1 Consideration of a special case.

We consider a system as depicted in Fig.33.

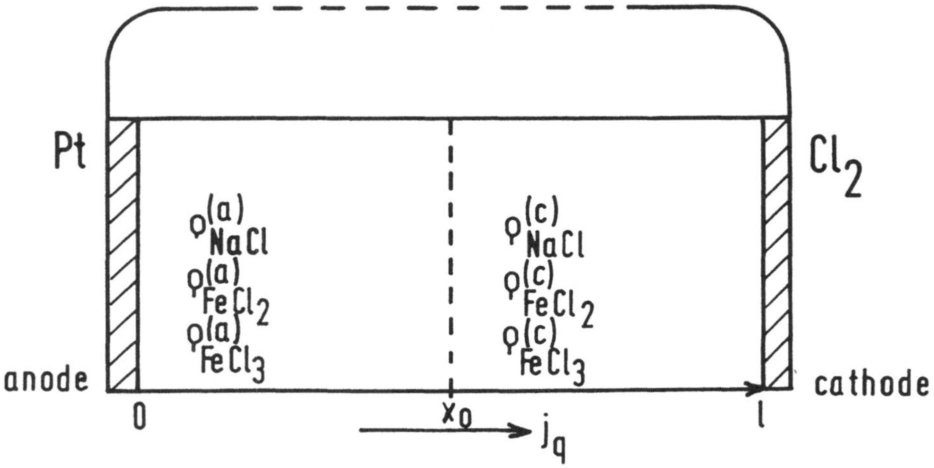

Fig.33 Schematic representation of a galvanic cell involving a redox electrode.

The chemical reaction occurring in this cell is

$$FeCl_2 + 1/2\ Cl_2 \longrightarrow FeCl_3$$

The first step in the treatment is to decide in which property space the system should be described. As an example and as an illustration of the general procedure adopted in this book, we begin using the total solute + two constituents representation. We choose Fe(II) and Fe(III) to be the two constituents. In order to derive suitable transformation relations we first consider a uniform system which is constructed from the components $FeCl_2$, $FeCl_3$, NaCl, and H_2O. The respective masses are $\mathfrak{m}_{FeCl_2}$ $\mathfrak{m}_{FeCl_3}$, $\mathfrak{m}_{NaCl}$, and $\mathfrak{m}_{H_2O}$. We wish to calculate the NaCl mass as a function of the total solute mass and the two independent constituent masses which we denote by m_2 and m_3, respectively.

We start from the three constituent masses $\mathfrak{m}_2, \mathfrak{m}_3, \mathfrak{m}_{Na}$; we have :

$$m_2 = m_{FeCl_2} \frac{M_{Fe}}{M_2}$$

$$m_3 = m_{FeCl_3} \frac{M_{Fe}}{M_3} \quad , \quad m_{Na} = m_{NaCl} \frac{M_{Na}}{M_{NaCl}}$$

with

$$M_2 = M_{FeCl_2} \quad , \quad M_3 = M_{FeCl_3}$$

Then of course we also have

$$\begin{aligned} m_{FeCl_2} &= m_2 \frac{M_2}{M_{Fe}} \\ m_{FeCl_3} &= m_3 \frac{M_3}{M_{Fe}} \\ m_{NaCl} &= m_{Na} \frac{M_{NaCl}}{M_{Na}} \end{aligned} \tag{329}$$

These are the three construction parameters m_{FeCl_2}, m_{FeCl_3}, and m_{NaCl} as a function of the three constituent masses. Now the total solute mass is:

$$m_s = m_{FeCl_2} + m_{FeCl_3} + m_{NaCl}$$

which gives

$$m_s = m_2 \frac{M_2}{M_{Fe}} + m_3 \frac{M_3}{M_{Fe}} + m_{Na} \frac{M_{NaCl}}{M_{Na}}$$

$$m_{Na} \frac{M_{NaCl}}{M_{Na}} = m_s - m_2 \frac{M_2}{M_{Fe}} - m_3 \frac{M_3}{M_{Fe}}$$

and consequently:

$$\begin{aligned} m_{NaCl} &= m_s - m_2 \frac{M_2}{M_{Fe}} - m_3 \frac{M_3}{M_{Fe}} \\ m_{FeCl_2} &= m_2 \frac{M_2}{M_{Fe}} \\ m_{FeCl_3} &= m_3 \frac{M_3}{M_{Fe}} \end{aligned} \tag{330}$$

When divided by the volume these are the transformation relations from the ρ_s, ρ_2, ρ_3 space to the construction space $\rho^*_{FeCl_2}$, $\rho^*_{FeCl_3}$, ρ^*_{NaCl}

The Gibbs free energy is:

$$G = m_{FeCl_2} \mu^*_{FeCl_2} + m_{FeCl_3} \mu^*_{FeCl_3} + m_{NaCl} \mu^*_{NaCl}$$

$$= m_{FeCl_2} \mu^*_{FeCl_2} + m_{FeCl_3} \mu^*_{FeCl_3} + \left(m_s - m_2 \frac{M_2}{M_{Fe}} - m_3 \frac{M_3}{M_{Fe}}\right) \mu^*_{NaCl}$$

$$= m_2 \frac{M_2}{M_{Fe}} (\mu^*_{FeCl_2} - \mu^*_{NaCl}) + m_3 \frac{M_3}{M_{Fe}} (\mu^*_{FeCl_3} - \mu^*_{NaCl}) + \mu^*_{NaCl}\, m_s$$

thus it follows:

$$G = m_2 \mu_2^* + m_3 \mu_3^* + m_s \mu_s^* \tag{331}$$

with

$$\begin{aligned} \mu_2^* &= \frac{M_2}{M_{Fe}} (\mu^*_{FeCl_2} - \mu^*_{NaCl}) \\ \mu_3^* &= \frac{M_3}{M_{Fe}} (\mu^*_{FeCl_3} - \mu^*_{NaCl}) \\ \mu_s^* &= \mu^*_{NaCl} \end{aligned} \tag{332}$$

After these formal preparations we turn to the system depicted schematically in Fig.32. We first calculate the change of the Gibbs-free energy at the electrodes.

a) anode:

In general we have

$$\frac{(dG)^{(a)}_{ef}}{A} = \mu_s^{*(a)} dm_s^{(a)} + \mu_2^{*(a)} dm_2^{(a)} + \mu_3^{*(a)} dm_3^{(a)}$$

Using the fundamental law of electrochemistry we find after the application of eqs.(175)-(177) and (332)

$$\frac{(dG)^{(a)}_{ef}}{A} = -\Big\{ \mu^{*(a)}_{NaCl} \Big(M_{Fe} \big[\tfrac{1}{2} t_2^{(a)} + \tfrac{1}{3} t_3^{(a)}\big] + M_{Na} t_{Na}^{(a)} - M_{Cl} t_{Cl}^{(a)} \Big) + (1 + \tfrac{1}{2} t_2^{(a)}) M_{Fe} \mu_2^* - (1 - \tfrac{1}{3} t_3^{(a)}) M_{Fe} \mu_3^* \Big\} \frac{j_e}{F} dt$$

$$\begin{aligned} = -\Big\{ & \mu^{*(a)}_{NaCl} M_{Fe} \tfrac{1}{2} t_2^{(a)} + \mu^{*(a)}_{NaCl} \tfrac{1}{3} M_{Fe} t_3^{(a)} + \mu^{*(a)}_{NaCl} M_{Na} t_{Na}^{(a)} - \mu^{*(a)}_{NaCl} M_{Cl} t_{Cl}^{(a)} \\ & + M_2 \mu^{*(a)}_{FeCl_2} - M_2 \mu^{*(a)}_{NaCl} + \tfrac{1}{2} t_2^{(a)} M_2 \mu^{*(a)}_{FeCl_2} - \tfrac{1}{2} t_2^{(a)} M_2 \mu^{*(a)}_{NaCl} \\ & - M_3 \mu^{*(a)}_{FeCl_2} + M_3 \mu^{*(a)}_{NaCl} + \tfrac{1}{3} t_3^{(a)} M_3 \mu^{*(a)}_{FeCl_3} - \tfrac{1}{3} t_3^{(a)} M_3 \mu^{*(a)}_{NaCl} \Big\} \frac{j_e\, dt}{F} \end{aligned}$$

$$\begin{aligned} = -\Big\{ & M_2 \mu^{*(a)}_{FeCl_2} (1 + \tfrac{1}{2} t_2^{(a)}) - M_3 \mu^{*(a)}_{FeCl_3} (1 - \tfrac{1}{3} t_3^{(a)}) \\ & - M_{Cl} \mu^{*(a)}_{NaCl} (t_2^{(a)} + t_3^{(a)} + t_{Cl}^{(a)} - 1) + \mu^{*(a)}_{NaCl} M_{Na} t_{Na}^{(a)} \Big\} \frac{j_e\, dt}{F} \end{aligned}$$

$$= -\Big\{ M_2 \mu^{*(a)}_{FeCl_2} \big(1 + \tfrac{1}{2} t_2^{(a)}\big) - M_3 \mu^{*(a)}_{FeCl_3} \big(1 - \tfrac{1}{3} t_3^{(a)}\big)$$
$$+ M_{Cl} \mu^{*(a)}_{NaCl} t_{Na}^{(a)} + M_{Na} \mu^{*(a)}_{NaCl} t_{Na}^{(a)} \Big\} \frac{j_2}{\mathcal{F}}$$

Thus our final result is:

$$\frac{(dG)_{ef}^{(a)}}{A} = -\Big\{ \mu^{(a)}_{FeCl_2} \big(1 + \tfrac{1}{2} t_2^{(a)}\big) - \mu^{(a)}_{FeCl_3} \big(1 - \tfrac{1}{3} t_3^{(a)}\big) + \mu^{(a)}_{NaCl} t_{Na}^{(a)} \Big\} \frac{j_2}{\mathcal{F}} dt \tag{334}$$

b) cathode:

For the computation of the free energy change connected with the cathode fluxes we use eqs.(186)-(190)

$$\frac{(dG)_{ef}^{(c)}}{A} = \mu_s^{*(c)} dm_s^{(c)} + \mu_2^{*(c)} dm_2^{(c)} + \mu_3^{*(c)} dm_3^{(c)}$$

$$= \Big\{ \mu^{*(c)}_{NaCl} M_{Fe} \tfrac{1}{2} t_2^{(c)} + \mu^{*(c)}_{NaCl} M_{Fe} \tfrac{1}{3} t_3^{(c)} + \mu^{*(c)}_{NaCl} M_{Na} t_{Na}^{(c)} + M_{Cl} \mu^{*(c)}_{NaCl}$$
$$- M_{Cl} \mu^{*(c)}_{NaCl} t_{Cl}^{(c)} + \mu_2^{*(c)} M_{Fe} \tfrac{1}{2} t_2^{(c)} + \mu_3^{*(c)} M_{Fe} \tfrac{1}{3} t_3^{(c)} \Big\} \frac{j_2}{\mathcal{F}} dt$$

$$= \Big\{ \mu^{*}_{NaCl} M_{Fe} \tfrac{1}{2} t_2^{(c)} + \mu^{*(c)}_{NaCl} M_{Fe} \tfrac{1}{3} t_3^{(c)} + \mu^{*(c)}_{NaCl} M_{Na} t_{Na}^{(c)}$$
$$+ \mu^{*(c)}_{NaCl} M_{Cl} - \mu^{*(c)}_{NaCl} M_{Cl} t_{Cl}^{(c)} + \mu^{*(c)}_{FeCl_2} M_2 \tfrac{1}{2} t_2^{(c)} - \mu^{*(c)}_{NaCl} M_2 \tfrac{1}{2} t_2^{(c)}$$
$$+ \mu^{*(c)}_{FeCl_3} M_3 \tfrac{1}{3} t_3^{(c)} - \mu^{*(c)}_{NaCl} M_3 \tfrac{1}{3} t_3^{(c)} \Big\} \frac{j_2}{\mathcal{F}} dt$$

which after some rearrangements yields:

$$\frac{(dG)_{ef}^{(c)}}{A} = \Big\{ \frac{1}{2} \mu^{(c)}_{FeCl_2} t_2^{(c)} + \frac{1}{3} \mu^{(c)}_{FeCl_3} t_3^{(c)} + \mu^{(c)}_{NaCl} t_{Na}^{(c)} \Big\} \frac{j_2}{\mathcal{F}} dt \tag{335}$$

Next we consider the contribution coming from the bulk of the solution.

The local excess increase of solute mass density per unit time is (see eq.(171) and subtract the term $j_2/\mathcal{F} \cdot M_{Cl}/\partial x$)

$$\frac{\partial \rho_s'}{\partial t} = -\frac{j_2}{\mathcal{F}} \Big[\frac{1}{2} (2M_{Cl} + M_{Fe}) \frac{dt_2}{dx} + \frac{1}{3} (3M_{Cl} + M_{Fe}) \frac{dt_3}{dx} + (M_{Cl} + M_{Na}) \frac{dt_{Na}}{dx} \Big] \tag{336}$$

and correspondingly, the excess local increases of the two constituent densities are:

$$\frac{\partial \varrho_2'}{\partial t} = -\frac{j_q}{2\mathcal{F}} M_{Fe} \frac{dt_2}{dx} \;; \quad \frac{\partial \varrho_3'}{\partial t} = -\frac{j_q}{3\mathcal{F}} M_{Fe} \frac{dt_3}{dx}$$

This gives then the total bulk increase of Gibbs-free energy:

$$\begin{aligned}\frac{(dG)_s}{A} &= -\Big[\int_0^l \mu_s^* \Big(\tfrac{1}{2}(2M_{Cl}+M_{Fe})\frac{dt_2}{dx} + \tfrac{1}{3}(3M_{Cl}+M_{Fe})\frac{dt_3}{dx} \\ &\quad + (M_{Na}+M_{Cl})\frac{dt_{Na}}{dx}\Big)dx + \frac{1}{2}\int_0^l \mu_2^* M_{Fe}\frac{dt_2}{dx}dx + \frac{1}{3}\int_0^l \mu_3^* M_{Fe}\frac{dt_3}{dx}dx\Big]\frac{j_q dt}{\mathcal{F}} \\ &= -\Big[\frac{1}{2}\int_0^l \mu_{FeCl_2}\frac{dt_2}{dx}dx + \frac{1}{3}\int_0^l \mu_{FeCl_3}\frac{dt_3}{dx}dx + \int_0^l \mu_{NaCl}\frac{dt_{Na}}{dx}dx\Big]\frac{j_q}{\mathcal{F}}dt\end{aligned}$$

To arrive at the last equation, first the chemical potentials μ_s^*, μ_2^* μ_3^* have been replaced by μ_{NaCl}^*, $\mu_{FeCl_2}^*$, and $\mu_{FeCl_3}^*$ (see eq.(332)) then the specific chemical potentials have been replaced by the corresponding molar quantities.

Now as before we set in the interval $x_0 \leq x \leq x_0 + \delta x$

$$\begin{aligned} t_2 &= t_2^{(a)} + \frac{\Delta t_2}{\delta x}(x - x_0) \\ t_3 &= t_3^{(a)} + \frac{\Delta t_3}{\delta x}(x - x_0) \\ t_{Na} &= t_{Na}^{(a)} + \frac{\Delta t_{Na}}{\delta x}(x - x_0) \end{aligned}$$

and with these equations we obtain:

$$\begin{aligned}\frac{(dG)_s}{A} &= -\Big[\frac{1}{2}\Delta t_2 \mu_{FeCl_2}^0 + \frac{1}{2}\frac{RT}{\delta x}\Delta t_2 \int_{x_0}^{x_0+\delta x} \ln a_{FeCl_2} dx \\ &\quad + \frac{1}{3}\Delta t_3 \mu_{FeCl_3}^0 + \frac{1}{3}\frac{RT}{\delta x}\Delta t_3 \int_{x_0}^{x_0+\delta x} \ln a_{FeCl_3} dx \\ &\quad + \Delta t_{Na}\mu_{NaCl}^0 + \frac{RT}{\delta x}\Delta t_{Na}\int_{x_0}^{x_0+\delta x} \ln a_{NaCl} dx\Big]\frac{j_q}{\mathcal{F}}dt\end{aligned}$$

Also for the activities we introduce the linear approximations. This gives the final result:

$$\frac{(dG)_s}{A} = \Big\{ -\frac{1}{2}\Delta t_2 \mu^o_{FeCl_2} - \frac{1}{3}\Delta t_3 \mu^o_{FeCl_3} - \Delta t_{Na} \mu^o_{NaCl}$$

$$-\frac{1}{2}\Delta t_2 RT \left[\frac{a_2^{(c)} \ln a_2^{(c)} - a_2^{(a)} \ln a_2^{(a)}}{\Delta a_2} - 1 \right]$$

$$-\frac{1}{3}\Delta t_3 RT \left[\frac{a_3^{(c)} \ln a_3^{(c)} - a_3^{(a)} \ln a_3^{(a)}}{\Delta a_3} - 1 \right]$$

$$- \Delta t_{Na} RT \left[\frac{a_{NaCl}^{(c)} \ln a_{NaCl}^{(c)} - a_{NaCl}^{(a)} \ln a_{NaCl}^{(a)}}{\Delta a_{Na}} - 1 \right] \Big\} \frac{j_q dt}{\mathcal{F}} \quad (337)$$

with

$$a_2^{(i)} = a_{FeCl_2}^{(i)} \; ; \quad a_3^{(i)} = a_{FeCl_3}^{(i)} \; , \quad \Delta a_j = a_j^{(c)} - a_j^{(a)}$$

In order to get the total increase of Gibbs-free energy we have to add the three contributions eqs.(334), (335) and (337). We find after some algebraic manipulations that

$$\frac{(dG)_{ef+s}}{A} = \Big\{ \mu_{FeCl_2}^{(a)} - \mu_{FeCl_2}^{(a)} + \frac{1}{3} RT t_3^{(c)} \ln a_{FeCl_2}^{(c)} - \frac{1}{3} RT t_3^{(a)} \ln a_{FeCl_3}^{(a)}$$

$$- \frac{1}{3} \Delta t_3 RT [\qquad] + \frac{1}{2} RT t_2^{(c)} \ln a_{FeCl_2}^{(c)} - \frac{1}{2} RT t_2^{(a)} \ln a_{FeCl_2}^{(a)}$$

$$- \frac{1}{2} \Delta t_2 RT [\qquad] + RT t_{Na}^{(c)} \ln a_{NaCl}^{(c)} - RT t_{Na}^{(a)} \ln a^{(a)} - \Delta t_{Na}$$

$$\cdot RT [\qquad] \Big\} \frac{j_q}{\mathcal{F}} dt$$

with the parentheses [ ] as given in eq.(337).
Then our final result is:

$$E = E_o + \frac{RT}{\mathcal{F}} \ln \frac{a_{FeCl_3}^{(a)}}{a_{FeCl_2}^{(a)}} + E_{diff} \quad (338)$$

with

$$E_o = \frac{1}{\mathcal{F}} \left(\mu^o_{FeCl_3} - \mu^o_{FeCl_2} - \frac{1}{2} \mu_{Cl_2} \right)$$

and

$$E_{diff} = \frac{RT}{\mathcal{F}} \left\{ \frac{t_3^{(c)}}{3} \left[\frac{a_3^{(a)} \ln a_3^{(a)}/a_3^{(c)}}{a_3^{(c)} - a_3^{(a)}} + 1 \right] - \frac{t_3^{(a)}}{3} \left[\frac{a_3^{(c)} \ln a_3^{(a)}/a_3^{(c)}}{a_3^{(c)} - a_3^{(a)}} + 1 \right] \right.$$

$$+ \frac{t_2^{(c)}}{2} \left[\frac{a_2^{(a)} \ln a_2^{(a)}/a_2^{(c)}}{a_2^{(c)} - a_2^{(a)}} + 1 \right] - \frac{t_2^{(a)}}{2} \left[\frac{a_2^{(c)} \ln a_2^{(a)}/a_2^{(c)}}{a_2^{(c)} - a_2^{(a)}} + 1 \right]$$

$$+ t_{Na}^{(c)} \left[\frac{a_{NaCl}^{(a)} \ln a_{NaCl}^{(a)}/a_{NaCl}^{(c)}}{a_{NaCl}^{(c)} - a_{NaCl}^{(a)}} + 1 \right]$$

$$\left. - t_{Na}^{(a)} \left[\frac{a_{NaCl}^{(c)} \ln a_{NaCl}^{(a)}/a_{NaCl}^{(c)}}{a_{NaCl}^{(c)} - a_{NaCl}^{(a)}} + 1 \right] \right\} \qquad (338a)$$

10.2 Modification of the galvanic cell involving a redox reaction

(a) The Cl_2 electrode is replaced by an Ag/AgCl electrode.
In this cell configuration the chemical reaction is:

$$FeCl_2 + AgCl \longrightarrow FeCl_3 + Ag \ .$$

Now the formulas (338)-(338a) remain the same except that the standard potential has changed

$$E_0 = \frac{1}{\mathcal{F}} (\mu^0_{FeCl_3} + \mu_{Ag} - \mu^0_{FeCl_2} - \mu_{AgCl})$$

(b) The cathode is a Na electrode instead of the Cl_2 electrode.
For this cell construction the reaction is:

$$FeCl_2 + NaCl \longrightarrow FeCl_3 + Na \ .$$

Now in eq.(186) we have only to replace t_{Na} by $-(1 - t_{Na})$ and $(1 - t_{Cl})$ by $-t_{Cl}$.
This gives for the change of Gibbs free energy due to the mass flux at the electrode (see derivation of eq.(335))

$$\frac{(dG)_{ef}^{(c)}}{A} = \Big\{ \mu^{*(c)}_{NaCl} M_{Fe} \frac{1}{2} t_2^{(c)} + \mu^{*(c)}_{NaCl} M_{Fe} \frac{1}{3} t_3^{(c)} - \mu^{*(c)}_{NaCl} M_{Na}$$

$$+ \mu^{*(c)}_{NaCl} M_{Na} t_{Na}^{(c)} - \mu^{*(c)}_{NaCl} M_{Cl} t_{Cl}^{(c)} + \mu_2^{*(c)} M_{Fe} \frac{t_2^{(c)}}{2}$$

$$+ \mu_3^{*(c)} M_{Fe} \frac{t_3^{(c)}}{3} \Big\} \frac{j_g}{F} dt$$

$$= \Big\{ \frac{1}{2} \mu^{(c)}_{FeCl_2} t_2^{(c)} + \frac{1}{3} \mu^{(c)}_{FeCl_3} t_3^{(c)} - \mu^{(c)}_{NaCl} (1 - t_{Na}^{(c)}) \Big\} \frac{j_g}{F} dt \qquad (339)$$

We see that eqs.(338) and (338a) remain unchanged, apart from the subtraction of $\mu^{(c)}_{NaCl}$.
Thus the final result is

$$E = E_o + \frac{RT}{F} \ln \frac{a^{(a)}_{FeCl_3}}{a^{(a)}_{FeCl_2} \, a^{(c)}_{NaCl}} + E_{diff}$$

with

$$E_o = \frac{1}{F} \{ \mu^o_{FeCl_3} + \mu_{Na} - \mu^o_{FeCl_2} - \mu^o_{NaCl} \}$$

where E_{diff} is given by eq.(338a).

(c) Addition of HCl to the system, and the cathode is a hydrogen electrode.

Now we have the chemical reaction

$$FeCl_2 + HCl \longrightarrow FeCl_3 + 1/2 \; H_2 \; .$$

The number of variables characterizing the electrolyte system has increased by one. We have to add the equation

$$m_{HCl} = m_H \frac{M_{HCl}}{M_H}$$

as a fourth equation to the set of eqs.(329).

At the anode the situation essentially remains as before only we have to add a term describing the removal of HCl from the solution:

$$\frac{(dG)_{ef}^{(a)}}{A} = - \Big\{ \mu^{(a)}_{FeCl_2} (1 + \tfrac{1}{2} t_2^{(a)}) - \mu^{(a)}_{FeCl_3} (1 - \tfrac{1}{3} t_3^{(a)}) + \mu^{(a)}_{NaCl} t_{Na}^{(a)} + \mu^{(a)}_{HCl} t_H^{(a)} \Big\} \frac{j_g \, dt}{F}$$

Correspondingly, in eq.(339) we have to add a term $-(1 - t_H^{(c)}) \mu^{(c)}_{HCl}$ and to remove a term $-\mu^{(c)}_{NaCl}$.

Thus the result is :

$$\frac{(dG)_{ef}^{(c)}}{A} = \left\{\frac{1}{2}\mu_{FeCl_2}^{(c)} t_2^{(c)} + \frac{1}{3}\mu_{FeCl_3}^{(c)} t_3^{(c)} + \mu_{NaCl}^{(c)} t_{Na}^{(c)} - \mu_{HCl}^{(c)}(1 - t_H^{(c)})\right\} \frac{j_e\,dt}{\mathcal{F}}$$

Then, considering that for the bulk electrolyte we have only to add a term

$$\int_0^{\ell} \mu_{HCl}^{*} M_{HCl} \frac{dt_H}{dx} dx$$

to the formula on page 200 . We arrive at the final result

$$E = E_0 + \frac{RT}{\mathcal{F}} \ln \frac{a_{FeCl_3}^{(a)}}{a_{FeCl_2}^{(a)}\, a_{HCl}^{(c)}} + E_{diff}$$

with

$$E_0 = \frac{1}{\mathcal{F}}\left(\mu_{FeCl_3}^{0} + \frac{1}{2}\mu_{H_2} - \mu_{FeCl_2}^{0} - \mu_{HCl}^{0}\right)$$

and

$$E_{diff} = \frac{RT}{\mathcal{F}}\left\{ \frac{t_3^{(c)}}{3}\left[\frac{a_3^{(a)}\cdots}{\cdots} + 1\right] - \frac{t_3^{(a)}}{3}\left[\frac{a_3^{(c)}\cdots}{\cdots} + 1\right] + \frac{t_2^{(c)}}{2}\left[\frac{a_2^{(a)}\cdots}{\cdots} + 1\right]\right.$$

$$- \frac{t_2^{(a)}}{2}\left[\frac{a_2^{(c)}\cdots}{\cdots} + 1\right] + t_{Na}^{(c)}\left[\frac{a_{NaCl}^{(a)}\cdots}{\cdots} + 1\right] - t_{Na}^{(a)}\left[\frac{a_{NaCl}^{(c)}\cdots}{\cdots} + 1\right]$$

$$\left. + t_H^{(c)}\left[\frac{a_{HCl}^{(a)}\cdots}{\cdots} + 1\right] - t_H^{(a)}\left[\frac{a_{HCl}^{(c)}\cdots}{\cdots} + 1\right]\right\}$$

(d) In the cathode compartment we have a HNO_3 solution; the cathode itself we choose to be a hydrogen electrode.

Now the total system comprising the anode and cathode compartment has 5 independent constituents: Fe(II); Fe(III); Na; NO_3; H.

<u>The construction of the system</u>: We consider the system to be constructed from the five components:

$FeCl_2$, $FeCl_3$, NaCl, $NaNO_3$, and HNO_3,

the respective masses are :

m_{FeCl_2}; m_{FeCl_3}, m_{NaCl}, m_{NaNO_3}, and m_{HNO_3}

Then we have the three chemical reactions :

1.) $HNO_3 + NaCl \longrightarrow NaNO_3 + HCl$

2.) $FeCl_3 + HNO_3 \longrightarrow FeCl_2NO_3 + HCl$

$+ 2\ HNO_3 \longrightarrow FeCl(NO_3)_2 + 2\ HCl$

$+ 3\ HNO_3 \longrightarrow Fe(NO_3)_3 + 3\ HCl$

3.) $FeCl_2 + HNO_3 \longrightarrow FeClNO_3 + HCl$

$+ 2\, HNO_3 \longrightarrow Fe(NO_3)_2 + 2\, HCl$

All the species occurring here are possible construction components; they could also be chosen if we so wished.

The transformation relations from the component construction space to the constituent space are given by

$$m_2 = m_{FeCl_2}\frac{M_{Fe}}{M_2}\,, \quad m_3 = m_{FeCl_3}\frac{M_{Fe}}{M_3}\,; \qquad m_H = m_{HNO_3}\frac{M_H}{M_{HNO_3}}$$

$$m_{Na} = m_{NaNO_3}\frac{M_{Na}}{M_{NaNO_3}} + m_{NaCl}\frac{M_{Na}}{M_{NaCl}}$$

$$m_{NO_3} = m_{HNO_3}\frac{M_{NO_3}}{M_{HNO_3}} + m_{NaNO_3}\frac{M_{NO_3}}{M_{NaNO_3}}$$

These relations are reversed to give the construction component masses in terms of the constituent masses:

$$m_{HNO_3} = M_{HNO_3}\frac{m_H}{M_H}$$

$$m_{NaNO_3} = M_{NaNO_3}\left(\frac{m_{NO_3}}{M_{NO_3}} - \frac{m_H}{M_H}\right)$$

$$m_{NaCl} = M_{NaCl}\left(\frac{m_{Na}}{M_{Na}} + \frac{m_H}{M_H} - \frac{m_{NO_3}}{M_{NO_3}}\right)$$

$$m_{FeCl_2} = m_2\frac{M_2}{M_{Fe}}$$

$$m_{FeCl_3} = m_3\frac{M_3}{M_{Fe}}$$

The condition for the representation of a given solution in terms of this set of construction parameters is that the second and third equations yield $m_{NaNO_3} \geqslant 0$ and $m_{NaCl} \geqslant 0$.

The Gibbs free energy of this system is:

$$G = m_{FeCl_2}\mu^*_{FeCl_2} + m_{FeCl_3}\mu^*_{FeCl_3} + m_{NaCl}\mu^*_{NaCl} + m_{HNO_3}\mu^*_{HNO_3} + m_{NaNO_3}\mu^*_{NaNO_3}$$

This quantity then can be given in the 5 constituents representation, or the ρ_s, Fe(II), Fe(III), H, NO_3, representations may also be chosen. As an example, we give the corresponding result:

$$G = m_s\mu_s^* + m_H\mu_H^* + m_{NO_3}\mu_{NO_3}^* + m_2\mu_2^* + m_3\mu_3^*$$

with

$$\begin{aligned}
\mu_s^* &= \mu_{NaCl}^* \\
\mu_{NO_3}^* &= (\mu_{NaNO_3}^* - \mu_{NaCl}^*)\frac{M_{NaNO_3}}{M_{NO_3}} \\
\mu_H^* &= \left[(\mu_{HNO_3}^* - \mu_{NaCl}^*)M_{HNO_3} - (\mu_{NaNO_3}^* - \mu_{NaCl}^*)M_{NaNO_3}\right]\frac{1}{M_H} \\
\mu_2^* &= (\mu_{FeCl_2}^* - \mu_{NaCl}^*)\frac{M_2}{M_{Fe}} \\
\mu_3^* &= (\mu_{FeCl_3}^* - \mu_{NaCl}^*)\frac{M_3}{M_{Fe}}
\end{aligned} \tag{340}$$

The free energy change contributions at the electrodes are:

(1) Anode: platinum electrode.

$$\frac{(dG)_{ef}^{(a)}}{A} = \mu_s^* dm_s + \mu_2^* dm_2 + \mu_3^* dm_3 + \mu_{NO_3}^* dm_{NO_3} + \mu_H^* dm_H$$

where we have omitted the superscript $^{(a)}$ for simplicity. When we introduce the constituent transport numbers, use eq.(340) and add the new terms to eq.(333), we obtain the result:

$$\frac{(dG)_{ef}^{(a)}}{A} = -\Big\{\mu_{FeCl_2}^{(a)}(1 + \tfrac{1}{2}t_2^{(a)}) - \mu_{FeCl_3}^{(a)}(1 - \tfrac{1}{3}t_3^{(a)}) + \mu_{NaCl}^{(a)}(t_{Na}^{(a)} + t_H^{(a)} + t_{NO_3}^{(a)}) + t_H^{(a)}\mu_{HNO_3}^{(a)} - (t_H^{(a)} + t_{NO_3}^{(a)})\mu_{NaNO_3}\Big\}\frac{j}{\mathcal{F}}dt \tag{341}$$

Note that $1 = t_2 + t_3 + t_{Na} + t_H + t_{NO_3} + t_{Cl}$.

(2) Cathode:

Correspondingly, at the cathode we have:

$$\frac{(dG)_{ef}^{(c)}}{A} = \mu_s^* dm_s + \mu_2^* dm_2 + \mu_3^* dm_3 + \mu_H^* dm_H + \mu_{NO_3}^* dm_{NO_3}$$

which after the necessary substitutions and rearrangements gives

$$\frac{(dG)_{ef}^{(c)}}{A} = \Big\{\frac{1}{2}\mu_{FeCl_2}^{(c)} t_2^{(c)} + \frac{1}{3}\mu_{FeCl_2}^{(c)} t_3^{(c)} - \mu_{NaCl}^{(c)}\big(1-(t_{Na}^{(c)} + t_H^{(c)} + t_{NO_3})\big)$$

$$-\mu_{HNO_3}^{(c)}(1-t_H^{(c)}) + \mu_{NaNO_3}^{(c)}\big(1-(t_H^{(c)} + t_{NO_3}^{(c)})\big)\Big\}\frac{jq}{\mathcal{F}}dt \qquad (342)$$

(3) The electrolyte solution.

We start from eq.(336) to which we have to add the two terms involved with the consitutents NO_3 and H. It follows:

$$\frac{\partial \rho_s'}{\partial t} = -\frac{jq}{\mathcal{F}}\Big[\frac{1}{2}(2M_{Cl} + M_{Fe})\frac{dt_2}{dx} + \frac{1}{3}(3M_{Cl} + M_{Fe})\frac{dt_3}{dx} + (M_{Cl} + M_{Na})\frac{dt_{Na}}{dx}$$
$$+ (M_{Cl} + M_H)\frac{dt_H}{dx} - (M_{NO_3} - M_{Cl})\frac{dt_{NO_3}}{dx}\Big]$$

Now for the first three terms in the brackets everything is the same as derived on pages 199-200. Thus we only need to consider the two added terms.

We have

$$\frac{(dG)_s}{A} = -\Big\{\ldots\ldots + \int_0^\ell \mu_s^*\Big((M_{Cl} + M_H)\frac{dt_H}{dx} - (M_{NO_3} - M_{Cl})\frac{dt_{NO_3}}{dx}\Big)dx$$
$$+ \int_0^\ell \mu_H^* M_H \frac{dt_H}{dx}dx - \int_0^\ell \mu_{NO_3}^* M_{NO_3}\frac{dt_{NO_3}}{dx}dx\Big\}\frac{jq}{\mathcal{F}}dt$$

$$= -\frac{jq\,dt}{\mathcal{F}}\Big\{\ldots + \int_0^\ell \mu_{NaCl}\Big(\frac{dt_H}{dx} + \frac{dt_{NO_3}}{dx}\Big)dx + \int_0^\ell \mu_{HNO_3}\frac{dt_H}{dx} - \int_0^\ell \mu_{NaNO_3}\Big(\frac{dt_H}{dx} + \frac{dt_{NO_3}}{dx}\Big)dx\Big\}$$

Taking this together with the results from pages 199-200 we obtain:

$$\frac{(dG)_s}{A} = -\Big\{\frac{1}{2}\int_0^\ell \mu_{FeCl_2}\frac{dt_2}{dx}dx + \frac{1}{3}\int_0^\ell \mu_{FeCl_3}\frac{dt_3}{dx}dx$$
$$+ \int_0^\ell \mu_{NaCl}\Big(\frac{dt_{Na}}{dx} + \frac{dt_H}{dx} + \frac{dt_{NO_3}}{dx}\Big)dx + \int_0^\ell \mu_{HNO_3}\frac{dt_H}{dx}dx$$
$$- \int_0^\ell \mu_{NaNO_3}\Big(\frac{dt_H}{dx} + \frac{dt_{NO_3}}{dx}\Big)dx\Big\}\frac{jq}{\mathcal{F}}dt \qquad (343)$$

Addition of eqs.(341), (342), and (343) yields the total change of Gibbs free energy, apart from the production of hydrogen. We consider two different cases:

α) The "general case" where all constitutents occur with finite concentrations everywhere in the system. In this case, by sorting out the terms without a transport number from those with one and applying the linear x dependence approximation for the t_i's and a_i's one finds:

$$\frac{dG}{A} = \frac{\dot{q}}{\mathcal{F}} dt \left\{ \mu_{FeCl_3}^{(a)} - \mu_{FeCl_2}^{(a)} - \mu_{HNO_3}^{(c)} + \mu_{NaNO_3}^{(c)} - \mu_{NaCl}^{(c)} + \frac{1}{2}\mu_{H_2} + \mathcal{F} E_{diff} \right\}$$

i.e.

$$E = E_0 + \frac{RT}{\mathcal{F}} \ln \frac{a_{FeCl_3}^{(a)}}{a_{FeCl_2}^{(a)} a_{HNO_3}^{(c)}} + \frac{RT}{\mathcal{F}} \ln \frac{a_{NaNO_3}^{(c)}}{a_{NaCl}^{(c)}} + E_{diff}$$

with

$$E_0 = \frac{1}{\mathcal{F}} \left[\mu_{FeCl_3}^{0} + \frac{1}{2}\mu_{H_2} + \mu_{NaNO_3}^{0} - \mu_{FeCl_2}^{0} - \mu_{HNO_3}^{0} - \mu_{NaCl}^{0} \right]$$

and

$$E_{diff} = \frac{RT}{\mathcal{F}} \Big[\frac{1}{3} t_3^{(c)} \left(a_{FeCl_3}^{(a)} \cdots \right) - \frac{1}{3} t_3^{(a)} \left(a_{FeCl_2}^{(c)} \cdots \right) + \frac{1}{2} t_2^{(c)} \cdot \left(a_{FeCl_2}^{(a)} \cdots \right) - \frac{1}{2} t_2^{(a)} \left(a_{FeCl_2}^{(c)} \cdots \right) + \left(t_{Na}^{(c)} + t_H^{(c)} + t_{NO_3}^{(c)} \right) \left(a_{NaCl}^{(a)} \cdots \right)$$
$$- \left(t_{Na}^{(a)} + t_H^{(a)} + t_{NO_3}^{(a)} \right) \left(a_{NaCl}^{(c)} \cdots \right) - \left(t_H^{(c)} + t_{NO_3}^{(c)} \right) \left(a_{NaNO_3}^{(a)} \cdots \right) + \left(t_H^{(a)} + t_{NO_3}^{(a)} \right) \left(a_{NaNO_3}^{(c)} \cdots \right) + t_H^{(c)} \left(a_{HNO_3}^{(a)} \cdots \right) - t_H^{(a)} \left(a_{HNO_3}^{(c)} \cdots \right) \Big]$$

where the expressions in the parentheses $(a_i^{(j)} \cdots)$ have the usual structure as given in eq.(338a). The chemical reaction, expressed in the construction space, is

$$FeCl_2 + HNO_3 + NaCl \longrightarrow FeCl_3 + 1/2\ H_2 + NaNO_3.$$

β) At the anode compartment we have only $FeCl_2$, $FeCl_3$ and NaCl, thus $t_H^{(a)} = 0$, $t_{NO_3}^{(a)} = 0$, we set $t_2^{(a)} = t_2$, $t_3^{(a)} = t_3$, $t_{Na}^{(a)} = t_{Na}$.

whereas at the cathode we have only nitric acid, i.e. $t_{Na}^{(c)} = t_2^{(c)} = t_3^{(c)} = 0$.

We set $t_H^{(c)} = t_H$; $t_{NO_3}^{(c)} = t_{NO_3}$.

In this situation at the liquid junction (at the coordinate $x = x_o$) we have the following linear approximations for the transport numbers:

$$
\begin{aligned}
t_{Na}(x) &= \frac{t_{Na}}{\delta x}(x_o + \delta x - x) \quad , \quad & \frac{dt_{Na}}{dx} &= -\frac{t_{Na}}{\delta x} \\
t_2(x) &= \frac{t_2}{\delta x}(x_o + \delta x - x) \quad , \quad & \frac{dt_2}{dx} &= -\frac{t_2}{\delta x} \\
t_3(x) &= \frac{t_3}{\delta x}(x_o + \delta x - x) \quad , \quad & \frac{dt_3}{dx} &= -\frac{t_3}{\delta x} \\
t_H(x) &= \frac{t_H}{\delta x}(x - x_o) \quad , \quad & \frac{dt_H}{dx} &= \frac{t_H}{\delta x} \\
t_{NO_3}(x) &= \frac{t_{NO_3}}{\delta x}(x - x_o) \quad , \quad & \frac{dt_{NO_3}}{dx} &= \frac{t_{NO_3}}{\delta x}
\end{aligned}
$$

This gives with eq.(343)

$$
\begin{aligned}
\frac{(dG)_s}{A} = \Big[& \frac{1}{2} t_2 \mu^o_{FeCl_2} + \frac{1}{2} t_2 RT \int_{x_o}^{x_o+\delta x} \ln a_{FeCl_2} dx + \frac{1}{3} t_3 \mu^o_{FeCl_3} + \frac{1}{3} t_3 RT \cdot \\
& \cdot \int_{x_o}^{x_o+\delta x} \ln a_{FeCl_3} dx + (t_{Na} - t_H - t_{NO_3}) \mu^o_{NaCl} + \frac{t_{Na} - t_H - t_{NO_3}}{\delta x} RT \int_{x_o} \ln a_{NaCl}\, dx \\
& + (t_H + t_{NO_3}) \mu^o_{NaNO_3} + \frac{t_H + t_{NO_3}}{\delta x} RT \int_{x_o}^{x_o+\delta x} \ln a_{NaNO_3} dx \\
& - t_H \mu^o_{HNO_3} - t_H RT \int_{x_o}^{x_o+\delta x} \ln a_{HNO_3} dx \Big] \frac{J_2}{F} dt
\end{aligned}
$$

Now the activities are approximated by ($x_o \leq x \leq x_o + \delta x$)

$$
\begin{aligned}
a_{FeCl_2}(x) &= \frac{a_{FeCl_2}}{\delta x}(x_o + \delta x - x) \\
a_{FeCl_3}(x) &= \frac{a_{FeCl_3}}{\delta x}(x_o + \delta x - x) \\
a_{NaCl}(x) &= \frac{a_{NaCl}}{\delta x}(x_o + \delta x - x) \\
a_{HNO_3}(x) &= \frac{a_{HNO_3}}{\delta x}(x - x_o)
\end{aligned}
$$

$$a_{NaNO_3}(x) = \frac{a_{NaNO_3}}{\delta x/2}(x - x_0) \quad \text{for} \quad x_0 \leq x \leq x_0 + \delta x/2$$

$$a_{NaNO_3}(x) = \frac{a_{NaNO_3}}{\delta x/2}(x_0 + \delta x - x) \quad \text{for} \quad x_0 + \delta x/2 \leq x \leq x_0 + \delta x$$

Note that the distribution of the activity of $NaNO_3$ forms a "triangular function". This is the product of the diffusion process, as was explained in sections 8.2 and 8.3.

With all these activity distributions it follows:

$$\frac{(dG)_s}{A} = \Big\{ \tfrac{1}{2} t_2 \mu^{o}_{FeCl_2} + \tfrac{1}{3} t_3 \mu^{o}_{FeCl_3} + t_{Na} \mu^{o}_{NaCl} + (t_H + t_{NO_3})(\mu^{o}_{NaNO_3}$$

$$- \mu^{o}_{NaCl}) - t_H \mu^{o}_{HNO_3} + t_{Na} RT(\ln a_{NaCl} - 1)$$

$$+ \tfrac{1}{2} t_2 RT(\ln a_{FeCl_2} - 1) + \tfrac{1}{3} t_3 RT(\ln a_{FeCl_3} - 1)$$

$$- (t_H + t_{NO_3}) RT(\ln a_{NaCl} - 1)$$

$$+ (t_H + t_{NO_3}) RT(\ln a_{NaNO_3} - 1) - t_H RT(\ln a_{HNO_3} - 1)\Big\} \frac{j_q}{\mathcal{F}} dt \quad (344)$$

We have now simply to write down the two electrode contributions and then to add them to eq.(344):

$$\frac{(dG)^{(a)}_{ef}}{A} = -\Big\{ \mu_{FeCl_2}(1 + \tfrac{1}{2} t_2) - \mu_{FeCl_3}(1 - \tfrac{1}{3} t_3) + \mu_{NaCl} t_{Na} \Big\} \frac{j_q}{\mathcal{F}} dt$$

$$\frac{(dG)^{(c)}_{ef}}{A} = -\mu_{HNO_3}(1 - t_H) \frac{j_q}{\mathcal{F}} dt$$

$$= \left[-\mu_{HNO_3} + \mu_{HNO_3}(1 - t_{NO_3}) \right] \frac{j_q}{\mathcal{F}} dt$$

The addition yields:

$$\frac{(dG)_{ef+s}}{A} = \frac{j_q dt}{\mathcal{F}} \Big\{ \mu^{o}_{FeCl_3} - \mu^{o}_{FeCl_2} - \mu^{o}_{HNO_3} + RT \ln \frac{a_{FeCl_3}}{a_{FeCl_2} a_{HNO_3}} - RT(\tfrac{1}{2} t_2 + \tfrac{1}{3} t_3)$$

$$+ RT t_H - RT t_{Na} + (t_H + t_{NO_3}) \left[\mu^{o}_{NaNO_3} - \mu^{o}_{NaCl} + RT \ln \frac{a_{NaNO_3}}{a_{NaCl}} \right] \Big\}$$

Then our final result is:

$$E = E_0 + \frac{RT}{\mathcal{F}} \ln \frac{a^{(a)}_{FeCl_3}}{a^{(a)}_{FeCl_2}\, a^{(c)}_{HNO_3}} + E_{diff}$$

with

$$E_0 = \frac{1}{\mathcal{F}} \left(\mu^{o}_{FeCl_3} + \frac{1}{2}\mu_{H_2} + \mu^{o}_{NaNO_3} - \mu^{o}_{FeCl_2} - \mu^{o}_{HNO_3} - \mu^{o}_{NaCl} \right)$$

and

$$E_{diff} = \frac{RT}{\mathcal{F}} \left[\ln \frac{a_{NaNO_3}}{a^{(a)}_{NaCl}} - \left(\frac{1}{2} t_2 + \frac{1}{3} t_3 + t_{Na} - t_H \right) \right]$$

In this formula the activity a_{NaNO_3} is not directly given by the construction of the system; it is the product of the diffusion process in the boundary between the anode and cathode compartment. It should also be recalled that all the other activities occurring in this formula refer to construction space.

Chapter 11

The galvanic cell with an oxygen electrode

11.1 The anode is a Na electrode.

Let us consider the following galvanic cell (Fig.34). In the presence of an electric current the chemical reaction is:

$$Na + 1/4\ O_2 + 1/2\ H_2O \longrightarrow NaOH$$

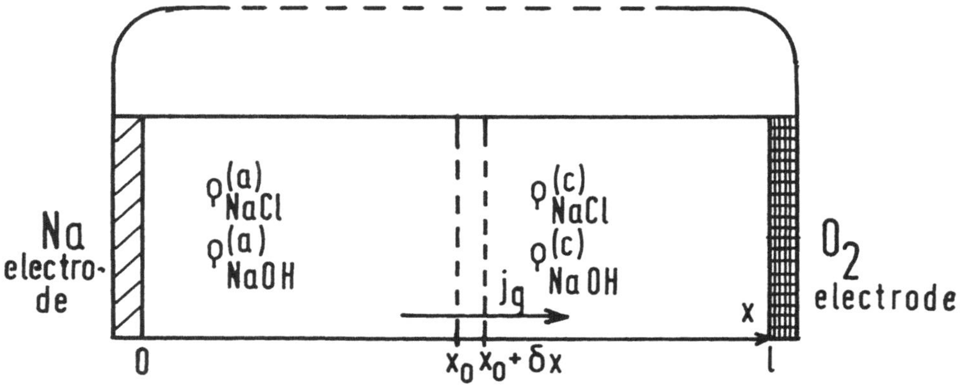

Fig. 34 Galvanic cell involving an oxygen electrode and a non-uniform electrolyte solution.

We begin with the electrode mass fluxes and choose the ρ_s, OH constituent representation.

There is only one kind of cation, Na, thus we eliminate Na by the aid of the relation:

$$c_{OH} + c_{Cl} = c_{Na}$$

The total solute density is:

$$\begin{aligned}\rho_s &= c_{OH} M_{OH} + c_{Cl} M_{Cl} + c_{Na} M_{Na} \\ &= c_{OH} M_{OH} + c_{Cl} M_{Cl} + M_{Na} (c_{OH} + c_{Cl}) \\ &= c_{OH} M_{NaOH} + c_{Cl} M_{NaCl}\end{aligned}$$

The anode is passive with respect to both the constituents OH and Cl. Thus we get the expressions for the fluxes:

$$(j_s)_{x=0} = \left(M_{NaCl}\, t_{Cl}^{(a)} + M_{NaOH}\, t_{OH}^{(a)}\right) \frac{j_q}{\mathcal{F}} dt \tag{345}$$

$$(j_{OH})_{x=0} = t_{OH} \frac{j_q}{\mathcal{F}} M_{OH} \tag{346}$$

The cathode is active with respect to the OH constituent, and there is a mass flux of H_2O towards the cathode. Thus we have:

$$(j_s)_{x=\ell} = \left(M_{NaCl}\, t_{Cl}^{(c)} - (1 - t_{OH}^{(c)}) M_{NaOH} + \frac{M_{H_2O}}{2}\right) \frac{j_q}{\mathcal{F}} \tag{347}$$

$$(j_{OH})_{x=\ell} = -(1 - t_{OH}^{(c)}) M_{OH} \frac{j_q}{\mathcal{F}} \tag{348}$$

$$(j_{H_2O})_{x=\ell} = \frac{1}{2} M_{H_2O} \frac{j_q}{\mathcal{F}} \tag{349}$$

We have now to formulate the mass and free energy relations.

The constituent masses as given by the construction masses are

$$m_{OH} = m_{NaOH} \frac{M_{OH}}{M_{NaOH}}$$

$$m_{Cl} = m_{NaCl} \frac{M_{Cl}}{M_{NaCl}}$$

thus, when these equations are reversed, we get

$$m_{NaOH} = M_{NaOH} \frac{m_{OH}}{M_{OH}}$$

$$m_{NaCl} = M_{NaCl} \frac{m_{Cl}}{M_{Cl}} \tag{350}$$

Next we write:

$$m_s = m_{NaOH} + m_{NaCl}$$

$$= M_{NaOH} \frac{m_{OH}}{M_{OH}} + M_{NaCl} \frac{m_{Cl}}{M_{Cl}}$$

if follows:

$$\frac{m_{Cl}}{M_{Cl}} = \frac{m_s}{M_{NaCl}} - \frac{M_{NaOH}}{M_{NaCl}} \cdot \frac{m_{OH}}{M_{OH}}$$

and the relation between the total solute-OH constituent mass and the construction description is:

$$m_{NaOH} = M_{NaOH} \cdot \frac{m_{OH}}{M_{OH}}$$

$$m_{NaCl} = m_s - M_{NaOH} \cdot \frac{m_{OH}}{M_{OH}} \tag{351}$$

The Gibbs free energy is:

$$\begin{aligned} G &= m_{NaCl}\mu^*_{NaCl} + m_{NaOH}\mu^*_{NaOH} \\ &= \mu^*_{NaCl}\left(m_s + M_{NaOH}\frac{m_{OH}}{M_{OH}}\right) + \mu^*_{NaOH} M_{NaOH}\frac{m_{OH}}{M_{OH}} \\ &= \mu^*_{NaCl} m_s + M_{NaOH}\frac{m_{OH}}{M_{OH}}\left(\mu^*_{NaOH} - \mu^*_{NaCl}\right) \\ &= \mu^*_s m_s + \mu^*_{OH} m_{OH} \end{aligned}$$

with

$$\mu^*_s = \mu^*_{NaCl} \; ; \; \mu^*_{OH} = \frac{M_{NaOH}}{M_{OH}}\left(\mu^*_{NaOH} - \mu^*_{NaCl}\right) \tag{352}$$

Now we can write down the free energy changes at the anode and cathode (see eqs.(345)-(350)):

$$\begin{aligned} \frac{(dG)^{(a)}_{ef}}{A} &= \mu^{*(a)}_s dm^{(a)}_s + \mu^{*(a)}_{OH} dm^{(a)}_{OH} \\ &= \left\{\mu^{*(a)}_{NaCl}\left(M_{NaCl} t^{(a)}_{Cl} + M_{NaOH} t^{(a)}_{OH}\right) \right. \\ &\quad \left. + \mu^{*(a)}_{OH} t^{(a)}_{OH} M_{OH}\right\} \frac{j_q}{\mathcal{F}} dt \end{aligned}$$

$$\begin{aligned} &= \left\{ \mu^{(a)}_{NaCl} t^{(a)}_{Cl} + M_{NaOH}\mu^{*(a)}_{NaCl} t^{(a)}_{OH} + M_{NaOH} t^{(a)}_{OH} \mu^{*(a)}_{NaOH} \right. \\ &\quad \left. - t^{(a)}_{OH} M_{NaOH} \mu^{*(a)}_{NaCl} \right\} \frac{j_q}{\mathcal{F}} dt \end{aligned} \tag{353}$$

where the fundamental law of electrochemistry and the relations (350)-(352) have been applied.

Likewise, the free energy change at the cathode is:

$$\frac{(dG)_{ef}^{(c)}}{A} = \Big\{\mu_s^*\big(-M_{NaCl}t_{Cl}^{(c)} + M_{NaOH}(1-t_{OH}^{(c)})\big) + M_{OH}\mu_{OH}^{*(c)}(1-t_{OH}^{(c)}) - \frac{1}{2}M_{H_2O}\mu_{H_2O}^*\Big\}\frac{j_2}{\mathcal{F}}dt$$

$$= \Big\{\mu_{NaOH}^{(c)} - t_{OH}^{(c)}\mu_{NaOH}^{(c)} - t_{Cl}^{(c)}\mu_{NaCl}^{(c)} - \frac{1}{2}\mu_{H_2O}^{(c)}\Big\}\frac{j_2}{\mathcal{F}}dt \quad (354)$$

As before, in the next step we have to consider the contribution of the bulk electrolyte solution:

$$\frac{(dG)_s}{A} = \Big\{\int_0^l \mu_s^*\frac{\partial m_s}{\partial t}dx + \int_0^l \mu_{OH}^*\frac{\partial m_{OH}}{\partial t}dx\Big\}\frac{j_2}{\mathcal{F}}dt$$

$$= \Big\{\int_0^l \mu_s^*\Big(\frac{dt_{Cl}}{dx}M_{NaCl} + \frac{dt_{OH}}{dx}M_{NaOH}\Big)dx + \int_0^l \mu_{OH}^* M_{OH}\frac{dt_{OH}}{dx}dx\Big\}\frac{j_2}{\mathcal{F}}dt$$

Now we set for the region of the junction:

$$\frac{dt_{Cl}}{dx} = \frac{\Delta t_{Cl}}{\delta x}, \quad \frac{dt_{OH}}{dx} = \frac{\Delta t_{OH}}{\delta x}, \quad x_0 \le x \le x_0 + \delta x$$

$$\Delta t_i = t_i^{(c)} - t_i^{(a)}$$

and we get:

$$\frac{(dG)_s}{A} = \Big\{\frac{\Delta t_{Cl}}{\delta x}M_{NaCl}\int_{x_0}^{x_0+\delta x}\mu_{NaCl}^* dx + \frac{\Delta t_{OH}}{\delta x}M_{NaOH}\int_{x_0}^{x_0+\delta x}\mu_{NaCl}^* dx$$

$$+ \frac{\Delta t_{OH}}{\delta x}M_{NaOH}\int_{x_0}^{x_0+\delta x}\mu_{NaOH}^* dx - \frac{\Delta t_{OH}}{\delta x}M_{NaOH}\int_{x_0}^{x_0+\delta x}\mu_{NaCl}^* dx\Big\}\frac{j_2}{\mathcal{F}}dt$$

$$= \frac{j\,q}{\mathcal{F}}\,dt\left\{\Delta t_{Cl}\,\mu^{o}_{NaCl} + \Delta t_{OH}\,\mu^{o}_{NaOH}\right.$$

$$+\Delta t_{Cl} RT\left[\frac{a^{(c)}_{NaCl}\ln a^{(c)}_{NaCl} - a^{(a)}_{NaCl}\ln a^{(a)}_{NaCl}}{\Delta a_{NaCl}} - 1\right]$$

$$\left.+\Delta t_{OH} RT\left[\frac{a^{(c)}_{NaOH}\ln a^{(c)}_{NaOH} - a^{(a)}_{NaOH}\ln a^{(a)}_{NaOH}}{\Delta a_{NaOH}} - 1\right]\right\} \qquad (355)$$

Now we add eqs.(353). (354) and (355) to get the total electrolyte free energy change:

$$\frac{(dG)_{ef+s}}{A} = \frac{j\,q}{\mathcal{F}}\,dt\left\{\mu^{(c)}_{NaOH} - \frac{1}{2}\mu^{(c)}_{H_2O}\right.$$

$$-t^{(c)}_{Cl} RT\ln a^{(c)}_{NaCl} + t^{(a)}_{Cl} RT\ln a^{(a)}_{NaCl}$$

$$-t^{(c)}_{OH} RT\ln a_{NaOH} + t^{(c)}_{OH} RT\ln a^{(a)}_{NaOH}$$

$$\left.+\Delta t_{Cl} RT\left[\qquad\right] + \Delta t_{OH} RT\left[\qquad\right]\right\}$$

$$= \frac{j\,q}{\mathcal{F}}\,dt\left\{\mu^{(c)}_{NaOH} - \frac{1}{2}\mu^{(c)}_{H_2O} + RT\left[t^{(a)}_{Cl}\left(a^{(c)}_{NaCl}\ldots\right)\right.\right.$$

$$\left.\left.-t^{(c)}_{Cl}\left(a^{(a)}_{NaCl}\ldots\right) + t^{(a)}_{OH}\left(a^{(c)}_{NaOH}\ldots\right) - t^{(c)}_{OH}\left(a^{(a)}_{NaOH}\ldots\right)\right]\right\}$$

Then our final result is:

$$E = E_o + \frac{RT}{\mathcal{F}}\ln\frac{a^{(c)}_{NaOH}}{\left(a^{(c)}_{H_2O}\right)^{1/2}} + E_{diff} \qquad (356)$$

with

$$E_o = \frac{1}{\mathcal{F}}\left(\mu^{o}_{NaOH} - \frac{1}{4}\mu_{O_2} - \frac{1}{2}\mu^{o}_{H_2O} - \mu_{Na'}\right)$$

and

$$E_{diff} = \frac{RT}{\mathcal{F}}\left[t_{Cl}^{(a)}\left(\frac{a_{NaCl}^{(c)} \ln a_{NaCl}^{(a)}/a_{NaCl}^{(c)}}{a_{NaCl}^{(c)} - a_{NaCl}^{(a)}} + 1\right)\right.$$

$$- t_{Cl}^{(c)}\left(\frac{a_{NaCl}^{(a)} \ln a_{NaCl}^{(a)}/a_{NaCl}^{(c)}}{a_{NaCl}^{(c)} - a_{NaCl}^{(a)}} + 1\right) + t_{OH}^{(a)}\left(\frac{a_{NaOH}^{(c)} \ln a_{NaOH}^{(a)}/a_{NaOH}^{(c)}}{a_{NaOH}^{(c)} - a_{NaOH}^{(a)}} + 1\right)$$

$$\left. - t_{OH}^{(c)}\left(\frac{a_{NaOH}^{(c)} \ln a_{NaOH}^{(a)}/a_{NaOH}^{(c)}}{a_{NaOH}^{(c)} - a_{NaOH}^{(a)}} + 1\right)\right] \tag{356a}$$

It is easy to write down the formulas for some simpler cases. If NaCl is absent, then the expression for the diffusion potential has only the terms involving t_{OH}, and if the transport number t_{OH} and t_{Cl} are practically equal in both compartments, then we have

$$E_{diff} = \frac{RT}{\mathcal{F}}\left(t_{Cl} \ln a_{NaCl}^{(a)}/a_{NaCl}^{(c)} + t_{OH} \ln a_{NaOH}^{(a)}/a_{NaOH}^{(c)}\right)$$

Also, if the concentrations are equal in both compartments, then

$$E_{diff} = 0$$

and we have a cell without transference.

11.2 A cell with a chloride and an oxygen electrode.

The arrangement we now wish to describe is the same as that shown in Fig.34, only at x = 0 the Na(metal) electrode (the anode) is replaced by a Cl_2(gas) electrode. In this situation the chemical reaction in the cell is:

$$NaCl + 1/4\ O_2 + 1/2\ H_2O \longrightarrow NaOH + 1/2\ Cl_2$$

and we have to replace the flux expression at the anode, eq.(345), by the formula

$$(j_s)_{x=0} = \left\{-(1 - t_{Cl}^{(a)})M_{NaCl} + M_{NaOH}\, t_{OH}^{(a)}\right\}\frac{I_2}{\mathcal{F}}\,dt \tag{357}$$

because now the anode is electrochemically active with respect to the constituent Cl. This has the consequence that in eq.(353) we have to subtract $\mu^{(a)}_{NaCl}$, and likewise in the final result eq.(356) we have to subtract $\mu^{(a)}_{NaCl}$ and to replace $-\mu_{Na}$ by $1/2\,\mu_{Cl_2}$ in the standard potential expression. Thus the result is

$$E = E_o + \frac{RT}{\mathcal{F}} \ln \frac{a^{(c)}_{NaOH}}{a^{(a)}_{NaCl}(a^{(c)}_{H_2O})^{1/2}} + E_{diff}$$

with

$$E_o = \frac{1}{\mathcal{F}}\left(\mu^o_{NaCl} + \frac{1}{2}\mu_{Cl_2} - \frac{1}{4}\mu_{O_2} - \frac{1}{2}\mu^o_{H_2O} - \mu^o_{NaCl}\right)$$

and E_{diff} the same as given by eq.(356a).

11.3 The replacement of the oxygen electrode by a hydrogen electrode.

Although this chapter is mainly devoted to cells involving an oxygen electrode, the general situation remains very similar if we modify the cell as described in the preceding section by replacing the oxygen electrode by a hydrogen electrode. So we may be allowed to discuss the the cell configuration shown in Fig. 35 here.

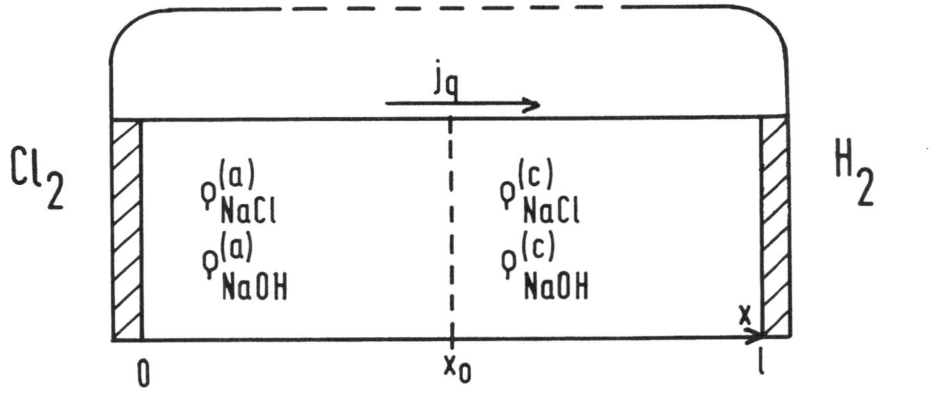

Fig. 35 A cell in which H_2 and Cl_2 is produced from NaCl(aq.) and H_2O.

Now the chemical reaction is:

$$NaCl + H_2O \longrightarrow NaOH + 1/2\ H_2 + 1/2\ Cl_2 \ .$$

The anode fluxes remain the same as given in eq.(357). However at the cathode, when we consider eq.(347) we have to replace $M_{H_2O}/2$ by M_{H_2O} . Apart from this, everything remains unchanged (only E_o is modified, as shown in eq.(358). Thus we have

$$E = E_o + \frac{RT}{\mathcal{F}} \ln \frac{a_{NaOH}^{(c)}}{a_{NaCl}^{(a)}\, a_{H_2O}^{(c)}} + E_{diff} \tag{358}$$

with

$$E_o = \frac{1}{\mathcal{F}} \left(\mu_{NaOH}^{o} + \frac{1}{2}\mu_{Cl_2} + \frac{1}{2}\mu_{H_2} - \mu_{NaCl}^{o} - \mu_{H_2O}^{o} \right)$$

and E_{diff} is the same as given in eq.(356a).

Chapter 12

The salt bridge.

12.1 General

As is well-known the salt bridge is a device which has the purpose of eliminating or at least reducing the diffusion or liquid junction potential. The set-up is simple. The two half cells from which the galvanic cell is to be constructed, are not brought into direct contact, rather they are connected via an electrolyte phase which in most cases consists of a concentrated KCl solution. Experimental details need not be given here. The conventional explanation of the observed effect, namely a drastic reduction of the diffusion potential, is based on the fact that the transport numbers in the bridge electrolyte, t_{Cl} and t_K are almost equal. As a consequence of the equal mobilities of the K^+ and Cl^- ions an electrical potential difference presumably inherent in any ionic diffusion process does not arise. In the present chapter we shall give the theory of the salt bridge within the framework of our treatment. It will be seen that the physical reason for the strong reduction of the liquid junction potential differs from that given in the classical treatment.

12.2 KCl-salt bridge connecting two chloride solutions.

We begin with the consideration of the "open" system shown in Fig. 36. The total galvanic cell may be a concentration cell, however

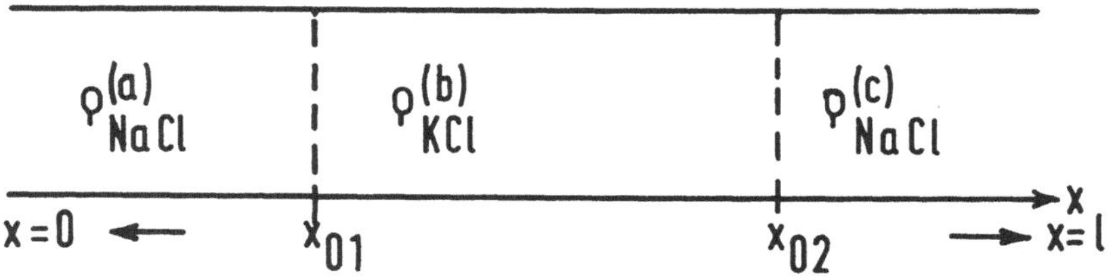

Fig.36 Schematic representation of a salt bridge.

the electrode processes themselves are not of direct interest for the present treatment; we have only to consider the anode and cathode com-

partments irrespective of the nature of the boundaries at x = 0 and x = ℓ . In the present chapter the superscript $^{(b)}$ has the meaning "boundary", as before, the superscripts $^{(a)}$ and $^{(c)}$ indicate that the anode and cathode compartments, respectively, are concerned.
As before we write:

$$\frac{(dG)_s}{A} = -\left\{ \int_0^\ell \mu_{NaCl} \frac{dt_{Na}}{dx} dx + \int_0^\ell \mu_{KCl} \frac{dt_K}{dx} dx \right\} \frac{j_q}{\mathcal{F}} dt$$

All our dt_i/dx are of the form

$$\frac{dt_i}{dx} = \frac{\Delta t_i}{\delta x}$$

with

$$\Delta t_{i1} = t_i^{(b)} - t_i^{(a)}$$

and

$$\Delta t_{i2} = t_i^{(c)} - t_i^{(b)}$$

Then, as has been derived in the previous chapters, we have

$$\begin{aligned} \frac{(dG)_s}{A} = \Big\{ & -\Delta t_{Na1} \overset{o}{\mu}_{NaCl} - \Delta t_{K1} \overset{o}{\mu}_{KCl} \\ & - \Delta t_{Na1} RT \left[\frac{a_{NaCl}^{(b)} \ln a_{NaCl}^{(b)} - a_{NaCl}^{(a)} \ln a_{NaCl}^{(a)}}{a_{NaCl}^{(b)} - a_{NaCl}^{(a)}} - 1 \right] \\ & - \Delta t_{K1} RT \left[\frac{a_{KCl}^{(b)} \ln a_{KCl}^{(b)} - a_{KCl}^{(a)} \ln a_{KCl}^{(a)}}{a_{KCl}^{(b)} - a_{KCl}^{(a)}} - 1 \right] \\ & - \Delta t_{Na2} \overset{o}{\mu}_{NaCl} - \Delta t_{K2} \overset{o}{\mu}_{KCl} \\ & - \Delta t_{Na2} RT \left[\frac{a_{NaCl}^{(c)} \ln a_{NaCl}^{(c)} - a_{NaCl}^{(b)} \ln a_{NaCl}^{(b)}}{a_{NaCl}^{(c)} - a_{NaCl}^{(b)}} - 1 \right] \\ & - \Delta t_{K2} RT \left[\frac{a_{KCl}^{(c)} \ln a_{KCl}^{(c)} - a_{KCl}^{(b)} \ln a_{KCl}^{(b)}}{a_{KCl}^{(c)} - a_{KCl}^{(b)}} - 1 \right] \end{aligned}$$

with

$$\Delta t_{Na1} = -t_{Na}^{(a)} \qquad a_{NaCl}^{(b)} = 0$$

$$\Delta t_{Na2} = t_{Na}^{(c)} \qquad a_{KCl}^{(a)} = a_{KCl}^{(c)} = 0$$

$$\Delta t_{K1} = t_K^{(b)}$$

$$\Delta t_{K2} = -t_K^{(b)}$$

it follows:

$$\frac{(dG)_s}{A} = \Big\{ \mu^{o}_{NaCl} t^{(a)}_{Na} - \mu^{o}_{KCl} t^{(b)}_{K} - t^{(c)}_{Na} \mu^{o}_{NaCl} + \mu^{o}_{KCl} t^{(b)}_{K}$$

$$+ t^{(a)}_{Na} \left[\frac{a^{(a)}_{NaCl} \ln a^{(a)}_{NaCl}}{a^{(a)}_{NaCl}} - 1 \right] RT$$

$$- t^{(b)}_{K} \left[\frac{a^{(b)}_{KCl} \ln a^{(b)}_{KCl}}{a^{(b)}_{KCl}} - 1 \right] RT$$

$$- t^{(c)}_{Na} \left[\frac{a^{(c)}_{NaCl} \ln a^{(c)}_{NaCl}}{a^{(c)}_{NaCl}} - 1 \right] RT$$

$$+ t^{(b)}_{K} \left[\frac{a^{(b)}_{KCl} \ln a^{(b)}_{KCl}}{a^{(b)}_{KCl}} - 1 \right] RT \Big\} \frac{j_2}{\mathcal{F}} dt$$

$$= \Big\{ \mu^{o}_{NaCl} t^{(a)}_{Na} - \mu^{o}_{NaCl} t^{(c)}_{Na} + RT t^{(a)}_{Na} \ln a^{(a)}_{NaCl}$$

$$- RT t^{(c)}_{Na} \ln a^{(c)}_{NaCl} \Big\} \frac{j_2}{\mathcal{F}} dt + \frac{RT}{\mathcal{F}} \left(t^{(c)}_{Na} - t^{(a)}_{Na} \right) j_2 \, dt$$

The first part on the right-hand side of this equation always cancels with the corresponding terms from the electrode fluxes. So we have for the diffusion potentials:

$$E_{diff} = \frac{RT}{\mathcal{F}} \left(t^{(c)}_{Na} - t^{(a)}_{Na} \right) j_2 \approx 0$$

The extension of this formula to the case where there is a different cation in the cathode compartment is obvious. For instance, in Fig.36 let us replace $\rho^{(c)}_{NaCl}$ by $\rho^{(c)}_{LiCl}$ for $x > x_{o2}$. Now we have

$$E_{diff} = \frac{RT}{\mathcal{F}} \left(t^{(c)}_{Li} - t^{(a)}_{Na} \right)$$

These formulas are valid in all cases, also if the chlorine electrode is used.

12.3 The general case of a salt bridge application.

Assume that we have two different binary electrolyte solutions, one containing NaI, and the other one HNO_3, say. These solutions are separated by a KCl salt bridge. The electrode reactions do not enter into the treatment explicitly and therefore, at the present stage need not be considered. The arrangement is sketched in Fig. 37; the dashed lines are drawn arbitrarily. They only ensure that the finite masses written

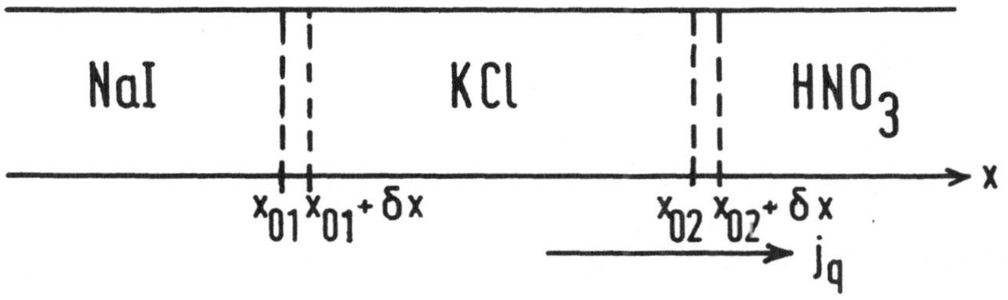

Fig. 37 KCl salt bridge between two electrolyte solutions of different kind and containing constituents different from K and Cl.

down in the following relations are meaningful quantities. Then we consider the system in question (Fig.37) to be constructed from the components NaI, KCl, NaCl, HNO_3, and KNO_3. Let us for a moment consider the masses of a homogeneous system. Then the construction masses are: m_{NaI}, m_{KCl}, m_{NaCl}, m_{HNO_3}, and m_{KNO_3}. It follows that the constituent masses are:

$$m_{Na} = m_{NaI}\frac{M_{Na}}{M_{NaI}} + m_{NaCl}\frac{M_{Na}}{M_{NaCl}}$$

$$m_{K} = m_{KCl}\frac{M_{K}}{M_{KCl}} + m_{KNO_3}\frac{M_{K}}{M_{KNO_3}}$$

$$m_{Cl} = m_{KCl}\frac{M_{Cl}}{M_{KCl}} + m_{NaCl}\frac{M_{Cl}}{M_{NaCl}}$$

$$m_{NO_3} = m_{HNO_3}\frac{M_{NO_3}}{M_{HNO_3}} + m_{KNO_3}\frac{M_{NO_3}}{M_{KNO_3}}$$

$$m_{H} = m_{HNO_3}\frac{M_{H}}{M_{HNO_3}}$$

These relations may be converted to give the construction masses in terms of the constituent masses:

$$m_{NaI} = M_{NaI}\left(\frac{m_{Na}}{M_{Na}} - \frac{m_{Cl}}{M_{Cl}} + \frac{m_K}{M_K} - \frac{m_{NO_3}}{M_{NO_3}} + \frac{m_H}{M_H}\right) \tag{359a}$$

$$m_{NaCl} = M_{NaCl}\left(\frac{m_{Cl}}{M_{Cl}} + \frac{m_{NO_3}}{M_{NO_3}} - \frac{m_K}{M_K} - \frac{m_H}{M_H}\right) \tag{359b}$$

$$m_{KCl} = M_{KCl}\left(\frac{m_K}{M_K} + \frac{m_H}{M_H} - \frac{m_{NO_3}}{M_{NO_3}}\right) \tag{359c}$$

$$m_{KNO_3} = M_{KNO_3}\left(\frac{m_{NO_3}}{M_{NO_3}} - \frac{m_H}{M_H}\right) \tag{359d}$$

$$m_{HNO_3} = M_{HNO_3}\frac{m_H}{M_H} \tag{359e}$$

It is instructive also to present the transformation relations from the total-solute plus four constituents representation to the construction representation. The total solute mass m_s is:

$$m_s = m_{NaI} + m_{KCl} + m_{NaCl} + m_{HNO_3} + m_{KNO_3}$$

and with the above equations this gives

$$m_s = M_{NaI}\frac{m_{Na}}{M_{Na}} + M_{KI}\frac{m_K}{M_K} + M_{HI}\frac{m_H}{M_H} + (M_{Cl}-M_I)\frac{m_{Cl}}{M_{Cl}} + (M_{NO_3}-M_I)\frac{m_{NO_3}}{M_{NO_3}} \tag{360}$$

Now we eliminate $\frac{m_{Na}}{M_{Na}}$ from eqs.(359) and (360) and find

$$m_{NaI} = m_s + \frac{m_K}{M_K}(M_{Na}-M_K) + \frac{m_H}{M_H}(M_{Na}-M_K) - \frac{m_{Cl}}{M_{Cl}}M_{NaCl} - \frac{m_{NO_3}}{M_{NO_3}}M_{KNO_3} \tag{361}$$

which replaces eq.(359a) in order to give the construction parameters in terms of the total solute and four constituent masses.

The Gibbs free energy in the construction representation is:

$$G = m_{NaI}\mu^*_{NaI} + m_{NaCl}\mu^*_{NaCl} + m_{KCl}\mu^*_{KCl} + m_{HNO_3}\mu^*_{KNO_3} + m_{HNO_3}\mu^*_{HNO_3}$$

With eqs.(359b) - (359e) and eq.(361) one derives the same quantity in terms of the total solute plus four constituents language:

$$G = \mu^*_{NaI} m_s + \frac{m_K}{M_K}\left(\left(M_{Na} - M_K\right)\mu^*_{NaI} - M_{NaCl}\mu^*_{NaCl} + M_{KNO_3}\mu^*_{KNO_3}\right)$$

$$+ \frac{m_H}{M_H}\left(\left(M_{Na} - M_H\right)\mu^*_{NaI} - M_{NaCl}\mu^*_{NaCl} + M_{KCl}\mu^*_{KCl}\right.$$

$$\left. - M_{KNO_3}\mu^*_{KNO_3} + M_{HNO_3}\mu^*_{HNO_3}\right) + \frac{m_{Cl}}{M_{Cl}}\left(M_{NaCl}\mu^*_{NaCl} - M_{NaCl}\mu^*_{NaI}\right)$$

$$+ \frac{m_{NO_3}}{M_{NO_3}}\left(M_{NaCl}\mu^*_{NaCl} - M_{NaCl}\mu^*_{NaI} - M_{KCl}\mu^*_{KCl} + M_{KNO_3}\mu^*_{KNO_3}\right)$$

$$= \mu^*_s m_s + \mu^*_K m_K + \mu^*_H m_H + \mu^*_{Cl} m_{Cl} + \mu^*_{NO_3} m_{NO_3} \quad (362)$$

with

$$\mu^*_s = \mu^*_{NaI} \quad (363a)$$

$$\mu^*_K = \frac{1}{M_K}\left\{\left(M_{Na} - M_K\right)\mu^*_{NaI} - M_{NaCl}\mu^*_{NaCl} + M_{KCl}\mu^*_{KCl}\right\} \quad (363b)$$

$$\mu^*_H = \frac{1}{M_H}\left\{\left(M_{Na} - M_H\right)\mu^*_{NaI} - M_{NaCl}\mu^*_{NaCl} + M_{KCl}\mu^*_{KCl} - M_{KNO_3}\mu^*_{KNO_3} + M_{HNO_3}\mu^*_{HNO_3}\right\} \quad (363c)$$

$$\mu^*_{Cl} = \frac{1}{M_{Cl}}\left\{M_{NaCl}\mu^*_{NaCl} - M_{NaCl}\mu^*_{NaI}\right\} \quad (363d)$$

$$\mu^*_{NO_3} = \frac{1}{M_{NO_3}}\left\{M_{NaCl}\mu^*_{NaCl} - M_{NaCl}\mu^*_{NaI} - M_{KCl}\mu^*_{KCl} + M_{KNO_3}\mu^*_{KNO_3}\right\} \quad (363e)$$

Next we write down the local mass productions in an inhomogeneous system. First consider the total solute mass density:

$$\rho_s = M_{Na} c_{Na} + M_K c_K + M_H c_H + M_{NO_3} c_{NO_3} + M_{Cl} c_{Cl} + M_I c_I$$

then with

$$c_{Na} + c_K + c_H - c_{Cl} - c_{NO_3} = c_I$$

$$\rho_s = (M_{Na} + M_I) c_{Na} + (M_K + M_I) c_K + (M_H + M_I) c_H + (M_{NO_3} - M_I) c_{NO_3} + (M_{Cl} - M_I) c_{Cl}$$

Using the fundamental law of electrochemistry this gives

$$\frac{\partial \rho_s'}{\partial t} = \Big\{ - M_{NaI} \frac{dt_{Na}}{dx} - M_{KI} \frac{dt_K}{dx} - M_{HI} \frac{dt_H}{dx} + (M_{NO_3} - M_I) \frac{dt_{NO_3}}{dx}$$

$$+ (M_{Cl} - M_I) \frac{dt_{Cl}}{dx} \Big\} \frac{j_2}{\mathcal{F}}$$

$$\frac{\partial \rho_K'}{\partial t} = - \frac{M_K}{\mathcal{F}} j_2 \frac{dt_K}{dx} \quad , \quad \frac{\partial \rho_H'}{\partial t} = - \frac{M_H}{\mathcal{F}} j_2 \frac{dt_H}{dx} \tag{364}$$

$$\frac{\partial \rho_{Cl}'}{\partial t} = \frac{M_{Cl}}{\mathcal{F}} j_2 \frac{dt_{Cl}}{dx} \quad , \quad \frac{\partial \rho_{NO_3}'}{\partial t} = \frac{M_{NO_3}}{\mathcal{F}} j_2 \frac{dt_{NO_3}}{dx}$$

Then the change of free energy in the bulk electrolyte (see eqs.(363a) -(363e) and (364) is:

$$\frac{(dG)_s}{A} = \frac{1}{A} \int_0^l \Big\{ \mu_s^* \frac{dm_s}{dx} + \mu_K^* \frac{dm_K}{dx} + \mu_H^* \frac{dm_H}{dx} + \mu_{Cl}^* \frac{dm_{Cl}}{dx} + \mu_{NO_3}^* \frac{dm_{NO_3}}{dx} \Big\} dx$$

$$= \Big[\int_0^l \Big\{ - M_{NaI} \mu_{NaI}^* \frac{dt_{Na}}{dx} - M_{KI} \mu_{NaI}^* \frac{dt_K}{dx} - M_{HI} \mu_{NaI}^* \frac{dt_H}{dx}$$

$$+ (M_{NO_3} - M_I) \mu_{NaI}^* \frac{dt_{NO_3}}{dx} + (M_{Cl} - M_I) \mu_{NaI}^* \frac{dt_{Cl}}{dx}$$

$$- M_{Na} \mu_{NaI}^* \frac{dt_K}{dx} + M_K \mu_{NaI}^* \frac{dt_K}{dx} + M_{NaCl} \mu_{NaCl}^* \frac{dt_K}{dx}$$

$$- M_{KCl} \mu_{KCl}^* \frac{dt_K}{dx} - M_{Na} \mu_{NaI}^* \frac{dt_H}{dx} + M_H \mu_{NaI}^* \frac{dt_H}{dx}$$

$$+ M_{NaCl} \mu_{NaCl}^* \frac{dt_H}{dx} - M_{KCl} \mu_{KCl}^* \frac{dt_H}{dx} + M_{KNO_3} \mu_{KNO_3}^* \frac{dt_H}{dx}$$

$$- M_{HNO_3} \mu_{HNO_3}^* \frac{dt_H}{dx} + M_{NaCl} \mu_{NaCl}^* \frac{dt_{Cl}}{dx} - M_{NaCl} \mu_{NaI}^* \frac{dt_{Cl}}{dx}$$

$$+ M_{NaCl} \mu_{NaCl}^* \frac{dt_{NO_3}}{dx} - M_{NaNO_3} \mu_{NaNO_3}^* \frac{dt_{NO_3}}{dx} - M_{KCl} \mu_{KCl}^* \frac{dt_{NO_3}}{dx} + M_{KNO_3} \mu_{KNO_3}^* \frac{dt_{NO_3}}{dx}$$

$$\Big\} dx \Big] \frac{j_2}{\mathcal{F}} dt \tag{365}$$

$$= \Big[\int_0^{\ell} \Big\{ -\mu_{NaI}\frac{dt_{Na}}{dx} - \mu_{NaI}\frac{dt_K}{dx} + \mu_{NaCl}\frac{dt_K}{dx} - \mu_{KCl}\frac{dt_K}{dx} - \mu_{NaI}\frac{dt_H}{dx}$$

$$+ \mu_{NaCl}\frac{dt_H}{dx} - \mu_{KCl}\frac{dt_H}{dx} + \mu_{KNO_3}\frac{dt_H}{dx} - \mu_{HNO_3}\frac{dt_H}{dx} - \mu_{NaI}\frac{dt_{Cl}}{dx} + \mu_{NaCl}\frac{dt_{Cl}}{dx}$$

$$- \mu_{NaI}\frac{dt_{NO_3}}{dx} + \mu_{NaCl}\frac{dt_{NO_3}}{dx} - \mu_{KCl}\frac{dt_{NO_3}}{dx} + \mu_{KNO_3}\frac{dt_{NO_3}}{dx} \Big\} dx \Big] \frac{j_2}{F} dt$$

$$= \Big[-\int_0^{\ell} \mu_{NaI}\Big(\frac{dt_{Na}}{dx} + \frac{dt_K}{dx} + \frac{dt_H}{dx} + \frac{dt_{Cl}}{dx} + \frac{dt_{NO_3}}{dx}\Big) dx$$

$$+ \int_0^{\ell} \mu_{NaCl}\Big(\frac{dt_K}{dx} + \frac{dt_H}{dx} + \frac{dt_{Cl}}{dx} + \frac{dt_{NO_3}}{dx}\Big) dx$$

$$- \int_0^{\ell} \mu_{KCl}\Big(\frac{dt_K}{dx} + \frac{dt_H}{dx} + \frac{dt_{NO_3}}{dx}\Big) dx + \int_0^{\ell} \mu_{KNO_3}\Big(\frac{dt_H}{dx} + \frac{dt_{NO_3}}{dx}\Big) dx$$

$$- \int_0^{\ell} \mu_{HNO_3}\frac{dt_H}{dx} dx \Big] \frac{j_2}{F} dt \qquad (366)$$

Now we begin the detailed treatment for the left-hand boundary x_{o1}. Here we have

$$t_{NO_3} = 0 \quad \text{and} \quad t_H = 0$$

and furthermore:

$$\frac{dt_{Na1}}{dx} = \frac{\Delta t_{Na1}}{\delta x} \; ; \quad \frac{dt_{K1}}{dx} = \frac{\Delta t_{K1}}{\delta x} \; ; \quad \frac{dt_{Cl1}}{dx} = \frac{\Delta t_{Cl1}}{\delta x}$$

Thus we find from eq.(366)

$$\frac{(dG)_s^{(1)}}{A} = \Big\{ -\Delta t_{Na1}\mu^o_{NaI} - \Delta t_{K1}\mu^o_{NaI} - \Delta t_{Cl1}\mu^o_{NaI}$$

$$+ \Delta t_{K1}\mu^o_{NaCl} + \Delta t_{Cl1}\mu^o_{NaCl} - \Delta t_{K1}\mu^o_{KCl}$$

$$- RT(\Delta t_{Na1} + \Delta t_{K1} + \Delta t_{Cl1})\Big[\frac{a^{(b)}_{NaI}\ln a^{(b)}_{NaI} - a^{(a)}_{NaI}\ln a^{(a)}_{NaI}}{a^{(b)}_{NaI} - a^{(a)}_{NaI}} - 1\Big]$$

$$+ RT(\Delta t_{K1} + \Delta t_{Cl1})\Big[\frac{a^{(b)}_{NaCl}\ln a^{(b)}_{NaCl} - a^{(a)}_{NaCl}\ln a^{(a)}_{NaCl}}{a^{(b)}_{NaCl} - a^{(a)}_{NaCl}} - 1\Big]$$

$$+ R\Delta t_{K1}\Big[\frac{a^{(b)}_{KCl}\ln a^{(b)}_{KCl} - a^{(a)}_{KCl}\ln a^{(a)}_{KCl}}{a^{(b)}_{KCl} - a^{(a)}_{KCl}} - 1\Big]\Big\} \frac{j_2}{F} dt$$

Considering the situation as shown in Fig.37 we see that we have to set

$$\Delta t_{Na1} = -t_{Na}^{(a)}, \qquad t_K^{(b)} + t_{Cl}^{(b)} = 1$$
$$\Delta t_{K1} = t_K^{(b)}, \qquad a_{KCl}^{(a)} = a_{NaCl}^{(a)} = 0$$
$$\Delta t_{Cl1} = t_{Cl}^{(b)}, \qquad a_{NaI}^{(b)} = 0.$$

In this way we obtain:

$$\begin{aligned}\frac{(dG)_s^{(1)}}{A} = \Big[& \underline{t_{Na}^{(a)}\mu_{NaI}^{o}} - \mu_{NaI}^{o} + \mu_{NaCl}^{o} - t_K^{(b)}\mu_{KCl} \\ & + RT\,\underline{t_{Na}^{(a)}(\ln a_{NaI}^{(a)}} - 1) - RT\ln a_{NaI}^{(a)}/a_{NaCl}^{(b)} \\ & - RT\,t_K^{(b)}(\ln a_{KCl}^{(b)} - 1)\Big]\frac{j_q}{\mathcal{F}}\,dt \qquad (367)\end{aligned}$$

Next we turn to the detailed treatment of the right-hand boundary at $x = x_{o2}$. Here we have $t_{Na} = 0$ and application of eq.(366) yields:

$$\begin{aligned}\frac{(dG)_s^{(2)}}{A} = \Big\{ & -(\Delta t_{K2} + \Delta t_{NO_3 2} + \Delta t_{H2})\mu_{NaI}^{o} \\ & + (\Delta t_{K2} + \Delta t_{NO_3 2} + \Delta t_{H2})\mu_{NaCl}^{o} \\ & - (\Delta t_{K2} + \Delta t_{NO_3 2} + \Delta t_{H2})\mu_{KCl}^{o} \\ & + (\Delta t_{H2} + \Delta t_{NO_3 2})\mu_{KNO_3}^{o} - \Delta t_{H2}\mu_{HNO_3}^{o} \\ & - (\Delta t_{K2} + \Delta t_{H2} + \Delta t_{NO_3 2})RT\left[\frac{a_{KCl}^{(c)}\ln a_{KCl}^{(c)} - a_{KCl}^{(b)}\ln a_{KCl}^{(b)}}{a_{KCl}^{(c)} - a_{KCl}^{(b)}} - 1\right] \\ & + (\Delta t_{H2} + \Delta t_{NO_3 2})RT\left[\frac{a_{KNO_3}^{(c)}\ln a_{KNO_3}^{(c)} - a_{KNO_3}^{(b)}\ln a_{KNO_3}^{(b)}}{a_{KNO_3}^{(c)} - a_{KNO_3}^{(b)}} - 1\right] \\ & - \Delta t_{H2}RT\left[\frac{a_{HNO_3}^{(c)}\ln a_{HNO_3}^{(c)} - a_{HNO_3}^{(b)}\ln a_{HNO_3}^{(b)}}{a_{HNO_3}^{(c)} - a_{HNO_3}^{(b)}} - 1\right]\Big\}\frac{j_q}{\mathcal{F}}\,dt\end{aligned}$$

With the special configuration of the right-hand boundary we have

$$\Delta t_{K2} = -t_K^{(b)} \quad ; \quad t_H^{(c)} + t_{NO_3}^{(c)} = 1$$
$$\Delta t_{H2} = t_H^{(c)} \quad ; \quad a_{KCl}^{(c)} = a_{KNO_3}^{(c)} = 0$$
$$\Delta t_{NO_3 2} = t_{NO_3}^{(c)} \quad ; \quad a_{HNO_3}^{(b)} = 0$$

we find

$$\frac{(dG)_s^{(2)}}{A} = \Big\{ t_K^{(b)} \mu^{\circ}_{KCl} - \mu^{\circ}_{KCl} + \mu^{\circ}_{KNO_3} - \underline{t_H^{(c)} \mu^{\circ}_{HNO_3}}$$
$$+ t_K^{(b)} RT(\ln a_{KCl}^{(b)} - 1) - RT \ln a_{KCl}^{(b)} / a_{KNO_3}^{(b)}$$
$$- \underline{t_H^{(c)} RT(\ln a_{HNO_3}^{(c)}} - 1) \Big\} \frac{I}{\mathcal{F}} dt \qquad (368)$$

In each of the equations (367) and (368) we have underlined two terms. These terms cancel with the corresponding electrode contribution in any case (compare e.g. eqs.(268) and (269)) irrespective of whether the electrode is electrochemically active with respect to the constituent whose transport number appears in the formula. Now, if we take eqs.(367) and (368) together, we arrive at the final result:

$$E_{diff} = \frac{1}{\mathcal{F}} \left(\frac{(dG)_s^{(1)}}{A} + \frac{(dG)_s^{(2)}}{A} \right)$$

i.e.

$$E_{diff} = \frac{1}{\mathcal{F}} \Big\{ (t_H^{(c)} - t_{Na}^{(a)}) RT + \mu^{\circ}_{NaCl} + \mu^{\circ}_{KNO_3} - \mu^{\circ}_{NaI} - \mu^{\circ}_{KCl}$$
$$+ RT \ln \frac{a_{NaCl}^{(b)} a_{KNO_3}^{(b)}}{a_{NaI}^{(a)} a_{KCl}^{(b)}} \Big\} \qquad (369)$$

The activities $a_{KNO_3}^{(b)}$ and $a_{NaCl}^{(b)}$ are determined by the diffusion process at the boundary so the value of the diffusion potential connected with the salt bridge is not operationally given in the strict sense. We shall describe some details of the diffusion process below; it will turn out that it is justified to set:

$$a_{NaCl}^{(b)} \approx a_{NaI}^{(a)} \qquad (370)$$

The contribution $RT(t_H^{(c)} - t_{Na}^{(a)})$ appears in eq.(369) because we have assumed that the NaI and HNO_3 activities vary linearly with the coordinate x and have the same relative slope as the Na and H transport numbers, which were assumed also to vary linearily in the boundary range. If the activities vary more slowly than the transport numbers which, as a consequence of the high KCl concentration is a reasonable postulate, then the two quantities t_{Na} and t_H have to be replaced by numbers $\tilde{t}_{Na}$, $\tilde{t}_H$ both of which are smaller:

$$0 \leq \tilde{t}_{Na} \leq t_{Na} \quad ; \quad 0 \leq \tilde{t}_H \leq t_H \tag{371}$$

In order to discuss eq.(369) let us, as a first example, assume that the anode compartment contains a Na electrode. Then in eq.(369) μ^o_{NaI} and $RT\ln a^{(a)}_{NaI}$ drop out against the anode contribution. Next, for the comparison of our eq.(369) with the classical result, it is convenient to write the chemical potentials in the single ion activity representation. We have

$$\mu_{NaCl} = \mu^o_{Na^+} + \mu^o_{Cl^-} + RT\ln a_{Na^+}a_{Cl^-}$$

$$\mu_{KCl} = \mu^o_{K^+} + \mu^o_{Cl^-} + RT\ln a_{K^+}a_{Cl^-}$$

$$\mu_{HNO_3} = \mu^o_{H^+} + \mu^o_{NO_3^-} + RT\ln a_{H^+}a_{NO_3^-}$$

$$\mu_{KNO_3} = \mu^o_{K^+} + \mu^o_{NO_3^-} + RT\ln a_{K^+}a_{NO_3^-}$$

These equations we insert in eq.(369). It will be seen that many of the single ion terms cancel out. Furthermore, if one writes the expression for the total EMF including the electrode contribution one finds that $RT\ln a^{(b)}_{NO_3^-}$ in eq.(369) cancels with the corresponding term for the cathode compartment. Then, considering eq.(371) one arrives at the formula:

$$E \approx \frac{1}{\mathcal{F}}\left(\mu^o_{Na^+} + \frac{1}{2}\mu_{H_2} - \mu^o_{H^+} - \mu_{Na'}\right) + \frac{RT}{\mathcal{F}}\ln\frac{a_{Na^+}}{a_{H^+}} \tag{372}$$

which is the classical result. The chemical reaction is

$$Na + H^+ \longrightarrow Na^+ + 1/2\ H_2.$$

Likewise, if as a second example we assume that the anode is an iodine electrode, then with

$$\mu_{NaI} = \mu^o_{Na^+} + \mu^o_{I^-} + RT\ln a_{Na^+}a_{I^-}$$

and eq.(371) we obtain the result:

$$E \approx \frac{1}{\mathcal{F}}\left(\frac{1}{2}\mu_{I_2} + \frac{1}{2}\mu_{H_2} - \mu^{\circ}_{I^-} - \mu^{\circ}_{H^+}\right) - RT \ln a_{H^+} \cdot a_{I^-} \qquad (373)$$

which again is the classical result for the process

$$I^- + H^+ \rightarrow 1/2\ I_2 + 1/2\ H_2$$

12.4 Replacement of the KCl-salt bridge by NaCl and $NaNO_3$ salt bridges

In this section we wish to consider the configuration: electrode compartments as described in the preceding section, but with the KCl salt bridge replaced by a $NaNO_3$ salt bridge. Since we can use eq.(367) directly, we first derive the expression for the diffusion potential connected with the arrangement

	x_{01}	x_{02}
NaI	NaCl	KCl

and then replace Cl by NO_3 and K by H.

We refer now to eq.(367) on page 228 and as a first step we have to write down the appropriate expression for the Δt_i's.

At the left-hand boundary we have

$$\Delta t_{Na1} = t^{(b)}_{Na} - t^{(a)}_{Na} \;; \qquad t^{(b)}_{Na} + t^{(b)}_{Cl} = 1$$

$$\Delta t_{Cl1} = t^{(b)}_{Cl} \;; \qquad a^{(b)}_{NaI} = 0$$

$$\Delta t_{K1} = 0 \;;$$

Then it follows for the free energy change occurring at this side of the salt bridge:

$$\frac{(dG)^{(1)}_s}{A} = \frac{i_q}{\mathcal{F}}\Big[-t^{(b)}_{Na}\mu^{\circ}_{NaI} + t^{(a)}_{Na}\mu^{\circ}_{NaI} - t^{(b)}_{Cl}\mu^{\circ}_{NaI} + t^{(b)}_{Cl}\mu^{\circ}_{NaCl}$$
$$- RT\left(t^{(b)}_{Na} - t^{(a)}_{Na} + t^{(b)}_{Cl}\right)\left[\ln a^{(a)}_{NaI} - 1\right] + t^{(b)}_{Cl}RT\left(\ln a^{(b)}_{NaCl} - 1\right)\Big]dt$$
$$= \Big[\underline{t^{(a)}_{Na}\mu^{\circ}_{NaI}} - \mu^{\circ}_{NaI} + t^{(b)}_{Cl}\mu^{\circ}_{NaCl} + \underline{t^{(a)}_{Na}RT\ln a^{(a)}_{NaI}} - t^{(a)}_{Na}RT$$
$$- RT\left(\ln a^{(a)}_{NaI} - 1\right) + t^{(b)}_{Cl}RT\left(\ln a^{(b)}_{NaCl} - 1\right)\Big]\frac{i_q}{\mathcal{F}}dt \qquad (374)$$

The two underlined terms are those which cancel with the electrode flux contributions.

Now we turn to the right-hand boundary. Here we have

$$\Delta t_{Cl2} = t_{Cl}^{(c)} - t_{Cl}^{(b)} \quad ; \quad t_{Na}^{(b)} + t_{Cl}^{(b)} = 1$$

$$\Delta t_{K2} = t_K^{(c)} \quad ; \quad t_K^{(c)} + t_{Cl}^{(c)} = 1$$

$$\Delta t_{Na2} = -t_{Na}^{(b)} \quad ;$$

and we get from eq.(341)

$$\frac{(dG)_s^{(2)}}{A} = \Big[-(-t_{Na}^{(b)} - t_{Cl}^{(b)} + t_{Cl}^{(c)} + t_K^{(c)})\mu_{NaI}^{o}$$

$$+ (t_K^{(c)} + t_{Cl}^{(c)} - t_{Cl}^{(b)})\mu_{NaCl}^{o}$$

$$-(-t_{Na}^{(b)} + t_K^{(c)} + t_{Cl}^{(c)} - t_{Cl}^{(b)})RT(\ln a_{NaI}^{(b)} - 1)$$

$$+ (t_K^{(c)} + t_{Cl}^{(c)} - t_{Cl}^{(b)})RT(\ln a_{NaCl}^{(b)} - 1)$$

$$-t_K^{(c)}\mu_{KCl}^{o} - t_K^{(c)}RT\ln a_{KCl}^{(c)} + t_K^{(c)}RT\Big]\frac{i_q}{\mathcal{F}}dt$$

$$= \Big[-t_{Cl}^{(b)}\mu_{NaCl}^{o} + \mu_{NaCl}^{o} - t_{Cl}^{(b)}RT(\ln a_{NaCl}^{(b)} - 1) + RT\cdot$$

$$\cdot(\ln a_{NaCl}^{(b)} - 1) - \underline{t_K^{(c)}\mu_{KCl}^{o}} - \underline{t_K^{(c)}RT\ln a_{KCl}^{(c)}} + t_K^{(c)}RT\Big]\frac{i_q}{\mathcal{F}}dt \qquad (375)$$

Then the sum of (374) and (375) gives:

$$\frac{(dG)_s}{A} = \Big[\mu_{NaCl}^{o} - \mu_{NaI}^{o} + RT(t_K^{(c)} - t_{Na}^{(a)}) + RT\ln\frac{a_{NaCl}^{(b)}}{a_{NaI}^{(a)}}\Big]\frac{i_q}{\mathcal{F}}dt \qquad (376)$$

where the "underlined" terms have already been omitted.
The liquid junction potential of the salt bridge arrangement is:

$$E_{diff} = \frac{1}{F}\left(\mu^0_{NaCl} - \mu^0_{NaI} + RT(t_K^{(c)} - t_{Na}^{(a)}) + RT \ln \frac{a^{(b)}_{NaCl}}{a^{(a)}_{NaI}}\right)$$

Now we return to our original problem. The salt bridge consists of a concentrated $NaNO_3$ solution and is placed between a NaI and a nitric acid solution.

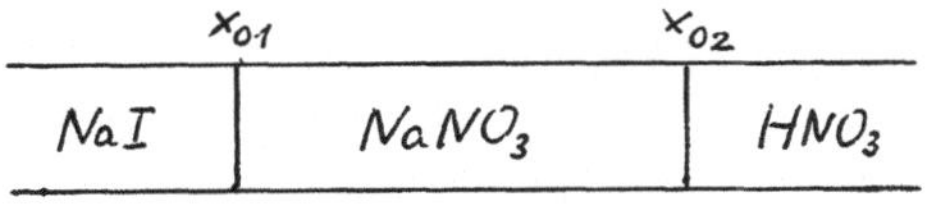

In order to treat this bridge configuration, in eq.(376) we have only to replace Cl by NO_3 and K by H. Thus, the result is:

$$\frac{(dG)_s}{A} = \left[\mu^0_{NaNO_3} - \mu^0_{NaI} + RT(t_H^{(c)} - t_{Na}^{(a)}) + RT \ln \frac{a^{(b)}_{NaNO_3}}{a^{(a)}_{NaI}}\right] \frac{i_s}{F}\, dt \tag{377}$$

or, rewritten as an electric potential contribution

$$E_{diff} = \frac{1}{F}\left[\mu^0_{NaNO_3} - \mu^0_{NaI} + RT(t_H^{(c)} - t_{Na}^{(a)}) + RT \ln \frac{a^{(b)}_{NaNO_3}}{a^{(a)}_{NaI}}\right] \tag{378}$$

When the anode is the Na electrode, then again, μ^0_{NaI} and $-RT \ln a^{(a)}_{NaI}$ drop out due to compensation with the anode processes. So we have to compare the two expressions (see eqs.(369) and (378)):

$$\mu^0_{NaCl} + \mu^0_{KNO_3} - \mu^0_{KCl} + RT \ln \frac{a^{(b)}_{NaCl}\, a^{(b)}_{KNO_3}}{a^{(b)}_{KCl}}$$

and

$$\mu^0_{NaNO_3} + RT \ln a^{(b)}_{NaNO_3}$$

It will be seen that one gets the same result if the conventional "single ion" description is used and again we arrive at eq.(372). The arrows indicate the sheme of cancellation of the various terms.

12.5 Another example: The KCl-salt bridge is placed between a NaI and a HI solution

We consider the liquid junction arrangement as shown in Fig.38.

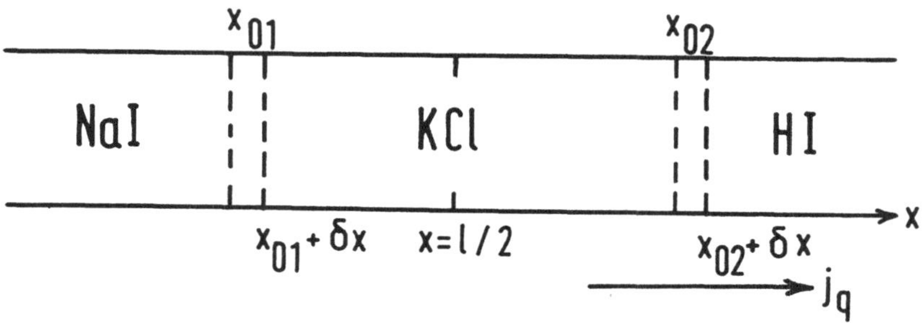

Fig. 38 KCl salt bridge between NaI and HI solutions.

This system contains four constituents which are independent: Na, K, H, and Cl. On the boundary x_{o1} NaCl is produced and on the boundary x_{o2} KI is produced. Thus we have to consider the system as constructed from the components: NaI, KCl, NaCl, KI and HI. These are more components than we have independent constituents; we require the possibility that KI and NaCl be produced. Thus the right-hand half of the system has to be built up from components other than the left-hand half. These are the components

HI, KCl, and KI,

leading to the "chemical reaction"

$$HI + KCl \longrightarrow KI + HCl .$$

Thus, in all our formulas so far written for the system constructed from NaI, KCl, and NaCl, we have to apply the symbol substitution:

$$Cl \longrightarrow I, \qquad Na \longrightarrow K,$$
$$K \longrightarrow H, \qquad I \longrightarrow Cl.$$

Then the constituent masses are

$$m_K = m_{KCl}\frac{M_K}{M_{KCl}} + m_{KI}\frac{M_K}{M_{KI}}$$
$$m_I = m_{KI}\frac{M_I}{M_{KI}} + m_{HI}\frac{M_I}{M_{HI}}$$

$$m_H = m_{HI}\frac{M_H}{M_{HI}}$$

or, when these relations are converted, the construction masses in terms of the constituent masses are:

$$m_{HI} = M_{HI}\frac{m_H}{M_H}$$

$$m_{KI} = M_{KI}\left(\frac{m_I}{M_I} - \frac{m_H}{M_H}\right)$$

$$m_{Kce} = M_{Kce}\left(\frac{m_K}{M_K} + \frac{m_H}{M_H} - \frac{m_I}{M_I}\right)$$

This gives the total solute mass:

$$m_s = M_{Kce}\frac{m_K}{M_K} + M_{Hce}\frac{m_H}{M_H} + (M_I - M_{ce})\frac{m_I}{M_I}$$

and the three construction components as functions of the set m_s, m_I, m_K:

$$m_{HI} = \frac{M_{HI}}{M_{Hce}}\left[m_s + (M_{ce} - M_I)\frac{m_I}{M_I} - M_{Kce}\frac{m_K}{M_K}\right]$$

$$m_{KI} = \frac{M_{KI}}{M_{Hce}}\left[M_{HI}\frac{m_I}{M_I} + M_{Kce}\frac{m_K}{M_K} - m_s\right]$$

$$m_{Kce} = \frac{M_{Kce}}{M_{Hce}}\left[m_s + \frac{m_K}{M_K}(M_H - M_K) - M_{HI}\frac{m_I}{M_I}\right]$$

The Gibbs free energy is:

$$\begin{aligned} G &= m_s\frac{1}{M_{Hce}}(\mu^*_{HI}M_{HI} - \mu^*_{KI}M_{KI} + \mu^*_{Kce}M_{Kce}) \\ &+ \frac{m_K}{M_K}\frac{M_{Kce}}{M_{Hce}}(\mu^*_{KI}M_{KI} - \mu^*_{HI}M_{HI} + \mu^*_{Kce}(M_H - M_K)) \\ &+ \frac{m_I}{M_I}\frac{M_{HI}}{M_{Hce}}(\mu^*_{KI}M_{KI} + \mu^*_{HI}(M_{ce} - M_I) - \mu^*_{Kce}M_{Kce}) \\ &= m_s\mu^*_s + m_K\mu^*_K + m_I\mu^*_I \end{aligned}$$

with

$$\mu^*_s = \frac{1}{M_{Hce}}(\mu^*_{HI}M_{HI} - \mu^*_{KI}M_{KI} + \mu^*_{Kce}M_{Kce})$$

$$\mu^*_K = \frac{1}{M_K}\frac{M_{Kce}}{M_{Hce}}\left[\mu^*_{KI}M_{KI} - \mu^*_{HI}M_{HI} + \mu^*_{Kce}(M_H - M_K)\right]$$

$$\mu_I^* = \frac{1}{M_I}\frac{M_{HI}}{M_{Hcl}}\left[\mu_{KI}^* M_{KI} + \mu_{HI}^*(M_{cl} - M_I) - \mu_{Kcl}^* M_{Kcl}\right]$$

Following the procedure given on pages 226-227 we can now write down the change of free energy in the right-hand part of the system $l/2 \leq x \leq l$, (see Fig.38):

$$\frac{(dG)_s^{(2)}}{A} = -\left\{\int_{l/2}^{l} \mu_{Kcl}\left(\frac{dt_K}{dx} + \frac{dt_H}{dx} + \frac{dt_I}{dx}\right)dx - \int_{l/2}^{l} \mu_{KI}\left(\frac{dt_H}{dx} + \frac{dt_I}{dx}\right)dx + \int_{l/2}^{l} \mu_{HI}\frac{dt_H}{dx}dx\right\}\frac{j_2}{F}\,dt$$

$$= -\Delta t_{K2}\mathring{\mu}_{Kcl} - \Delta t_{H2}\mathring{\mu}_{Kcl} - \Delta t_{I2}\mathring{\mu}_{Kcl} + \Delta t_{H2}\mathring{\mu}_{KI}$$

$$+ \Delta t_{I2}\mathring{\mu}_{KI} - \Delta t_{H2}\mathring{\mu}_{HI}$$

$$- RT(\Delta t_{K2} + \Delta t_{H2} + \Delta t_{I2})\left[\frac{a_{Kcl}^{(c)}\ln a_{Kcl}^{(c)} - a_{Kcl}^{(b)}\ln a_{Kcl}^{(b)}}{a_{Kcl}^{(c)} - a_{Kcl}^{(b)}} - 1\right]$$

$$+ RT(\Delta t_{H2} + \Delta t_{I2})\left[\frac{a_{KI}^{(c)}\ln a_{KI}^{(c)} - a_{KI}^{(b)}\ln a_{KI}^{(b)}}{a_{KI}^{(c)} - a_{KI}^{(b)}} - 1\right]$$

$$- RT\Delta t_{H2}\left[\frac{a_{HI}^{(c)}\ln a_{HI}^{(c)} - a_{HI}^{(b)}\ln a_{HI}^{(b)}}{a_{HI}^{(c)} - a_{HI}^{(b)}} - 1\right]\Bigg\}\frac{j_2}{F}\,dt \qquad (379)$$

Now we have

$$\Delta t_{K2} = -t_K^{(b)}, \quad \Delta t_{H2} = t_H^{(c)}, \quad \Delta t_{I2} = t_I^{(c)}$$

$$t_H^{(c)} + t_I^{(c)} = 1$$

Then we get from eq.(379)

$$\frac{(dG)_s^{(2)}}{A} = \Big[t_K^{(b)}\mathring{\mu}_{Kcl} - \mathring{\mu}_{Kcl} + \mathring{\mu}_{KI} - \underline{t_H^{(c)}\mathring{\mu}_{HI}}$$

$$+ t_K^{(b)} RT(\ln a_{KCl}^{(b)} - 1) - RT(\ln a_{KCl}^{(b)} - 1)$$

$$+ RT(\ln a_{KI}^{(b)} - 1) - \underline{t_H^{(c)} RT(\ln a_{HI}^{(c)} - 1)}\Big] \frac{I_q}{F} dt \qquad (380)$$

The underlined terms cancel with the corresponding electrode terms. We have to add this result to eq.(367) to get:

$$\frac{(dG)_s^{(1)} + (dG)_s^{(2)}}{A} = \Big\{ -\mu_{NaI}^{o} + \mu_{NaCl}^{o} - t_K^{(b)} \mu_{KCl}^{o} - RT\, t_{Na}^{(a)}$$

$$- RT \ln a_{NaI}^{(a)} / a_{NaCl}^{(b)} - RT\, t_K^{(b)} (\ln a_{KCl}^{(b)} - 1)$$

$$+ t_K^{(b)} \mu_{KCl}^{o} - \mu_{KCl}^{o} + \mu_{KI}^{o} + t_K^{(b)} RT(\ln a_{KCl}^{(b)} - 1)$$

$$- RT \ln a_{KCl}^{(b)} / a_{KI}^{(b)} + t_H^{(c)} RT \Big\} \frac{I_q}{F} dt$$

Thus the final result is:

$$E_{diff} = \frac{1}{F} \Big\{ RT(t_H^{(c)} - t_{Na}^{(a)}) + \mu_{NaCl}^{o} + \mu_{KI}^{o} - \mu_{NaI}^{o} - \mu_{KCl}^{o}$$

$$+ RT \ln \frac{a_{NaCl}^{(b)}\, a_{KI}^{(b)}}{a_{NaI}^{(a)}\, a_{KCl}^{(b)}} \Big\} \qquad (381)$$

This is the same formula as eq.(369); we have only to replace NO_3 by I. Thus we see that eq.(369) can also be derived from the addition of two separate half-bridge contributions.

12.6 The salt bridge (KCl) between a redox electrode and a hydrogen electrode.

We apply the method of half-bridge contributions which in the previous section turned out to be a valid and simple procedure when dealing with more complicated junction configurations. The salt bridge is built up as follows (Fig.39).

First we consider the left-hand part $x < \ell/2$. We have to use the modified eq.(343)

$$\frac{(dG)_s^{(1)}}{A} = -\Big\{ \frac{1}{2} \int_0^{\ell/2} \mu_{FeCl_2} \frac{dt_2}{dx} dx + \frac{1}{3} \int_0^{\ell/2} \mu_{FeCl_3} \frac{dt_3}{dx} dx$$

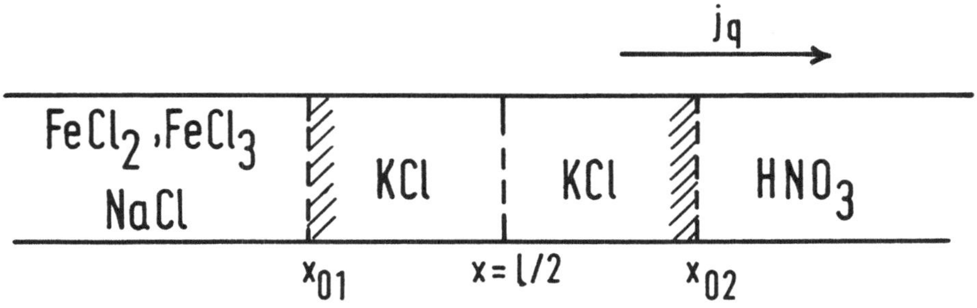

Fig.39 KCl salt bridge between half cells containing a redox electrode and a hydrogen electrode.

$$+\int_0^{l/2}\mu_{NaCl}\frac{dt_{Na}}{dx}dx + \int_0^{l/2}\mu_{KCl}\frac{dt_K}{dx}dx\Big\}\frac{j_q}{F}dt$$

$$= \Big\{-\tfrac{1}{2}\Delta t_2\,\mu^{\circ}_{FeCl_2} - \tfrac{1}{3}\Delta t_3\,\mu^{\circ}_{FeCl_3} - \Delta t_{Na}\,\mu^{\circ}_{NaCl} - \Delta t_K\,\mu^{\circ}_{KCl}$$

$$-\tfrac{1}{2}\Delta t_2 RT\left[\frac{a_2^{(b)}\ln a_2^{(b)} - a_2^{(a)}\ln a_2^{(a)}}{a_2^{(b)} - a_2^{(a)}} - 1\right] - \tfrac{1}{3}\Delta t_3 RT\,\cdot$$

$$\cdot\left[\frac{a_3^{(b)}\ln a_3^{(b)} - a_3^{(a)}\ln a_3^{(a)}}{a_3^{(b)} - a_3^{(a)}} - 1\right] - \Delta t_{Na} RT\left[\frac{a_{NaCl}^{(b)}\ln a_{NaCl}^{(b)} - a_{NaCl}^{(a)}\cdot}{a_{NaCl}^{(b)} -}\right.$$

$$\left.\frac{\cdot\ln a_{NaCl}^{(a)}}{- a_{NaCl}^{(a)}} - 1\right] - \Delta t_K RT\left[\frac{a_{KCl}^{(b)}\ln a_{KCl}^{(b)} - a_{KCl}^{(a)}\ln a_{KCl}^{(a)}}{a_{KCl}^{(b)} - a_{KCl}^{(a)}} - 1\right]\Big\}\frac{j_q}{F}dt$$

Now with our usual linear approximations for the t_i 's and a_i 's we find:

$$\frac{(dG)_s^{(1)}}{A} = \Big[\tfrac{1}{2}t_2^{(a)}\mu^{\circ}_{FeCl_2} + \tfrac{1}{3}t_3^{(a)}\mu^{\circ}_{FeCl_3} + t_{Na}^{(a)}\mu^{\circ}_{NaCl} - t_K^{(b)}\mu^{\circ}_{KCl}$$

$$+ \underline{\tfrac{1}{2}t_2^{(a)}RT(\ln a_{FeCl_2}^{(a)} - 1)} + \underline{\tfrac{1}{3}t_3^{(a)}RT(\ln a_{FeCl_3}^{(a)} - 1)}$$

$$+ \underline{t_{Na}^{(a)}RT(\ln a_{NaCl}^{(a)}} - 1) - t_K^{(b)}RT(\ln a_{KCl}^{(b)} - 1)\Big]\frac{j_q}{F}dt$$

$$= \Big[-RT\big(\tfrac{1}{2}t_2^{(a)} + \tfrac{1}{3}t_3^{(a)} + t_{Na}^{(a)}\big) - t_K^{(b)}RT(\ln a_{KCl}^{(b)} - 1)\Big]\frac{j_q}{F}dt \qquad (382)$$

This equation we have to add to eq.(379) where, however we are replacing the anionic constituent I by NO_3.

Then the final result is :

$$E_{diff} = \frac{1}{\mathcal{F}}\Big[RT\big(t_H^{(c)} - \tfrac{1}{2}t_2^{(a)} - \tfrac{1}{3}t_3^{(a)} - t_{Na}^{(a)}\big) + RT\ln\frac{a_{KNO_3}^{(b)}}{a_{KCl}^{(b)}}\Big] \qquad (383)$$

As was explained above (page 230), in the real system the term RT(...) containing the trasnport numbers will be closer to zero than given in eq.(383). Incorporating the electrode reactions and using the single ion description, we see that eq.(383) represents the classical result:

$$E \approx \frac{1}{\mathcal{F}}\Big(\mu^{o}_{Fe^{+++}} + \tfrac{1}{2}\mu_{H_2} - \mu^{o}_{Fe^{++}} - \mu^{o}_{H^+} + RT\ln\frac{a_{Fe^{+++}}}{a_{Fe^{++}}\cdot a_{H^+}}\Big) \qquad (384)$$

the "ionic" reaction is

$$Fe^{2+} + H^{+} \longrightarrow Fe^{3+} + 1/2\ H_2.$$

For ease of comparison eqs.(372), (373), and (384) have been given in terms of the "classical" single ion activities. It is important now to point out that we could also have derived the equivalents to eqs.(369), (377),(378). . . (383) through application of a regular coordinate system transformation from the construction representation to the full constituents mode of description as the starting point. The transformation relations for the chemical potentials are eqs.(281) to which, in the present situation, the two relations

$$\mu_{HNO_3} = \mu_H + \mu_{NO_3}$$

$$\mu_{KNO_3} = \mu_K + \mu_{NO_3}$$

have to be added. Then we would have obtained results which formally are the same as eqs.(372), (373), and (384), only the + and - signs would have to be dropped. It should be mentioned that the formal results like eq.(369) and the corresponding ones computed using the full constituents representation need not necessarily be exactly the same. In the former case it has been assumed that the $a_{ij} = a_i \cdot a_j$ are linear functions of the cell coordinate whereas in the latter case this assumption refers to each single constituent activity a_i.

12.7 The physical reason for the effectiveness of the salt bridge - an estimate of component concentration in the salt bridge boundary.

It will have been seen from the various examples given in this chapter that the classical statement giving the reason for the effectiveness of a salt bridge, namely the equality of the transport numbers for K^+ and Cl^-, does not play any role in our treatment. In fact, apart from minor modifications, the final result obtained for the electromotive force was independent of the salt used for the construction of the salt bridge. It is important only that the bridge contains a concentrated solution such that at the salt bridge boundary the transport numbers of the constitutents in the electrode compartments (or half-cells) go rapidly to zero as x increases. This reduces the contribution of the term RT(...) (see eqs.(369), (381), (382)) containing the difference of transport numbers in the two electrode compartments, as has been explained above. To the author's knowledge, a systematic experimental study to test the influence of the kind of electrolyte used in the salt bridge has never been performed because the hypotheses that the diffusion potential can be avoided if the two ionic constitutents of the bridge have the same mobility seemed to be entirely self-evident. On the other hand, salt bridges containing concentrated solutions of other electrolytes have successfully been used in those cases where for chemical or solubility requirements the presence of KCl had to be avoided. The concentration cell without transference is well-known in the classical treatments of electrochemistry. Two half-cells of different concentration are placed against an electrode which is of the same kind in both cases, for instance a metal electrode or a hydrogen electrode. Then these two latter electrodes are connected, the result is the concentration cell without transference. Now we have seen in chapter 4, 5, and 6 that any boundary between electrolyte solutions of varying composition can be considered as acting like an electrode. So the salt bridge which is placed between the two electrolyte solutions is equivalent to the two"central" metal or hydrogen electrodes just mentioned. Let AB be the substance being produced and depleted in a concentration cell without transference. Then the scheme of mass production and depletion is as shown in Fig.40.

One sees from Fig.40a that the result of the action of the electric current is the

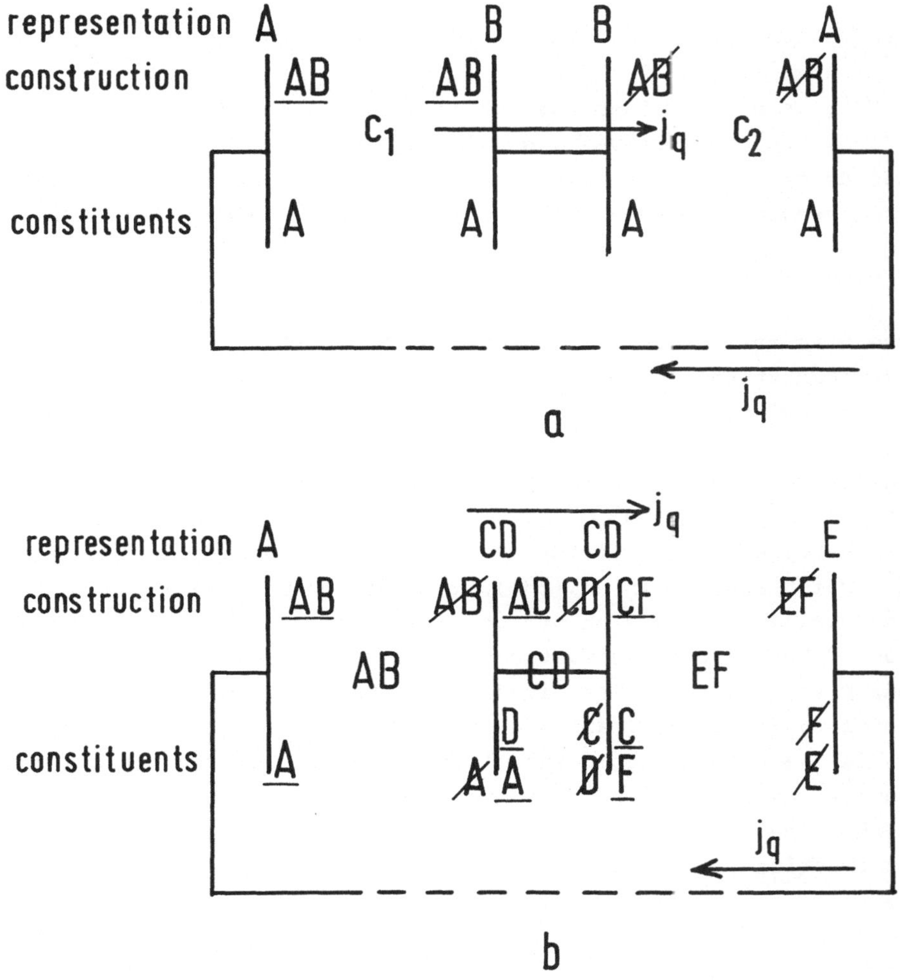

Fig.40: Comparison of the action of a concentration cell without transference with a galvanic cell equipped with a salt bridge

production of $\begin{cases} AB \\ A \end{cases}$ in C_1 and the depletion of $\begin{cases} AB \\ A \end{cases}$

in C_2. Here we have written the result in the construction and constituents language, respectively. Now in Fig. 40b the salt bridge is sketched according to the same scheme. As demonstrated in the figure, the total result connected with the action of the electric current is the

production of $\begin{cases} AD,\ CF \\ A \end{cases}$ and the depletion of $\begin{cases} CD,\ EF \\ E \end{cases}$

Again we have written this statement in the construction and constituents representation. The observed EMF is connected with the respective chemical reaction.

Finally we have to deal with the following problem. In eq.(369) the activities $a_{NaCl}^{(b)}$ and $a_{KNO_3}^{(b)}$ occur which are not directly fixed through the construction of the galvanic cell. The same is true for $a_{KI}^{(b)}$ which appears in eq.(381). In eq.(370) we have set $a_{NaCl}^{(b)} \approx a_{NaI}^{(a)}$ as an approximation. We wish now to give a slightly more detailed description of the physical situation at the boundaries x_{01} and x_{02}. This then may give a certain clarification of the nature of the approximation involved in eq.(370) and a similar relation for $\rho_{KNO_3}^{(b)}$ and $\rho_{KI}^{(b)}$. In Fig.41a we have depicted the distribution of the constituent concentration c_{Na} as a function of the cell coordinate x around the boundary NaI-KCl. The other curves represent the component concentrations in the construction coordinate system. They are given by the relations:

$$c_{NaI}^{*} = \frac{\rho_{NaI}^{*}}{M_{NaI}} \quad , \quad c_{NaCl}^{*} = \frac{\rho_{NaCl}^{*}}{M_{NaCl}} \quad , \quad c_{KCl}^{*} = \frac{\rho_{KCl}^{*}}{M_{KCl}}$$

The original plane of contact at t = 0 is at $x = x_{01}$. In the figure the process of mutual diffusion has already proceeded for a while. The salt bridge contains a very high KCl concentration; thus we have $c_{KCl}^{*} \gg c_{Na}$ and for $x > x_{01}'$ the sum of t_K and t_{Cl} is already virtually unity. Let us assume that the diffusion process is described in the full constituents property space and let $c_{Na}(x_{01}')$, $c_K(x_{01}')$ and $c_{Cl}(x_{01}')$ be the three respective constituent concentrations at x_{01}' and at a given time. Then according to eqs.(16)-(18) the construction component concentrations are

$$c_{KCl}^{*} = c_K(x_{01}')$$
$$c_{NaCl}^{*} = c_{Cl}(x_{01}') - c_K(x_{01}')$$

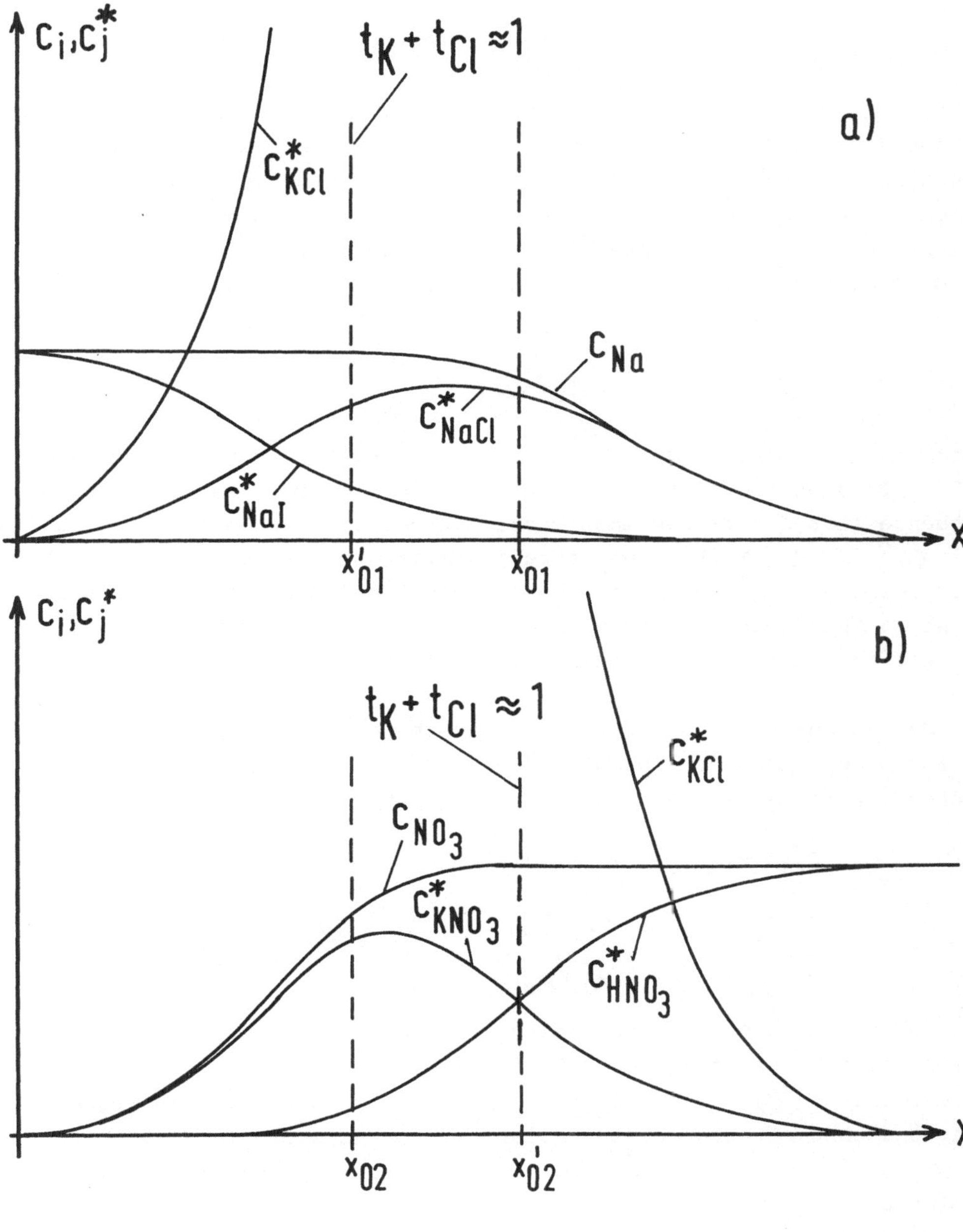

Fig. 41 Schematic representation of the concentration distribution in the two boundaries of a salt bridge. The starred quantities refer to the construction representation. The remaining concentrations are given in the constituents representation. For further details see text.

$$c^{*}_{NaI} = c_{Na}(x'_{01}) + c_{K}(x'_{01}) - c_{Cl}(x'_{01})$$

If the $c_i(x'_{01})$, i = K, Na, Cl, were known, then, assuming the linear x dependence approximation, we would have the desired component concentrations occurring in eq.(369). Since in most practical cases the multicomponent electrolyte solution mutual diffusion coefficients are not known, we have to rely on estimates. In a concentrated KCl solution (c ≈ 3 M) the ratio of the self-diffusion coefficients of K and Cl is almost unity i.e. $\mathcal{D}_K / \mathcal{D}_{Cl} = 0.97$ [1]. Thus if as an approximation we set:

$$c^{*}_{NaCl}(x'_{01}) = c_{Cl}(x'_{01})\left(1 - \frac{\mathcal{D}_K}{\mathcal{D}_{Cl}}\right) \approx c_{Cl}(x'_{01}) \cdot 0{,}03$$

with $c_{KCl} \approx 3$ M, we obtain $c^{*}_{NaCl} \approx 0.1$ M. Moreover, the ratio of the self-diffusion coefficients of Na and I is ≈ 0.7, which has the consequence that I diffuses much faster into the KCl solution than Na. Thus if $c^{(a)}_{NaI} \lesssim 0.1$ M it seems to be justified to assume that at the boundary coordinate x'_{01} we have a NaCl concentration which is virtually equal to the Na constituent concentration and consequently also equal to the NaI concentration in the anode compartment.

Now let us turn to Fig.41b. We use the eq.(359c) - (359e) in order to find the component concentrations (in the construction coordinate system) in terms of the constituent concentrations:

$$c^{*}_{KCl}(x'_{02}) = c_{K}(x'_{02}) + c_{H}(x'_{02}) - c_{NO_3}(x'_{02})$$

$$c^{*}_{KNO_3}(x'_{02}) = c_{NO_3}(x'_{02}) - c_{H}(x'_{02})$$

$$c^{*}_{HNO_3}(x'_{02}) = c_{H}(x'_{02})$$

Although we have no experimental knowledge of the mutual diffusion coefficient for the constituent H in the system in question here, it is expected that $\mathcal{D}_H$ is comparatively large. This has the consequence that the concentration distribution will be as shown in Fig.41b, that is $c_H(x'_{02}) \approx 0$ and $c^{*}_{KNO_3}(x'_{02}) \approx c^{(c)}_{HNO_3}$. For c_{KI} the situation is analogous.

Reference

1) H.G. Hertz and R. Mills, J.Chim.Phys., 1974, 71, 1355

Chapter 13

Membrane potentials

13.1 Simple membrane arrangements.

We begin with the simplest example. We have two NaCl solutions and these solutions are separated by a membrane in which some compound of the constituent Na is formed but which does not contain the constituent Cl. Then, in the left-hand and right-hand compartments we have Cl_2 electrodes or also Ag/AgCl - or calomel-electrodes -. Thus the system has the construction scheme as shown in Fig. 42.

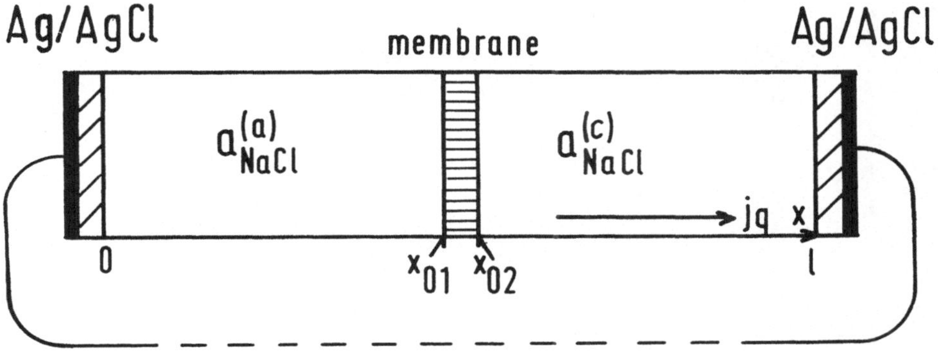

Fig. 42 A simple arrangement to measure the membrane potential

The Gibbs-free energy change localized at the anode is

$$\frac{(dG)_{ef}^{(a)}}{A} = -\left(1 - t_{Cl}^{(a)}\right)\mu_{NaCl}^{(a)} \frac{j_q}{F}\, dt \tag{385}$$

and correspondingly at the cathode we have

$$\frac{(dG)_{ef}^{(c)}}{A} = \left(1 - t_{Cl}^{(c)}\right)\mu_{NaCl}^{(c)} \frac{j_q}{F}\, dt \tag{386}$$

In the bulk electrolyte including the membrane we have:

$$\frac{(dG)_s}{A} = -\left\{ \int_{x_{01}}^{x_{01}+\delta x} \mu_{NaCl}^{(a)} \frac{dt_{Na}^{(1)}}{dx}\, dx + \int_{x_{02}}^{x_{02}+\delta x} \mu_{NaCl}^{(c)} \frac{dt_{Na}^{(2)}}{dx}\, dx \right\} \frac{j_q}{F}\, dt$$

For the transport numbers of Na we have at the two membrane boundaries:

$$t_{Na}^{(1)}(x) = t_{Na}^{(a)} + \frac{1 - t_{Na}^{(a)}}{\delta x}(x - x_{01}) , \quad \frac{dt_{Na}^{(1)}}{dx} = \frac{1 - t_{Na}^{(a)}}{\delta x}$$

$$t_{Na}^{(2)}(x) = 1 + \frac{t_{Na}^{(c)} - 1}{\delta x}(x - x_{02}) , \quad \frac{dt_{Na}^{(2)}}{dx} = - \frac{1 - t_{Na}^{(c)}}{\delta x}$$

Now the behaviour of μ_{NaCl} at the membrane boundary is essentially the same as it is at the electrode boundary, namely μ_{NaCl} = const., and then immediately dropping to zero. Then we obtain

$$\frac{(dG)_s}{A} = \left[-(1 - t_{Na}^{(a)})\mu_{NaCl}^{(a)} + (1 - t_{Na}^{(c)})\mu_{NaCl}^{(c)} \right] \frac{j}{\mathcal{F}} dt$$

$$= \left[- t_{Cl}^{(a)}\mu_{NaCl}^{(a)} + t_{Cl}^{(c)}\mu_{NaCl}^{(c)} \right] \frac{j}{\mathcal{F}} dt \qquad (387)$$

We add this to the eqs.(385) and (386) and find

$$\frac{(dG)_{ef}^{(a)} + (dG)_{ef}^{(c)} + dG_s}{A} = - \mu_{NaCl}^{(a)} + \mu_{NaCl}^{(c)}$$

$$= RT \ln \frac{a_{NaCl}^{(c)}}{a_{NaCl}^{(a)}} + \mu_{NaCl}^{o}(P^{(c)}) - \mu_{NaCl}^{o}(P^{(a)})$$

which gives the membrane potential:

$$E = \frac{RT}{\mathcal{F}} \ln \frac{a_{NaCl}^{(c)}}{a_{NaCl}^{(a)}} + \frac{1}{\mathcal{F}}\left[\mu_{NaCl}^{o}(P^{(c)}) - \mu_{NaCl}^{o}(P^{(a)}) \right] \qquad (388)$$

This is a well-known formula. $\mu_{NaCl}^{o}(P^{(c)})$ means the standard chemical potential of NaCl at an (osmotic) pressure $P^{(c)}$. If the osmotic pressure equilibrium is not attained, for instance, if both compartments are under atmospheric pressure, then we have only

$$E = \frac{RT}{\mathcal{F}} \ln \frac{a_{NaCl}^{(c)}}{a_{NaCl}^{(a)}}$$

What is the physical process involved in these arrangements? In the right-hand compartment NaCl is produced. It is produced at the cathode surface and at the membrane surface. In the anode compartment NaCl is removed; it is removed at the anode and at the left-hand surface of the membrane according to the chemical reaction:

Ag + NaCl + membrane ⟶ AgCl + Na(membrane)

In the right-hand compartment the process is:

AgCl + Na(membrane ⟶ Ag + NaCl + membrane.

Now let us use Na-electrodes instead of the Cl electrodes, then the electrode contributions are:

$$\frac{(dG)_{ef}^{(a)}}{A} = t_{Cl}^{(a)} \mu_{NaCl}^{(a)} \frac{i}{\mathcal{F}} dt$$

$$\frac{(dG)_{ef}^{(c)}}{A} = - t_{Cl}^{(c)} \mu_{NaCl}^{(c)} \frac{i}{\mathcal{F}} dt$$

and consequently we have

$$dG = (dG)_{ef}^{(a)} + (dG)_{ef}^{(c)} + (dG)_s = 0$$

The chemical process involved here is the following: There is production of NaCl at the surface of the anode and removal of NaCl at the left-hand surface of the membrane. At the right-hand surface of the membrane we have a production of NaCl and at the surface of the cathode the removal of NaCl. Thus, only Na metal is transported from the left-hand to the right-hand side of the system; such a process is not connected with a change of free energy.

Next it is of interest to consider the reverse situation. The membrane forms a compound with Cl, the anion constituent. Then the free energy change in the bulk + membrane is:

$$\frac{(dG)_s}{A} = \left[+ t_{Na}^{(a)} \mu_{NaCl}^{(a)} - t_{Na}^{(c)} \mu_{NaCl}^{(c)} \right] \frac{i}{\mathcal{F}} dt \qquad (389)$$

and consequently, the potential difference is only observed if Na electrodes are applied. For another configuration we may also apply hydrogen electrodes in the two respective compartments.

Now the chemical reactions are:
in the anode compartment

$1/2\ H_2$ + Cl(membrane) + NaOH ⟶ NaCl + H_2O

and in the cathode compartment

NaCl + H_2O ⟶ $1/2\ H_2$ + Cl(membrane) + NaOH .

In this set up the change of Gibbs free energy at the anode is obtained when eq.(354) is applied and then modified according to the description given on page 218 . One finds:

$$\frac{(dG)_{ef}^{(a)}}{A} = \left[-\mu_{NaOH}^{(a)} + t_{OH}^{(a)}\mu_{NaOH} + t_{Cl}^{(a)}\mu_{NaCl} + \mu_{H_2O}^{(a)}\right]\frac{I^2}{F}\,dt \tag{390}$$

Correspondingly, at the cathode we have:

$$\frac{(dG)_{ef}^{(c)}}{A} = \left[\mu_{NaOH}^{(c)} - t_{OH}^{(c)}\mu_{NaOH}^{(c)} - t_{Cl}^{(c)}\mu_{NaCl}^{(c)} - \mu_{H_2O}^{(c)}\right]\frac{I^2}{F}\,dt \tag{391}$$

Here the component NaOH is a construction component according to the scheme

$$NaCl + H_2O \longrightarrow NaOH + HCl$$

$$= NaCl + HOH \longrightarrow NaOH + HCl\ .$$

It is beyond the scope of this book to give more details. In eqs.(390) and (391) we assume for simplicity $\mu_{NaOH}^{(c)} \approx \mu_{NaOH}^{(a)}$; of course we have $t_{OH} \approx 0$. Then combination of eqs.(389), (390), and (391) yields:

$$\frac{dG}{A} = RT \ln \frac{a_{NaCl}^{(a)}}{a_{NaCl}^{(c)}} + \mu_{NaCl}^{o}(P^{(a)}) - \mu_{NaCl}^{o}(P^{(c)})$$

or

$$E = \frac{RT}{F} \ln \frac{a_{NaCl}^{(a)}}{a_{NaCl}^{(c)}} + \frac{1}{F}\left[\mu_{NaCl}^{o}(P^{(a)}) - \mu_{NaCl}^{o}(P^{(c)})\right]$$

Next we assume that we have <u>an iodine electrode</u> in the cathode compartment and the electrolyte solution is NaI. Then at the anode the reaction is as before; at the cathode the reaction is

$$1/2\ I_2 + Na(\text{membrane}) \longrightarrow NaI + \text{membrane}$$

and consequently the membrane potential is (apart from an additive quantity)

$$E = \frac{RT}{F} \ln \frac{a_{NaI}^{(c)}}{a_{NaCl}^{(a)}} + \ldots .$$

13.2 The Donnan potential

The arrangement we have now to discuss is as follows:

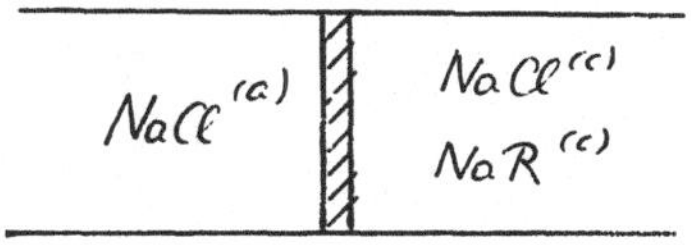

or

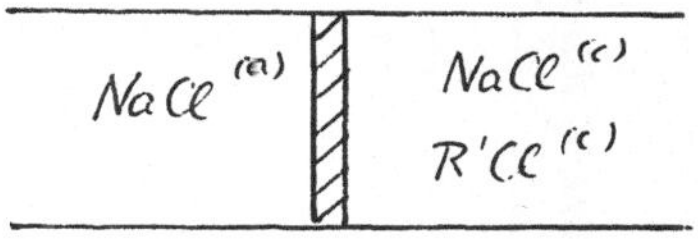

The substances NaR or R'Cl cannot pass through the membrane, that is, the mutual diffusion coefficients of the compounds NaR or R'Cl with respect to the material of the membrane are practically zero. However, NaCl can pass through the membrane. Thus, in fact, essentially we are dealing with an ordinary concentration cell with respect to NaCl. If the difference of the transport numbers on both sides of the membrane is neglected, then we have:

$$E = \frac{RT}{\mathcal{F}}(1-t_{Cl})\ln\frac{a_{NaCl}^{(c)}}{a_{NaCl}^{(a)}}$$

Here we have set $\mu_{NaCl} = \mu_{NaCl}^{o} + RT\ln a_{NaCl}$, thus the standard state is the same on both sides of the membrane; it is the standard state in the left-hand side, say. Then the activity may not change at all along the path x and we have a distribution equilibrium of NaCl, however, the concentration changes, $c_{NaCl}^{(a)} \neq c_{NaCl}^{(c)}$. This means that the activity coefficients vary strongly when the components NaR or R'Cl are present. Of course, if $a_{NaCl}^{(c)} = a_{NaCl}^{(a)} = f^{(c)}c_{NaCl}^{(c)} = f^{(a)}c_{NaCl}^{(a)}$, in the equilibrium situation no electric work can be delivered by the system. This has also been pointed out in KORTÜM's book [2]. Only if $a_{NaCl}^{(c)} \neq a_{NaCl}^{(a)}$ can we measure an electrical potential. Treatments of this subject are obscure in other textbooks.

13.3 The glass electrode.

We shall now consider the following arrangement (see Fig.43). First we consider the left-hand side (1) of this cell arrangement. We have the following contributions to the change of Gibbs free energy.
At the anode the relation holds

$$\frac{(dG)_{ef}^{(a)}}{A} = -(1-t_{Cl}^{(a)})\mu_{KCl}^{(a)}\frac{j_q}{\mathcal{F}}dt \qquad (392)$$

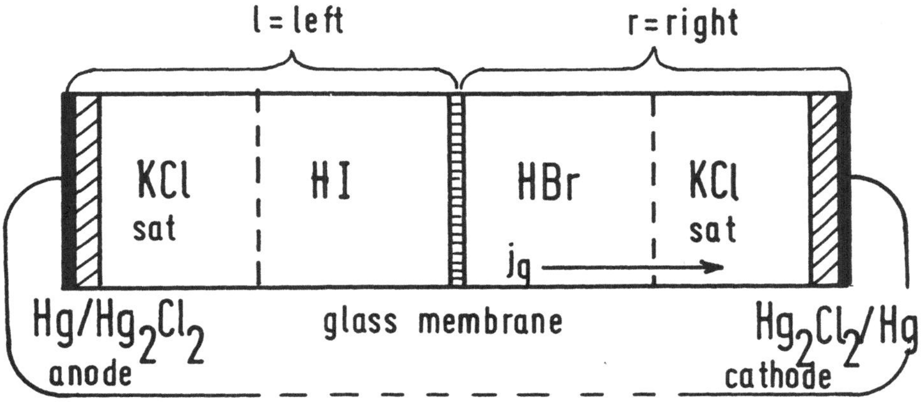

Fig. 43 Schematic representation of a galvanic cell arrangement involving a glass electrode.

For the electrolyte solution contribution, we take eq.(380) however replacing $^{(b)}$ by $^{(a)}$ and $^{(c)}$ by $^{(\ell)}$:

$$\frac{(dG)_s^{(\ell)}}{A} = \Big[t_K^{(a)}\mathring{\mu}_{KCl} - \mathring{\mu}_{KCl} + \mathring{\mu}_{KI} - t_H^{(\ell)}\mathring{\mu}_{HI}$$

$$+ t_K^{(a)} RT(\ln a_{KCl}^{(a)} - 1) - RT(\ln a_{KCl}^{(a)} - 1)$$

$$+ RT(\ln a_{KI}^{(\ell)} - 1) - t_H^{(\ell)} RT(\ln a_{HI}^{(\ell)} - 1)\Big] \frac{j_g}{F} dt$$

$$= \Big[-(1 - t_K^{(a)})\mu_{KCl}^{(a)} + (1 - t_K^{(a)})RT + \mu_{KI}^{(\ell)} - RT$$

$$- t_H^{(\ell)}\mu_{HI}^{(\ell)} + t_H^{(\ell)} RT \Big] \frac{j_g}{F} dt$$

$$= \Big[t_K^{(a)}\mu_{KCl}^{(a)} + \mu_{KI}^{(\ell)} - \mu_{KCl}^{(a)} - t_H^{(\ell)}\mu_{HI}^{(\ell)} + (t_H^{(\ell)} - t_K^{(a)})RT \Big] \frac{j_g}{F} dt \tag{393}$$

Addition of eqs.(392) and (393) gives the total left-hand contribution:

$$\frac{(dG)^{(\ell)}}{A} = \Big[\mu_{KI}^{(\ell)} - \mu_{KCl}^{(a)} - t_H^{(\ell)}\mu_{HI}^{(\ell)} + (t_H^{(\ell)} - t_K^{(a)})RT \Big] \frac{j_g}{F} dt \tag{394}$$

$$t_K^{(a)} = 1 - t_{Cl}^{(a)}$$

Then the corresponding expression for the right-hand side is:

$$\frac{(dG)^{(r)}}{A} = \left[-\mu_{KBr}^{(r)} + \mu_{KCl}^{(c)} + t_H^{(r)}\mu_{HBr}^{(r)} - (t_H^{(r)} - t_K^{(c)})RT\right]\frac{j_q}{\mathcal{F}}\,dt \qquad (395)$$

where we keep in mind that

$$\mu_{KCl}^{(c)} = \mu_{KCl}^{(a)} \quad , \quad t_K^{(a)} = t_K^{(c)} \qquad (396)$$

due to the saturated KCl solutions on both sides of the system. Next we write the expression for the free energy production on both sides of the glass membrane. Here we can simply take eq.(387), only t_{Na} has to be replaced by t_H. The result is:

$$\frac{(dG)^{(\text{both sides})}_{\text{membrane}}}{A} = \left[-(1 - t_H^{(l)})\mu_{HI}^{(l)} + (1 - t_H^{(r)})\mu_{HBr}^{(r)}\right]\frac{j_q}{\mathcal{F}}\,dt \qquad (397)$$

Now we add eqs.(394),(395), and (397) and taking into account eq.(396), we find after some cancellations:

$$\frac{dG}{A} = \left[\mu_{KI}^{(l)} - \mu_{KBr}^{(r)} + \mu_{HBr}^{(r)} - \mu_{HI}^{(l)} + (t_H^{(l)} - t_H^{(r)})RT\right]\frac{j_q}{\mathcal{F}}\,dt$$

$$= \left[\mu_{KI}^{\circ} - \mu_{KBr}^{\circ} + RT\ln\frac{a_{KI}^{(l)}}{a_{KBr}^{(r)}} + \mu_{HBr}^{\circ} - \mu_{HI}^{\circ} + RT\ln\frac{a_{HBr}^{(r)}}{a_{HI}^{(l)}} + (t_H^{(l)} - t_H^{(r)})RT\right]\frac{j_q}{\mathcal{F}}\,dt$$

Thus our final result is:

$$\frac{dG}{A} = \left[RT\ln\frac{a_{KI}^{(l)}}{a_{KBr}^{(r)}} + RT\ln\frac{a_{HBr}^{(r)}}{a_{HI}^{(l)}} + (t_H^{(l)} - t_H^{(r)})RT\right]\frac{j_q}{\mathcal{F}}\,dt$$

or rewritten to give the EMF:

$$E = \frac{RT}{\mathcal{F}}\left\{\ln\frac{a_{KI}^{(l)}}{a_{KBr}^{(r)}} + \ln\frac{a_{HBr}^{(r)}}{a_{HI}^{(l)}} + (t_H^{(l)} - t_H^{(r)})\right\}$$

The chemical reaction on the left-hand side is:

$$\text{Hg} + \text{KCl} + \text{membrane} + \text{HI} \longrightarrow 1/2\ \text{Hg}_2\text{Cl}_2 + \text{KI} + \text{H(membrane)}$$

and the chemical reation on the right-hand side is:

$$\text{H(membrane)} + \text{KBr} + 1/2\ \text{Hg}_2\text{Cl}_2 \longrightarrow \text{Hg} + \text{KCl} + \text{HBr(membrane)}$$

Thus the total chemical process is:

$$\text{HI} + \text{KBr} \longrightarrow \text{HBr} + \text{KI}\ .$$

Next we choose the left-hand compartment to contain NaOH instead of HI, such that the cell arrangement has the form depicted in Fig.44 .

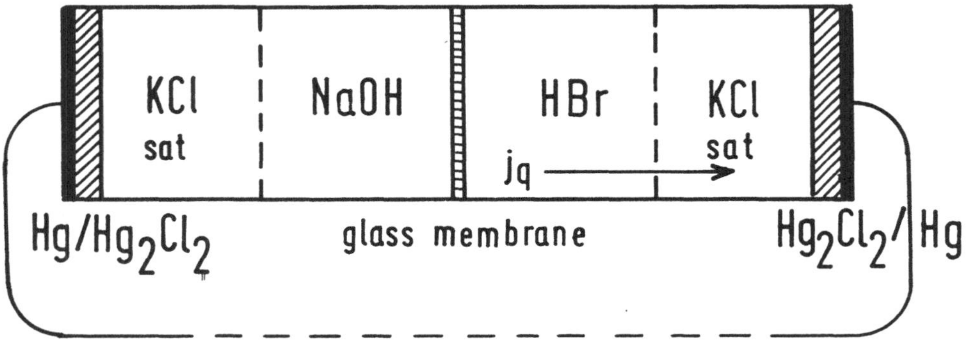

Fig. 44 Glas electrode arrangement in which a pH value >7 is measured.

Then in formula (393) we have to replace HI by NaOH and to add a contribution due to the flux of water. The result is:

$$\frac{(dG)_s^{(l)}}{A} = \Big[-(1-t_K^{(a)})\mu_{KCl}^{(a)} + (1-t_K^{(a)})RT + \mu_{KOH}^{(l)} - RT$$
$$- t_{Na}^{(l)}\mu_{NaOH}^{(l)} + t_{Na}^{(l)}RT - \mu_{H_2O}^{(l)}\Big]\frac{j_q}{F}dt$$
$$= \Big[t_K^{(a)}\mu_{KCl}^{(a)} + \mu_{KOH}^{(l)} - \mu_{KCl}^{(l)} - t_{Na}^{(l)}\mu_{NaOH}^{(l)}$$
$$+ (t_{Na}^{(l)} - t_K^{(a)})RT - \mu_{H_2O}^{(l)}\Big]\frac{j_q}{F}dt \qquad (398)$$

At the membrane we have

$$\frac{(dG)_{membrane}^{(both\ sides)}}{A} = \Big[t_{Na}^{(l)}\mu_{NaOH}^{(l)} + (1-t_H^{(r)})\mu_{HBr}^{(r)}\Big]\frac{j_q}{F}dt \qquad (399)$$

Now we add eqs.(398), (395) and (399). The result is:

$$\frac{dG}{A} = \Big\{\mu_{KOH}^{(l)} - t_{Na}^{(l)}\mu_{NaOH}^{(l)} + t_{Na}^{(l)}RT - t_K^{(a)}RT - \mu_{H_2O}^{(l)} - \mu_{KBr}^{(r)} + t_H^{(r)}\mu_{HBr}^{(r)}$$
$$- t_H^{(r)}RT + t_K^{(a)}RT + t_{Na}^{(l)}\mu_{NaOH}^{(l)} + \mu_{HBr}^{(r)} - t_H^{(r)}\mu_{HBr}^{(r)}\Big\}\frac{j_q}{F}dt$$

$$= \left\{ \mu_{KOH}^{(\ell)} - \mu_{KBr}^{(r)} + \mu_{HBr}^{(r)} - \mu_{H_2O}^{(\ell)} + (t_{Na}^{(\ell)} - t_{H}^{(r)})RT \right\} \frac{j_2}{\mathcal{F}} dt$$

$$= \left\{ \mu_{KOH}^{o} + RT \ln a_{KOH}^{(\ell)} - \mu_{KBr}^{o} - RT \ln a_{KBr}^{(r)} + \mu_{HBr}^{o} \right.$$

$$\left. + RT \ln a_{HBr}^{(r)} - \mu_{H_2O}^{o} - RT \ln a_{H_2O}^{(\ell)} + (t_{Na}^{(\ell)} - t_{H}^{(r)})RT \right\} \frac{j_2}{\mathcal{F}} dt$$

$$= \left\{ \mu_{K}^{o} + \mu_{OH}^{o} - \mu_{K}^{o} - \mu_{Br}^{o} + \mu_{H}^{o} + \mu_{Br}^{o} - \mu_{H}^{o} - \mu_{OH}^{o} \right.$$

$$\left. + RT \ln a_{KOH}^{(\ell)} a_{HBr}^{(r)} - RT \ln a_{KBr}^{(r)} a_{H_2O}^{(\ell)} + (t_{Na}^{(\ell)} - t_{H}^{(r)})RT \right\} \frac{j_2 dt}{\mathcal{F}}$$

and thus finally we arrive at the formula:

$$E = \frac{1}{\mathcal{F}} \left[RT \ln a_{KOH}^{(\ell)} a_{HBr}^{(r)} - RT \ln a_{H_2O}^{(\ell)} a_{KBr}^{(r)} + (t_{Na}^{(\ell)} - t_{H}^{(r)})RT \right] \tag{400}$$

The chemical reaction in the total system is:

$$KBr + H_2O \longrightarrow KOH + HBr,$$

which may be deduced from the addition of the left-hand side reaction:

$$\text{membrane} + H_2O + KCl + Hg \longrightarrow H(\text{membrane}) + KOH + 1/2\ Hg_2Cl_2$$

with the right-hand side reaction given on page 251 .

It will be seen that the constitutent Na does not enter in the total chemical reaction. The arrangement as shown in Fig.44 and described by eq.(400) has a limiting situation when $\rho_{NaOH} \rightarrow 0$ in which case there is only pure water on the left-hand side. The addition of other substances like NH_3 or NH_4Cl then opens the door wide to acid-base theory and to a general consideration which however will not be developed here.

Acknowledgement

The first thoughts which led to the writing of this book arose in the quiet environment of Canberra, Australia, when the author was a visiting fellow in the Diffusion Research Unit of the Australian National University. The author wishes to thank Dr. R. Mills for having made this stay possible. Furthermore, Dr. Mills was kind enough to take care of the English language in this book.

A. F. Williams

A Theoretical Approach to Inorganic Chemistry

1979. 144 figures, 17 tables. XII, 316 pages
ISBN 3-540-09073-8

This book outlines the application of simple quantum mechanics to the study of inorganic chemistry, and shows its potential for systematizing and understanding the structure, physical properties, and reactivities of inorganic compounds. The considerable strides made in inorganic chemistry in recent years necessitate the establishment of a theoretical framework if the student is to acquire a sound knowledge of the subject. A wide range of topics is covered, and the reader is encouraged to look for further extensions of the theories discussed. The book emphasizes the importance of the cirtical application of theory and, although it is chiefly concerned with molucular orbital theory, other approaches are discussed. This text is intended for students in the latter half of their undergraduate studies. (235 references)

From the Foreword by Prof. C. K. Jørgensen

"...Dr. Alan Williams has acquired a considerable experience in work with transition metal complexes at the Universities of Cambridge and Geneva. In this book he has tried to avoid the variety of ephemeral and often contradictory rationalisations encountered in this field, and has made a careful comparison of modern opinions about chemical bonding. In my opinion this effort is fruitful for all students and active scientists in the field of inorganic chemistry. The distant relations to group theory, atomic spectroscopy and epistemology are brought into daylight.
...The interdisciplinary approach of the book shows up in the careful consideration given to many experimental techniques such as vibrational (infra-red and Raman), electronic (visible and ultraviolet), Mössbauer, magnetic resonance, and photoelectron spectra, with data for gaseous and solid samples as well as selected facts about solution chemistry. The book could not have been written a few years ago, and is likely to remain a highly information survey of modern inorganic chemistry and chemical physics."

Springer-Verlag
Berlin
Heidelberg
New York

Lecture Notes in Chemistry

Vol. 1: G. H. Wagnière, Introduction to Elementary Molecular Orbital Theory and to Semiempirical Methods. V, 109 pages. 1976.

Vol. 2: E. Clementi, Determination of Liquid Water Structure. VI, 107 pages. 1976.

Vol. 3: S. R. Niketic and K. Rasmussen, The Consistent Force Field. IX, 212 pages. 1977.

Vol. 4: A. Graovac, I. Gutman and N. Trinajstić, Topological Approach to the Chemistry of Conjugated Molecules. IX, 123 pages. 1977.

Vol. 5: R. Carbo and J. M. Riera, A General SCF Theory. XII, 210 pages. 1978.

Vol. 6: I. Hargittai, Sulphone Molecular Structures. VIII, 175 pages. 1978.

Vol. 7: Ion Cyclotron Resonance Spectrometry. Edited by H. Hartmann and K.-P. Wanczek. VI, 326 pages. 1978.

Vol. 8: E. E. Nikitin and L. Zülicke, Selected Topics of the Theory of Chemical Elementary Processes. X, 175 pages. 1978.

Vol. 9: A. Julg, Crystals as Giant Molecules. VII, 135 pages. 1978.

Vol. 10: J. Ulstrup, Charge Transfer Processes in Condensed Media. VII, 419 pages. 1979.

Vol. 11: F. A. Gianturco, The Transfer of Molecular Energies by Collision: Recent Quantum Treatments. VIII, 328 pages. 1979.

Vol. 12: The Permutation Group in Physics and Chemistry. Edited by J. Hinze. VI, 230 pages. 1979.

Vol. 13: G. Del Re et al., Electronic States of Molecules and Atom Clusters. (In preparation).

Vol. 14: E. W. Thulstrup, Aspects of the Linear and Magnetic Circular Dichroism of Planar Organic Molecules. VI, 100 pages. 1980.

Vol. 15: A.T. Balaban et al, Steric Fit in Quantitative Structure-Activity Relations. VII, 178 pages. 1980.

Vol. 16: P. Čársky and M. Urban, Ab Initio Calculations. VI, 247 pages. 1980.

Vol. 17: H. G. Hertz, Electrochemistry. X, 254 pages. 1980.